EXERCISES IN MATHEMATICS

EXERCISES IN MATHEMATICS

J. BASS

École Nationale Supérieure de l'Aéronautique
Paris, France

Simple and multiple integrals. Series of functions.
Fourier series and Fourier integrals. Analytic functions.
Ordinary and partial differential equations.

Translated by SCRIPTA TECHNICA, INC.

1966

ACADEMIC PRESS New York and London

First published in the French language under the title
Exercises de Mathématiques
and copyrighted 1965 by Masson & Cie, Paris, France.

ACADEMIC PRESS INC.
111 Fifth Avenue, New York, New York 10003

United Kingdom Edition published by
ACADEMIC PRESS INC. (LONDON) LTD.
Berkeley Square House, London W.1

PRINTED IN THE UNITED STATES OF AMERICA

Preface

It is not without some hesitation that I have decided to publish this volume of solved mathematical problems. For while it is essential to offer the student a large choice of exercises, it is less evident that providing detailed models of solutions will be useful for him. More precisely, one might feel that a small number of such models, judiciously chosen, should normally form a part of the mathematical treatises themselves and that this should be sufficient. However, after several years of experience, I have come to feel that a collection of exercises, varied in nature and worked out in detail, can be beneficial, and that the student needs to learn how to work them out as much as he does to reason in connection with them and to find their solutions.

This collection is intended for students of applied mathematics, physics, and engineering. The spirit in which it is written corresponds approximately to that of the upper division level of instruction in courses on mathematical techniques. The subjects treated, however, are often found in courses on a somewhat lower level (e.g., integral calculus) as well as a somewhat higher level (e.g., mathematical methods in physics).

These exercises do not pretend in any way to take the place of a course of instruction. To help the reader, each chapter is preceded by a brief introduction in which certain essential formulas are recalled and in which the prerequisites are briefly outlined. Of course, such prerequisites should have already been acquired in oral or written instruction.

Although there are few books covering all the material, one can find the substance of it in quite a large number of works under the general headings of applied mathematics, advanced calculus, matrix and tensor calculus, etc., written for the upper undergraduate level and designed for engineers and physicists. Among the texts that are best suited we may cite:

R. Courant and D. Hilbert, "Methods of Mathematical Physics," Volume 1. Wiley (Interscience), New York, 1953.

This is a fundamental reference work which has served as a model for all advanced books in applied mathematics. One need not have studied it in detail to be able to do these exercises, but familiarity with certain chapters is essential.

R. V. Churchill, "Complex Variables and Applications," 2nd ed. McGraw-Hill, New York, 1960.

R. Courant and F. John, "Introduction to Calculus and Analysis." Wiley (Interscience), New York, 1965.

J. Irving and N. Mullineux, "Mathematics in Physics and Engineering." Academic Press, New York, 1959.

I. S. Sokolnikoff and R. M. Redheffer, "Mathematics of Physics and Modern Engineering." McGraw-Hill, New York, 1958.

It is difficult to state explicitly the sources of the exercises whose solutions are worked out in the present collection. Some of them are original. Many have been suggested in a more or less conscious way by analogous exercises that appear rather ubiquitously in mathematical literature. Some of them are of the nature of supplementary material to a course of instruction. With very few exceptions, the statements of the problems appear in the author's "Cours de mathématiques" (Masson, Paris, 1964). As a whole, the exercises chosen are of medium difficulty. The more difficult ones are indicated by an asterisk preceding the title of the exercise.

The subjects treated have been divided into six chapters which are devoted to simple integrals, uniform convergence and normed spaces, line integrals and multiple integrals, analytic functions, ordinary differential equations, and partial differential equations. This order is not happenstance. In each chapter the reader should be capable of using the material constituting the subject of the exercises of the preceding chapters. Although the beginning of Chapter 2 is devoted to uniform convergence, numerous exercises in the following chapters constitute a training in the techniques of continuity of series and integrals, and of differentiation and integration under the integral sign. Similarly, the residue theorem, which is the subject of the end of Chapter 4, is used in numerous exercises in Chapters 5 and 6. In Chapter 1 there are no exercises on the convergence of improper integrals. Such exercises inevitably constitute an initial problem in various exercises in Chapter 2 dealing with the uniform convergence of integrals. Orthogonal functions appear first in Chapter 2 and again in Chapter 5 in connection with problems on second-order linear differential equations. Very few topics from the calculus of variations have been treated. They appear at the end of Chapter 5 although they are sometimes more of an algebraic than a functional nature.

Here we do not propose to give exercises on probability theory. Such a subject requires special preparation. However, the relationships between probability theory and certain questions in algebra and integral calculus are so close that several exercises are worded in probability theory language. For the reader who may not be familiar with the elements of probability theory, the few words of introduction preceding the statement of the problems will enable him to make the necessary adjustment.

J. BASS

February, 1966

A Note about the French Edition

The original French edition of this volume contains all the material found here and, in addition, a chapter on linear algebra and a short chapter on probability. It corresponds very closely to the author's "Cours de mathématiques" (3rd ed., Masson, Paris, 1964) and, with only a few exceptions, the statements of the problems can be found there. Among the French texts suitable for background study are:

J. Bass, "Éléments de calcul des probabilités." Masson, Paris, 1964. An English translation, "Elements of Probability Theory," will be published by Academic Press in 1966.

A. Hocquenghem and P. Jaffard, "Mathématiques." Masson, Paris, 1964.

J. Kuntzmann, "Mathématiques de la physique et de la technique." Hermann, Paris.

L. Schwartz, "Méthodes mathématiques pour les sciences physiques." Hermann, Paris, 1961.

The rough drafts of these exercises were prepared by the assistants of l'École Nationale Superieure de l'Aéronautique between 1962 and 1964. MM. J. Azema, J. P. Bertrandias, J. Couot, M. Gatesoupe, M. Mendès-France, M. Pianko, J. Servant, and Vo Khac Khoan had the responsibility for individual chapters, and I am very grateful to them for their assistance. I also wish to thank MM. J. Dhombres, F. Dress, F. Hoffman, J-F. Mela, Pham Phu Hien, and Ph. Robba who reread the manuscripts and the proofs and the former assistants at l'École Nationale Supérieure de l'Aéronautique and l'École Nationale Supérieure des Mines — MM. P. Belayche, F. Bourion, F. Germain, J. Germain, J-P. Guiraud, G. Legrand, J. Stern, and R. Vallée — to whom I am indebted for many interesting comments.

J. BASS

Contents

CHAPTER 1

Sequences, Series, and Definite Integrals

This chapter contains several exercises on the fundamental concepts of analysis. These exercises do not require extensive preparation on the part of the reader. It will be sufficient for him to know various definitions that are part of almost all curricula or special mathematical courses and certain elementary properties of the objects in question. The most frequently encountered are the following:

Distance in general. The distance between two functions. The distance between two sequences.

Norms in a vector space. The norm of a function.

Classical properties of continuous and monotonic functions.

Convergence of numerical sequences and series. Absolute convergence. Alternating series.

Definite integrals. Integrable functions of a single variable. The first and second mean-value theorems. Schwarz' inequality.

The integrals, of course, are in the sense of Riemann. The exercises on Hölder's inequality would have quite extensive sequels if we were using the Lebesgue integral. However, since we are dealing here with purely algebraic properties of the *definite* integral, only the elementary axioms come into play: The integral is a linear functional and is additive as a function of the interval of integration; integrable functions constitute an algebra; the integral of a positive function is positive; the integral of the absolute value of a function is at least as great as the absolute value of the integral of the function; if f is nonnegative and its integral is zero, then f is zero almost everywhere. With these principles, it hardly matters what kind of integral one is using.

All the definite integrals are of bounded functions defined on a compact interval. The chapter contains no exercises on "generalized" (i.e., improper) integrals, either integrals over an infinite interval or integrals of unbounded functions. Integrals of this type will appear frequently in Chapter 2 in connection with exercises on uniform convergence. Among the questions posed, one frequently encountered will be to prove the convergence of an integral, and it has not seemed advisable to isolate these exercises. We therefore refer the reader to Chapter 2. Also in Chapter 2 are exercises on the Bolzano-Weierstrass theorem, in connection with uniform convergence.

1. Examples of the distance between two functions and between two sequences

Let us give an example of a distance between two functions that is not a norm.

Problems. A. Suppose that a and b are positive numbers. Show that

$$\frac{a+b}{1+a+b} \leqslant \frac{a}{1+a} + \frac{b}{1+b}.$$

B. Let $f(x)$ be an integrable function on the interval $\alpha \leqslant x \leqslant \beta$ and set

$$L(f) = \int_\alpha^\beta \frac{|f(x)|}{1+|f(x)|}\,dx.$$

Show that

$$L(f+g) \leqslant L(f) + L(g).$$

Can we define the distance between two functions with the aid of the functional L?

C. Let u and v be any two sequences of real numbers the values of which are $u_1, \dots, u_n, \dots$ and $v_1, \dots, v_n \dots$. Let a_n be the general term of a convergent sequence of positive numbers. Set

$$L(u) = \sum_n a_n \frac{|u_n|}{1+|u_n|}, \qquad L(v) = \sum_n a_n \frac{|v_n|}{1+|v_n|}.$$

Denote by $u + v$ the sequence whose general term is $u_n + v_n$. Show that

$$L(u + v) \leqslant L(u) + L(v).$$

Solutions. **A.** Since a and b are positive numbers, the inequality

$$\frac{a + b}{1 + a + b} \leqslant \frac{a}{1 + a} + \frac{b}{1 + b} \tag{1}$$

follows from the obvious inequalities

$$\frac{a}{1 + a + b} \leqslant \frac{a}{1 + a}, \qquad \frac{b}{1 + a + b} \leqslant \frac{b}{1 + b}.$$

Equality can hold only if a or b is zero.

B. Since f is an integrable function on the interval $[\alpha, \beta]$, the functional L maps f into the positive number

$$L(f) = \int_\alpha^\beta \frac{|f(x)|}{1 + |f(x)|} \, dx \qquad (\alpha < \beta).$$

Let us show that the functional L verifies the triangle inequality

$$L(f + g) \leqslant L(f) + L(g). \tag{2}$$

First of all, we have

$$|f + g| \leqslant |f| + |g|.$$

Since the function $x/(1 + x)$ is an increasing function, we obtain

$$\frac{|f + g|}{1 + |f + g|} \leqslant \frac{|f| + |g|}{1 + |f| + |g|}$$
$$\leqslant \frac{|f|}{1 + |f|} + \frac{|g|}{1 + |g|}$$

on the basis of inequality (1). This proves inequality (2).

Let us now recall the definition of the distance between two functions: The distance $l(f, g)$ between two functions f and g is a real number that satisfies the following axioms:

(a) $l(f, g) > 0$ if $f \neq g$,

(b) $l(f, f) = 0$,

(c) $l(f, g) = l(g, f)$,

(d) $l(f, g) \leqslant l(f, h) + l(h, g)$ for arbitrary functions f, g, h.

If we agree to consider two functions f and g equal when $L(f - g) = 0$, we see that the functional L enables us to define the distance between two functions f and g. We simply set

$$l(f, g) = L(f - g).$$

This definition is valid for integrable functions defined on the interval $[\alpha, \beta]$. The functional $L(f)$ may be used to define the distance between two functions, but it is not a *norm* in the vector space of functions that are integrable on $[\alpha, \beta]$. To see this, consider the vector space of real functions. Here, if m denotes a real number, the requirement of homogeneity $L(mf) = |m| L(f)$ is not verified.

C. We have seen that

$$\frac{|u_n + v_n|}{1 + |u_n + v_n|} \leqslant \frac{|u_n|}{1 + |u_n|} + \frac{|v_n|}{1 + |v_n|}.$$

Since $|u_n|/(1 + |u_n|) \leqslant 1$, the series $\sum a_n |u_n|/(1 + |u_n|)$ converges.
If we multiply the two sides of the above inequality by a_n and sum over n, we obtain

$$L(u + v) \leqslant L(u) + L(v),$$

so that

$$L(u) = \sum a_n \frac{|u_n|}{1 + |u_n|}.$$

The functional $L(u - v)$ can be considered as the *distance between the two series* u and v. It is positive except when $u_n = v_n$ for every n, in which case it is zero. It is symmetric with respect to u and v. Finally, it satisfies the triangle inequality. It is interesting to note that this definition of distance is valid for infinite sequences *irrespective of their convergence or divergence*.

2. Extreme values of continuous functions

Problems. A. Let $f(x)$ be a continuous function that is positive everywhere in the closed interval $[a, b]$, where $a < b$. Then $f(x)$ attains a maximum value M at a point x_0 in $[a, b]$. Show that

$$M = \lim_{n \to +\infty} \left[\frac{1}{b - a} \int_a^b [f(x)]^n \, dx\right]^{1/n}.$$

B. Show that, if $f(x)$ is strictly positive, the minimum of $f(x)$ is equal to

$$\lim_{n\to-\infty}\left[\frac{1}{b-a}\int_a^b [f(x)]^n\,dx\right]^{1/n}.$$

Solutions. **A.** By hypothesis, the function f is positive and continuous on the closed interval $[a, b]$. Therefore, it attains its maximum M at least once on $[a, b]$, at some point x_0. Let us suppose first that x_0 is different from a and b. For any $\epsilon > 0$, there exists a number α less than $x_0 - a$ and $b - x_0$ such that the inequality

$$|x - x_0| < \alpha$$

implies that

$$M - \epsilon \leqslant f(x) \leqslant M.$$

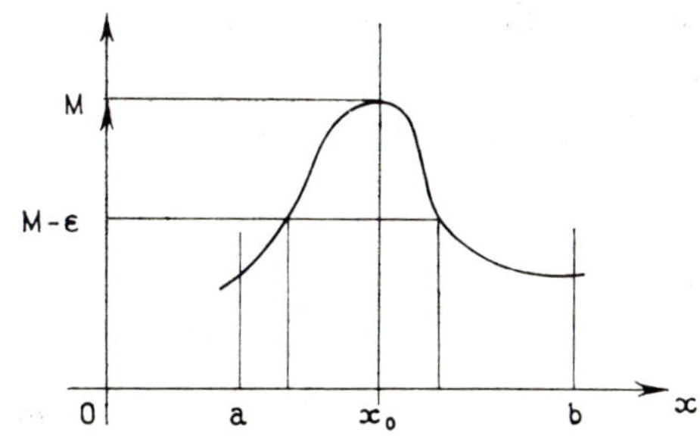

FIGURE 1

Let us denote by E the union of the two intervals $(a, x_0 - \alpha)$ and $(x_0 + \alpha, b)$. Then

$$\frac{1}{b-a}\int_a^b [f(x)]^n\,dx = \frac{1}{b-a}\int_{x_0-x}^{x_0+\alpha} [f(x)]^n\,dx + \frac{1}{b-a}\int_E [f(x)]^n\,dx.$$

Consequently,

$$\frac{1}{b-a}\int_a^b [f(x)]^n\,dx \geqslant \frac{1}{b-a}\,2\alpha(M-\epsilon)^n. \tag{1}$$

On the other hand, the inequality $f(x) \leqslant M$ implies that

$$\frac{1}{b-a}\int_a^b [f(x)]^n\,dx \leqslant M^n,$$

so that

$$\left(\frac{2\alpha}{b-a}\right)^{1/n}(M-\epsilon) \leqslant \left[\frac{1}{b-a}\int_a^b [f(x)]^n\,dx\right]^{1/n} \leqslant M. \tag{2}$$

As n increases without bound, the factor $[2\alpha/(b - a)]^{1/n}$ tends to 1. Therefore, there exists a number N_ϵ such that the inequality $n > N_\epsilon$ implies the double inequality

$$M - 2\epsilon \leqslant \left[\frac{1}{b - a} \int_a^h [f(x)]^n \, dx\right]^{1/n} \leqslant M.$$

This double inequality means that

$$\lim_{n \to +\infty} \left[\frac{1}{b - a} \int_a^b [f(x)]^n \, dx\right]^{1/n} = M. \tag{3}$$

Remark. In the above, the number α was chosen sufficiently small that the interval $(x_0 - \alpha, x_0 + \alpha)$ was contained in the interval (a, b). In the event that x_0 is, let us say, the point a, it will be sufficient to replace the interval $(x_0 - \alpha, x_0 + \alpha)$ with $(a, x_0 + \alpha)$.

B. Let us use inequality (3) to show that

$$\lim_{n \to -\infty} \left[\frac{1}{b - a} \int_a^b [f(x)]^n \, dx\right]^{1/n} = \min f(x). \tag{4}$$

We set $n' = -n$. Then

$$\lim_{n \to -\infty} \left[\frac{1}{b - a} \int_a^b [f(x)]^n \, dx\right]^{1/n} = \lim_{n' \to +\infty} \left[\frac{1}{b - a} \int_a^b \frac{1}{[f(x)]^{n'}} \, dx\right]^{-1/n'}$$

$$= \left(\max \frac{1}{f(x)}\right)^{-1} = \min f(x).$$

The function $1/f(x)$ verifies the conditions of problem A, since $f(x)$ does not vanish on the closed interval $[a, b]$ and has a positive minimum in that interval.

C. These formulas express the extreme values of a continuous function on an interval in terms of a limit without requiring that the function be differentiable. Therefore, this procedure is more general (although less practical) than that of finding the points at which the derivative of $f(x)$ (which is also the result of a limiting process) vanishes.

Example. $f(x) = 1 - x$ for $0 \leqslant x \leqslant 1$.
The maximum of $f(x)$ is given by

$$\lim_{n \to \infty} \left[\int_0^1 (1 - x)^n \, dx\right]^{1/n} = \lim_{n \to \infty} \left(\frac{1}{n + 1}\right)^{1/n} = 1.$$

This maximum appears at the point [well defined since $f(x)$ is continuous] at which $f(x) = 1$. This is the point $x = 0$.

Remark. We denote by L^n the space of functions $f(x)$ for which the integral $\int_a^b |f(x)|^n \, dx$ exists. It is basically a question of the existence of the integral in the Lebesgue sense. This space contains all functions that are integrable in the Riemann sense over the closed interval $[a, b]$.

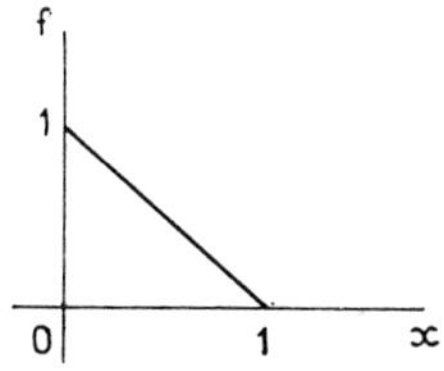

FIGURE 2

It can be shown (see Exercise 3) that L^n is a vector space and that

$$\left[\int_a^b |f(x)|^n \, dx\right]^{1/n}$$

has all the properties of a norm in L^n.

We see that the maximum of a continuous function is the limit of its norm in the L^n sense as $n \to \infty$. The maximum of $|f|$ also possesses the properties of a norm in the space of functions that are bounded in absolute value on the closed interval $[a, b]$.

3. The Schwarz, Hölder, and Minkowski inequalities

Schwarz' inequality is well known. For integrable functions, it takes the form

$$\left|\int_\alpha^\beta fg \, dx\right| \leqslant \left(\int_\alpha^\beta f^2 \, dx\right)^{1/2} \left(\int_\alpha^\beta g^2 \, dx\right)^{1/2} \qquad (\alpha < \beta).$$

To prove it, we assume that the second-degree real trinomial

$$\int_\alpha^\beta (f + \lambda g)^2 \, dx$$

is positive, that is, that its discriminant is negative.

This inequality implies that the functional

$$\left(\int_\alpha^\beta f^2 \, dx\right)^{1/2}$$

possesses the properties of a norm in the vector space of functions that are integrable on the interval $[a, b]$.

In this exercise we shall present a new proof of Schwarz' inequality and a generalization of it known as Hölder's inequality, which leads to a more general form for the norm of an integrable function.

In the remaining exercises, we shall see new proofs of Schwarz' inequality and proofs of Hölder's inequality that employ the properties of convex functions or multiple integrals.

Problems. A. Suppose that $f(x) = P(x) + iQ(x)$ is a complex integrable function of a real variable x. Show that

$$\left| \int_\alpha^\beta f\,dx \right| \leqslant \int_\alpha^\beta |f|\,dx.$$

Replace f with f^2 and prove Schwarz' inequality

$$\left| \int_\alpha^\beta PQ\,dx \right|^2 \leqslant \int_\alpha^\beta P^2\,dx \int_\alpha^\beta Q^2\,dx.$$

B. Let p and q denote two numbers both greater than 1 such that

$$\frac{1}{p} + \frac{1}{q} = 1.$$

Show that, for any two positive numbers a and b,

$$ab \leqslant \frac{a^p}{p} + \frac{b^q}{q}.$$

How may we choose a and b for equality to hold? Consider the case in which $p = q = 2$.

C. Let $f(x)$ and $g(x)$ denote two functions that are integrable on an interval $[\alpha, \beta]$ where $\alpha < \beta$. Use the inequality of Problem B to prove the inequality

$$\left| \int_\alpha^\beta fg\,dx \right| \leqslant \left(\int_\alpha^\beta |f|^p\,dx \right)^{1/p} \left(\int_\alpha^\beta |g|^q\,dx \right)^{1/q}.$$

D. Suppose that $a_1, \ldots, a_n$ and $b_1, \ldots, b_n$ are two sequences of real numbers. Prove the inequality

$$\sum_{k=1}^n a_k b_k \leqslant \left(\sum_{k=1}^n a_k^p \right)^{1/p} \left(\sum_{k=1}^n b_k^q \right)^{1/q}.$$

E. Prove that the functional

$$\left(\int_\alpha^\beta |f|^p\,dx \right)^{1/p}$$

is a norm for the function f in the space of functions that are integrable over the interval $[\alpha, \beta]$.

(N.B.: The first of these problems is independent of the three following t.)

Solutions. **A.** Consider the complex function $f(x) = P(x) + iQ(x)$ of the real variable x. The functions $P(x)$ and $Q(x)$ are assumed to be integrable over the interval $[\alpha, \beta]$. This ensures the existence of the two integrals

$$\int_\alpha^\beta f\,dx \qquad \text{and} \qquad \int_\alpha^\beta |f|\,dx = \int_\alpha^\beta (P^2 + Q^2)^{1/2}\,dx.$$

The inequality

$$\left| \int_\alpha^\beta f(x)\,dx \right| \leqslant \int_\alpha^\beta |f(x)|\,dx \qquad (\alpha < \beta) \tag{1}$$

follows from passage to the limit in the analogous inequality for Riemann sums,

$$\left| \sum (x_j - x_{j-1}) f(\xi_i) \right| \leqslant \sum (x_j - x_{j-1}) \,|f(\xi_j)|.$$

Let us recall that the x_j are the points of partition of the interval $[\alpha, \beta]$ and that ξ_j is a point in the interval $[x_{j-1}, x_j]$.

To prove Schwarz' inequality, we apply inequality (1) to the function $f^2 = P^2 - Q^2 + 2iPQ$, the absolute value of which is $P^2 + Q^2$. We then have

$$\left| \int_\alpha^\beta (P^2 - Q^2 + 2iPQ)\,dx \right| \leqslant \int_\alpha^\beta (P^2 + Q^2)\,dx.$$

After taking the square of both sides and performing some elementary manipulations, we obtain

$$\left[\int_\alpha^\beta (P^2 - Q^2)\,dx\right]^2 + 4\left[\int_\alpha^\beta PQ\,dx\right]^2 \leqslant \left[\int_\alpha^\beta (P^2 + Q^2)\,dx\right]^2,$$

from which we get

$$\begin{aligned}\left[\int_\alpha^\beta PQ\,dx\right]^2 &\leqslant \tfrac{1}{4}\left[\int_\alpha^\beta (P^2 + Q^2)\,dx\right]^2 - \tfrac{1}{4}\left[\int_\alpha^\beta (P^2 - Q^2)\,dx\right]^2 \\ &\leqslant \int_\alpha^\beta P^2\,dx \int_\alpha^\beta Q^2\,dx.\end{aligned}$$

This is Schwarz' inequality.

B. Throughout the remainder of this exercise, p and q denote numbers exceeding 1 such that

$$\frac{1}{p} + \frac{1}{q} = 1 \qquad \text{or} \qquad q = \frac{p}{p-1}.$$

If a and b are positive numbers, they satisfy the inequality

$$ab \leqslant \frac{a^p}{p} + \frac{b^q}{q}.$$

To see this, let us set

$$\varphi(b) = \frac{b^q}{q} + \frac{a^p}{p} - ab.$$

Let us observe the behavior of this function $\varphi(b)$. It is positive for $b = 0$ and it tends to $+\infty$ as $b \to \infty$. On the other hand,

$$\varphi'(b) = b^{q-1} - a,$$

and $\varphi'(b)$ vanishes at $b = a^{1/(q-1)}$, that is, at $b^q = a^{q/(q-1)} = a^p$, or

$$ab = a^{1+(1/(q-1))} = a^{q/(q-1)} = a^p.$$

At this value,

$$\varphi(b) = \frac{a^p}{p} + \frac{a^p}{q} - a^p = 0.$$

Thus $\varphi(b)$ has a zero minimum, which it attains when $a^p = b^q$. From this it follows that

$$ab \leqslant \frac{a^p}{p} + \frac{b^q}{q}.$$

Remark 1. If $p = q = 2$, this inequality becomes the obvious inequality

$$ab < \frac{a^2 + b^2}{2}.$$

Remark 2. The inequality $ab \leqslant a^p/p + b^q/q$ lends itself to a simple geometric interpretation. Let $f(x)$ be a continuous monotonic function on the interval $[0, b]$. Let g denote the inverse function of f; that is, if $y = f(x)$, then $x = g(y)$. Let a denote a positive number.

To compare the areas, we note that

$$ab \leqslant \int_0^b f(x)\,dx + \int_0^a g(x)\,dx,$$

equality holding if $a = f(b)$.

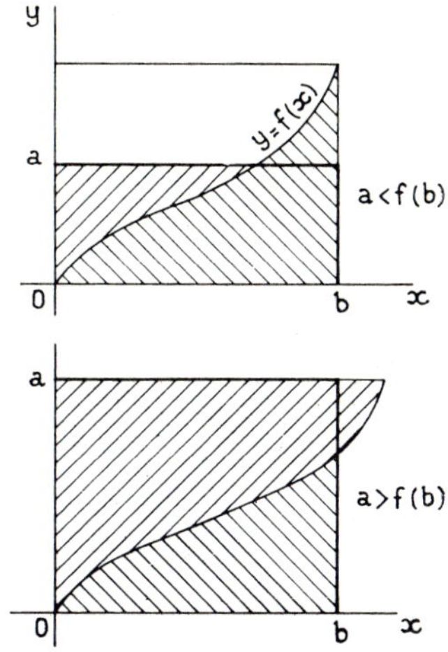

FIGURE 3

Let us take in particular

$$f(x) = x^{q-1} \qquad (q > 1).$$

Then

$$g(x) = x^{p-1},$$

where

$$\frac{1}{p} + \frac{1}{q} = 1.$$

Then

$$ab \leqslant \int_0^b x^{q-1}\,dx + \int_0^a x^{p-1}\,dx = \frac{b^q}{q} + \frac{a^p}{p}.$$

C. Suppose that $f(x)$ and $g(x)$ are two positive integrable functions defined on the interval $[\alpha, \beta]$. Let us set

$$a = f(x) \Big/ \left(\int_\alpha^\beta f^p\,dx \right)^{1/p}, \qquad b = g(x) \Big/ \left(\int_\alpha^\beta g^q\,dx \right)^{1/q}.$$

The inequality of Problem B becomes

$$fg\Big/\Big(\int_\alpha^\beta f^p\,dx\Big)^{1/p}\Big(\int_\alpha^\beta g^q\,dx\Big)^{1/q} \leqslant \Big(f^p\Big/p\int_\alpha^\beta f^p\,dx\Big) + \Big(g^q\Big/q\int_\alpha^\beta g^q\,dx\Big).$$

Let us integrate this inequality from α to β. Since $1/p + 1/q = 1$, the right-hand member is equal to 1. Thus we have the inequality

$$\int_\alpha^\beta fg\,dx \leqslant \Big(\int_\alpha^\beta f^p\,dx\Big)^{1/p}\Big(\int_\alpha^\beta g^q\,dx\Big)^{1/q}.$$

This is Hölder's inequality. A modified form of it,

$$\Big|\int_\alpha^\beta fg\,dx\Big| \leqslant \Big(\int_\alpha^\beta |f|^p\,dx\Big)^{1/p}\Big(\int_\alpha^\beta |g|^q\,dx\Big)^{1/q},$$

can be used for functions of varying sign. It becomes Schwarz' inequality when $p = q = 2$.

Equality holds in the above inequality if, in the notation of the preceding problem,

$$a^p = b^q\,.$$

This means that the ratio $|f(x)|^p/|q(x)|^q$ has a constant value independent of x except possibly for isolated values of x that do not change the value of the integrals.

D. Let

$$a_1\,, \ldots\,, a_n \qquad \text{and} \qquad b_1\,, \ldots\,, b_n$$

be two sequences of positive numbers. We define

$$a = \frac{a_k}{(a_1{}^p + \cdots + a_n{}^p)^{1/p}}\,, \qquad b = \frac{b_k}{(b_1{}^q + \cdots + b_n{}^q)^{1/q}}\,.$$

The inequality of Problem B becomes

$$\frac{a_k b_k}{(a_1{}^p + \cdots + a_n{}^p)^{1/p}(b_1{}^q + \cdots + b_n{}^q)^{1/q}} \leqslant \frac{a_k{}^p}{p(a_1{}^p + \cdots + a_n{}^p)} + \frac{b_k{}^q}{q(b_1{}^q + \cdots + b_n{}^q)}\,.$$

Let us sum these inequalities for all values of k from 1 to n. The right-hand member is equal to 1. Thus we have the inequality

$$\sum_{k=1}^n a_k b_k \leqslant \Big(\sum_{k=1}^n a_k{}^p\Big)^{1/p}\Big(\sum_{k=1}^n b_k{}^q\Big)^{1/q}.$$

This is the "discrete" form of Hölder's inequality. For $p = q = 2$, it reduces to the so-called *Cauchy-Schwarz inequality*.

E. The expression

$$\|f\| = \left(\int_\alpha^\beta |f(x)|^p \, dx\right)^{1/p}$$

is one possible form of the *norm* of a function f in the vector space of real integrable functions on the interval $[\alpha, \beta]$.

To see this, note first that $\|f\| \geqslant 0$ with equality holding only when $f = 0$ almost everywhere. If m is a real number, we have

$$\|mf\| = |m| \, \|f\|.$$

It remains to prove the triangle inequality

$$\|f + g\| \leqslant \|f\| + \|g\|.$$

This inequality is known as *Minkowski's inequality*. We write

$$\int_\alpha^\beta |f + g|^p \, dx = \int_\alpha^\beta |f + g|^{p-1} |f| \, dx + \int_\alpha^\beta |f + g|^{p-1} |g| \, dx.$$

If we apply Hölder's inequality to each of the terms in the right-hand member, we obtain

$$\int_\alpha^\beta |f + g|^{p-1} |f| \, dx \leqslant \left(\int_\alpha^\beta |f|^p \, dx\right)^{1/p} \left(\int_\alpha^\beta |f + g|^{(p-1)q} \, dx\right)^{1/q},$$
$$\int_\alpha^\beta |f + g|^{p-1} |g| \, dx \leqslant \left(\int_\alpha^\beta |g|^p \, dx\right)^{1/p} \left(\int_\alpha^\beta |f + g|^{(p-1)q} \, dx\right)^{1/q}$$

Now, $(p - 1)\, q = p$, Therefore, the last integral is equal to

$$\left(\int_\alpha^\beta |f + g|^p \, dx\right)^{1/q}$$

Therefore,

$$\int_\alpha^\beta |f + g|^p \, dx \leqslant \left[\left(\int_\alpha^\beta |f|^p \, dx\right)^{1/p} + \left(\int_\alpha^\beta |g|^p \, dx\right)^{1/p}\right]\left(\int_\alpha^\beta |f + g|^p \, dx\right)^{1/p}$$

If we divide this inequality by the last integral on the right, we obtain, remembering that $1 - 1/q = 1/p$,

$$\left(\int_\alpha^\beta |f + g|^p \, dx\right)^{1/p} \leqslant \left(\int_\alpha^\beta |f|^p \, dx\right)^{1/p} + \left(\int_\alpha |g|^p \, dx\right)^{1/p}$$

or

$$\|f+g\| \leqslant \|f\| + \|g\|.$$

This is the desired inequality.

4. Convex functions

The concept of a convex function is almost as important as that of a monotonic function. In the present exercise, we shall present several properties of convex functions. In the case of functions that are twice differentiable, the statement that $f'' \geqslant 0$ or that f' is an increasing function, which is equivalent to the statement that f is convex, establishes a relationship between the terms "convex," "increasing," and "positive." The concept of convexity may also be used to establish Hölder's inequality, which we shall prove in a different manner in Exercise 3. We shall establish and employ this inequality for the case of sequences. By passing to the limit, we shall then show easily that the corresponding inequality holds for functions of a continuous argument.

Problems. A continuous function $f(x)$ is said to be convex in the open interval (a, b) if, for arbitrary x and y in that interval,

$$f\left(\frac{x+y}{2}\right) \leqslant \frac{f(x)+f(y)}{2}. \tag{1}$$

A. Show that if the m numbers $x_1, x_2, \ldots, x_m$ belong to the interval (a, b), then

$$f\left(\frac{\sum x_i}{m}\right) \leqslant \frac{1}{m} \sum f(x_i). \tag{2}$$

B. Let $\lambda_1, \lambda_2, \ldots, \lambda_m$ denote m nonnegative constants whose sum is 1. Show that

$$f\left(\sum \lambda_i x_i\right) \leqslant \sum \lambda_i f(x_i). \tag{3}$$

Give a geometric interpretation.

C. If $f(x)$ has a continuous second derivative, a necessary and sufficient condition for $f(x)$ to be convex is that $f'' \geqslant 0$.

D. Show that the function $f(x) = x^p$ (where $p > 1$) is convex for $x \geqslant 0$. By applying (3) to this function, prove Hölder's inequality

$$\sum a_i b_i \leqslant \left[\sum (a_i)^p\right]^{1/p} \left[\sum (b_i)^q\right]^{1/q}, \tag{4}$$

where p and q are numbers greater than 1 satisfying the equation

$$\frac{1}{p} + \frac{1}{q} = 1$$

and the numbers $a_1, a_2, \ldots, a_n, b_1, b_2, \ldots, b_n$ are positive.

E. Under the same conditions, prove Minkowski's inequality,

$$\left[\sum (a_i + b_i)^p\right]^{1/p} \leqslant \left[\sum (a_i)^p\right]^{1/p} + \left[\sum (b_i)^p\right]^{1/p}. \tag{5}$$

Hint for Problem C. Define a function

$$u(x) = f(x) - 2f\left(\frac{x+y}{2}\right) + f(y).$$

For the converse, set $\lambda(z - y)/(z - x)$ and $\mu = (y - x)/(z - x)$ and then majorize $f(\lambda x + \mu z)$.

Solutions. **A.** Let $f(x)$ denote a continuous function. It is said to be *convex* on an interval if, for every x and y belonging to that interval,

$$f\left(\frac{x+y}{2}\right) \leqslant \frac{f(x) + f(y)}{2}. \tag{1}$$

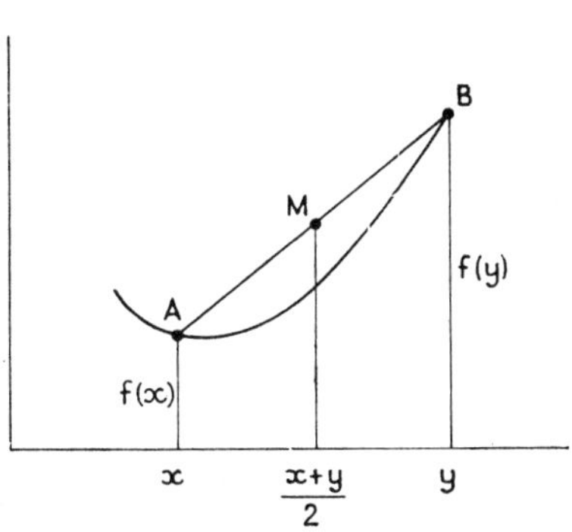

FIGURE 4

Let A and B be points with abscissas x and y on the curve representing f.

Inequality (1) states that the midpoint M of the segment AB lies above the curve.

If the numbers $x_1, x_2, \ldots, x_m, \ldots$ belong to the interval (a, b) then inequality (1) implies the inequality

$$f\left(\frac{x_1 + \cdots + x_4}{4}\right) \leqslant \frac{1}{2}\left[f\left(\frac{x_1 + x_2}{2}\right) + f\left(\frac{x_3 + x_4}{2}\right)\right]$$
$$\leqslant \frac{1}{4}[f(x_1) + \cdots + f(x_4)].$$

More generally, we obtain by induction

$$f\left(\frac{1}{2^N}\sum_{i=1}^{2^N} x_i\right) \leqslant \frac{1}{2^N}\sum_{i=1}^{2^N} f(x_i).$$

If in this inequality we set

$$x_{m+1} = x_{m+2} = \cdots = x_{2^N} = \frac{1}{m}(x_1 + x_2 + \cdots + x_m),$$

we see that

$$f\left(\frac{1}{m}\sum_{i=1}^{m} x_i\right) \leqslant \frac{1}{m}\sum_{i=1}^{m} f(x_i). \tag{2}$$

B. Let $p_1, p_2, \ldots, p_n$ denote positive integers such that $\sum p_k = m$. In inequality (2), we choose the numbers x_i in such a way that

$$x_1 = x_2 = \cdots = x_{p_1} = y_1$$
$$x_{p_1+1} = \cdots = x_{p_1+p_2} = y_2$$
$$\cdots\cdots\cdots\cdots\cdots\cdots\cdots\cdots\cdots$$
$$x_{p_1+\cdots+p_{n-1}+1} = \cdots = x_{p_1+p_2+\cdots+p_n} = y_n .$$

Then inequality (2) can be written in the form

$$f\left[\frac{p_1}{m}y_1 + \frac{p_2}{m}y_2 + \cdots + \frac{p_n}{m}y_n\right] \leqslant \frac{p_1}{m}f(y_1) + \cdots + \frac{p_n}{m}f(y_n).$$

Now, let $r_1, r_2, \ldots, r_n$ denote n positive rational numbers such that

$$\sum r_i = 1.$$

It is always possible to choose integers $m, p_1, p_2, \ldots, p_n$ in such a way that

$$r_i = \frac{p_i}{m} \qquad (i = 1, 2, 3, \ldots, n).$$

Thus if the rational numbers r_i satisfy the two conditions

$$r_i > 0, \qquad \sum r_i = 1,$$

we have the inequality

$$f\left(\sum r_i y_i\right) \leqslant \sum r_i f(y_i).$$

Now using the continuity of the function f, we see that the above inequality can be extended to nonnegative real numbers.

This proves the inequality

$$f\left(\sum \lambda_i y_i\right) \leqslant \sum \lambda_i f(y_i), \tag{3}$$

where

$$\lambda_i \geqslant 0, \qquad \sum \lambda_i = 1.$$

This inequality means that the center of mass of n points possessing masses $\lambda_1, \ldots, \lambda_n$ lies *above* the curve.

C. Let us show that, if the function f has a continuous nonnegative second derivative on an open interval (a, b), then f is a convex function on that interval. Suppose that $f'' \geqslant 0$. Let us define

$$u(x) = f(x) - 2f\left(\frac{x+y}{2}\right) + f(y),$$

where y plays the role of a parameter. We see that $u(y) = 0$. Furthermore,

$$u'(x) = f'(x) - f'\left(\frac{x+y}{2}\right).$$

Since $f'' \geqslant 0$, f' is a strictly increasing function.

Therefore, if $y \leqslant x$, so that $(x + y)/2 \leqslant x$, we have

$$u'(x) \geqslant 0,$$

so that $u(x)$ is an increasing function that vanishes for $x = y$. It follows that

$$f(x) - 2f\left(\frac{x+y}{2}\right) + f(y) \geqslant 0$$

for $a < y < x < b$.

But the above expression is symmetric in x and y so that, for every pair (x, y) in the interval (a, b),

$$f\left(\frac{x+y}{2}\right) \leqslant \tfrac{1}{2}[f(x) + f(y)].$$

Therefore, $f(x)$ is convex.

Conversely, suppose that f is convex. Consider three numbers x, y, z such that

$$x < y < z.$$

Let us set

$$\lambda = \frac{z - y}{z - x}, \qquad \mu = \frac{y - x}{z - x},$$

where λ and μ are two positive numbers such that $\lambda + \mu = 1$. Then

$$f(\lambda x + \mu z) \leqslant \lambda f(x) + \mu f(z).$$

Now, $\lambda x + \mu z = y$. Therefore,

$$f(y) \leqslant \frac{z - y}{z - x} f(x) + \frac{y - x}{z - x} f(z)$$

and hence

$$\frac{f(y) - f(y)}{y - x} \leqslant \frac{f(z) - f(y)}{z - y}.$$

This inequality states that, for three consecutive points A, B, and C on the curve representing f, the slope of the chord BC is greater than the slope of the chord AB.

Consider a fourth point D lying to the right of C on the curve. As D approaches C, the slope of the chord CD tends by hypothesis to a limit equal to $f'(z)$. If α is the slope of the chord BC, we have

$$\alpha \leqslant f'(z).$$

Similarly, as A tends to B, we have in the limit

$$f'(y) \leqslant \alpha.$$

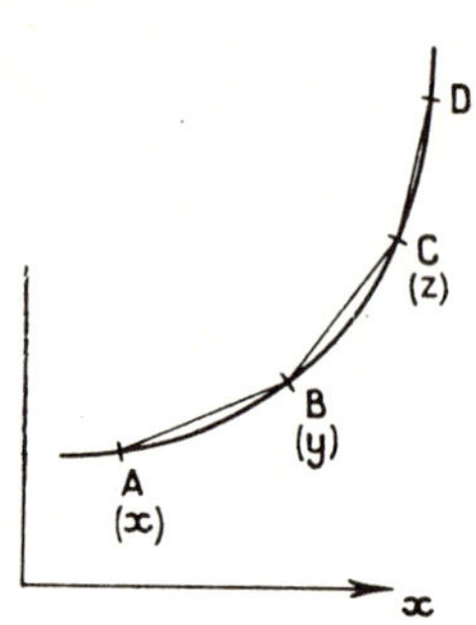

FIGURE 5

From this it follows that $f'(y) \leqslant f'(z)$ whenever $y \leqslant z$; that is, f' is an increasing function. Therefore, $f''(x) \geqslant 0$.

D. The function $f(x) = x^p$ (where $p > 1$) is an example of a convex function for $x > 0$, since $f''(x) = p(p-1)\, x^{p-2} \geqslant 0$. Therefore, inequality (3) implies the inequality

$$\left(\sum \lambda_i x_i\right)^p \leqslant \sum \lambda_i (x_i)^p \qquad \left(\lambda_i \geqslant 0,\ \sum \lambda_i = 1\right).$$

Let us denote by $\mu_1, \mu_2, \ldots, \mu_i \ldots$ a finite sequence of nonnegative numbers. Let us replace λ_i in the preceding inequality with $u_i / \Sigma \mu_i$. We then have

$$\left(\sum \mu_i x_i\right)^p \leqslant \left(\sum \mu_i x_i{}^p\right)\left(\sum \mu_i\right)^{p-1}$$

Let us make a change of notation by setting

$$q = \frac{p}{p-1}, \qquad \mu_i = b_i{}^q, \qquad x_i = a_i b_i{}^{1-q}.$$

We then obtain Hölder's inequality,

$$\sum a_i b_i \leqslant \left[\sum a_i{}^p\right]^{1/p} \left[\sum b_i{}^q\right]^{1/q}, \qquad \frac{1}{p} + \frac{1}{q} = 1, \qquad p > 1. \tag{4}$$

This is the sequence form of the inequality. In Exercise 3 we proved the analogous form of the inequality for functions of a continuous variable.

E. Finally, let us prove Minkowski's inequality. First we write

$$\sum (a_i + b_i)^p = \sum (a_i + b_i)^{p-1} a_i + \sum (a_i + b_i)^{p-1} b_i.$$

If we apply Hölder's inequality to each of the terms in the second member, we obtain

$$\sum (a_i + b_i)^p \leqslant \left[\sum (a_i + b_i)^{q(p-1)}\right]^{1/q} \left[\sum a_i{}^p\right]^{1/p} + \left[\sum (a_i + b_i)^{q(p-1)}\right]^{1/q} \left[\sum b_i{}^p\right]^{1/p}.$$

Now,

$$q(p-1) = p.$$

Therefore,

$$\sum (a_i + b_i)^p \leqslant \left[\sum (a_i + b_i)^p\right]^{1/q} \left[\left(\sum a_i{}^p\right)^{1/p} + \left(\sum b_i{}^p\right)^{1/p}\right],$$

from which we derive the inequality

$$\left[\sum (a_i + b_i)^p\right]^{1/p} \leqslant \left[\sum (a_i)^p\right]^{1/p} + \left[\sum (b_i)^p\right]^{1/p} \tag{5}$$

An application. Let u denote a finite sequence of m real terms $(u_1, u_2, \ldots, u_m)$. The set of all such m-tuples u defines a vector space over the field of real numbers. The sum of two sequences u and u' is the sequence whose ith term is $u_i + u_i'$. The product of a sequence u and a real number λ is the sequence whose ith term is λu_i.

This vector space can be normed by taking

$$\| u \| = \left[\sum_{i=1}^{m} | u_i |^p\right]^{1/p}$$

The norm $\| u \|$ satisfies the defining properties of a norm:

(1) $\| \lambda u \| = | \lambda | \, \| u \|$.

(2) $\| u \| \geqslant 0$.

(3) $\| u \| = 0$ and only if all the u_i are 0.

(4) According to Minkowski's inequality, $\| u \|$ satisfies the triangle inequality.

In the case in which $p = 2$, $\| u \|$ is the ordinary length of the vector whose components in an orthonormal basis are $u_1, \ldots, u_m$.

5. Two applications of the second mean-value theorem

The second mean-value theorem (Bonnet's theorem) is as follows: *Let $f(x)$ and $g(x)$ denote two integrable functions defined on an interval $[\alpha, \beta]$. Suppose that $f(x)$ is a nonincreasing function.*

(a) *If $f(x)$ is nonnegative there exists a number ξ lying between α and β such that*

$$\int_\alpha^\beta f(x)g(x)\,dx = f(\alpha) \int_\alpha^\xi g(x)\,dx.$$

(b) *If no assumptions are made on the sign of $f(x)$, there exists a number ξ' between α and β such that*

$$\int_\alpha^\beta f(x)g(x)\,dx = f(\alpha) \int_\alpha^{\xi'} g(x)\,dx + f(\beta) \int_{\xi'}^\beta g(x)\,dx.$$

Problems. **A.** Let $f(x)$ be a positive nonincreasing integrable function defined on the interval $[0, 1]$. *Set* $f(0) = a$ and $f(1) = b$. Let c denote a number between 0 and b. Show that there exists a number ξ between 0 and 1 depending on c such that

$$\int_0^1 f(x)\,dx = a\xi + c(1 - \xi).$$

B. Show that, for $m < 2$,

$$\lim_{x\to 0} \frac{1}{x^m} \int_0^x \sin\frac{1}{t}\,dt = 0.$$

Solutions. **A.** We set

$$a = f(0), \qquad b = f(1).$$

Let $F(x)$ denote the function that is equal to $f(x)$ for $0 \leqslant x < 1$ and equal to c at $x = 1$. Since

$$0 \leqslant c \leqslant b,$$

$F(x)$ is positive and nonincreasing. The functions F and f coincide everywhere except at $x = 1$, at which point

$$f(x) = b, \qquad F(x) = c.$$

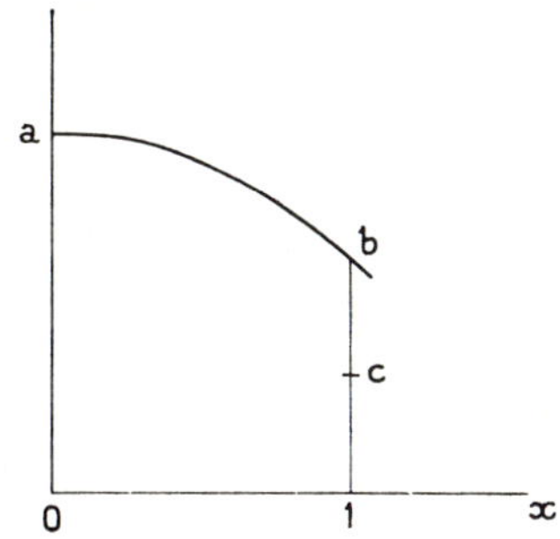

FIGURE 6

From the second mean-value theorem [form (b)], we have

$$\int_0^1 F(x)\,dx = F(0)\int_0^\xi dx + F(1)\int_\xi^1 dx,$$

so that

$$\int_0^1 f(x)\,dx = a\xi + c(1-\xi),$$

where ξ is a number between 0 and 1 depending on c.

If we choose $c = 0$, we obtain form (a) of the second mean-value theorem for $f(x)$ [with $g(x)$ taken equal to 1]. On the other hand, if $c \neq 0$, we obtain an expression of the integral of $f(x)$ in accordance with form (b) but with ξ depending on the arbitrary constant c.

It is easy to verify directly that ξ remains between 0 and 1 as c varies from 0 to b. Specifically,

$$\xi = \int_0^1 f(x)\,dx - c/a - c.$$

From the first mean-value theorem, the integral $\int_0^1 f(x)\,dx$ lies between the greatest lower and least upper bounds of $f(x)$ on [0, 1], that is, between b and a:

$$b < \int_0^1 f(x)\,dx < a.$$

Now, ξ is a homographic decreasing function of c. For $c = 0$,

$$\xi = \frac{1}{a}\int_0^1 f(x)\,dx.$$

Therefore, as c varies from 0 to b, ξ decreases and is always less than

$$\frac{1}{a}\int_0^1 f(x)\,dx,$$

which is less than 1. The minimum value of ξ is

$$\int_0^1 f(x)\,dx - \frac{b}{a-b},$$

which is positive. Finally, we see that

$$0 < \int_0^1 f(x)\,dx - \frac{b}{a-b} \leqslant \xi \leqslant \frac{1}{a}\int_0^1 f(x)\,dx < 1.$$

B. In the integral we make the change of variable $1/t = u$:

$$\frac{1}{x^m}\int_0^x \sin\frac{1}{t}\,dt = \frac{1}{x^m}\int_{1/x}^{+\infty} \sin u\,\frac{du}{u^2}. \tag{1}$$

Since the function $1/u^2$ is a decreasing function, the second mean-value theorem enables us to write

$$\int_{1/x}^{A} \sin u \frac{du}{u^2} = x^2 \int_{1/x}^{\xi} \sin u \, du,$$

where ξ is a number between $1/x$ and A. When we perform the integration, the right-hand member becomes

$$x^2 \left(-\cos \xi + \cos \frac{1}{x} \right).$$

Therefore,

$$\frac{1}{x^m} \left| \int_{1/x}^{A} \sin u \frac{du}{u^2} \right| \leqslant 2 \cdot x^{2-m}.$$

For every x, this inequality holds whenever $A > 1/x$. Therefore,

$$\frac{1}{x^m} \left| \int_{1/x}^{+\infty} \sin u \frac{du}{u^2} \right| \leqslant 2 \cdot x^{2-m}.$$

Therefore, (1) and the inequality $m < 2$ show that

$$\lim_{x \to +0} \frac{1}{x^m} \int_0^x \sin \frac{1}{t} \, dt = 0.$$

Remark. It is not possible to apply the second mean-value theorem directly to the improper integral

$$\int_{1/x}^{\infty} \sin u \frac{du}{u^2}.$$

An improved method. The above procedure does not enable us to draw our conclusion if $m = 2$. The following, more powerful, method brings into play the oscillatory nature of the function

$$f(x) = \frac{1}{x^m} \int_0^x \sin \frac{1}{t} \, dt.$$

If we integrate the expression $\int_{1/x}^{+\infty} \sin u(du/u^2)$ by parts, we see that

$$f(x) = \frac{1}{x^m} \left(x^2 \cos \frac{1}{x} - 2 \int_{1/x}^{+\infty} \cos u \frac{du}{u^3} \right).$$

Application of the second mean-value theorem to the integral on the right leads, just as above, to the inequality

$$\left| \int_{1/x}^{+\infty} \cos u \frac{du}{u^3} \right| < 2x^3.$$

Therefore, if we assume that $m < 3$, we have

$$\lim_{x \to +0} \left[f(x) - x^{2-m} \cos \frac{1}{x} \right] = 0.$$

We conclude that $f(x)$ tends to 0 if $2 - m > 0$ and that $f(x)$ has no limit if $m \geqslant 2$.

6. An example of a linear integral equation

This exercise is a very simple example of the solution, accompanied by a discussion of a linear integral equation of the Fredholm type. The general Fredholm equation is of the form

$$y(x) - \lambda \int_a^b k(x, s) y(s) \, ds = g(x),$$

where $k(x, s)$ is a given "kernel," $g(x)$ is a given "right-hand member," and λ is a parameter. If $g(x) = 0$, the equation is homogeneous. It always has the trivial solution $y = 0$, and it may, for certain values of λ, have nonzero solutions.

The theory and the results depend on the class of functions to which k and g belong and on the function space in which we are seeking solutions. In the present case, we consider only continuous functions, and the solution amounts to a study of a first-degree equation in λ.

Later, we shall see more complicated examples of Fredholm integral equations.

Problems. This is an exercise in finding continuous functions $y(x)$ that satisfy the linear integral equation

$$\lambda y(x) = \int_0^1 xsy(s) \, ds + g(x),$$

where $g(x)$ is a given continuous function and λ is a real parameter.

A. Let us suppose first that $g(x) = 0$. Show that, for $\lambda \neq 0$, the only possible solutions are of the form $y = mx$ and that m must be zero if $\lambda \neq \frac{1}{3}$. What solutions exist when $\lambda = \frac{1}{3}$? (The number $\frac{1}{3}$ is called an eigenvalue of the equation.)

B. Suppose now that $g(x) \neq 0$ and that $\lambda \neq 0$. Prove the following:

(a) If $\lambda \neq \frac{1}{3}$, the equation has one and only one solution.

(b) If $\lambda \neq \frac{1}{3}$, the equation has no solution unless $\int_0^1 xg(x) \, dx = 0$.

C. Treat the different cases when

$$g(x) = x + \alpha.$$

D. Examine the particular case in which $\lambda = 0$.

Solutions. **A.** Consider the set of functions $y(x)$ that are continuous on the closed interval $[0, 1]$. This set constitutes a vector space of functions. On this space we define an operator E by

$$E[y](x) = \int_0^1 xsy(s)\, ds. \tag{1}$$

This operator is linear, since

$$E[y_1 + y_2] = E[y_1] + E[y_2] \qquad \text{and} \qquad E[\alpha y] = \alpha E[y],$$

where α is any real number.

We wish to solve the linear integral equation

$$E[y] = \lambda y,$$

that is, the equation

$$\lambda y(x) = x \int_0^1 sy(s)\, ds. \tag{2}$$

If $\lambda \neq 0$, we see that y must necessarily be of the form $y = mx$, where m is a constant independent of x although it depends on the form of the function y.

Substituting this value in (2), we obtain

$$\lambda mx = x \int_0^1 ms^2\, ds,$$

so that

$$(\lambda - \tfrac{1}{3})m = 0.$$

If $\lambda \neq 0$ and $\lambda \neq \frac{1}{3}$, (2) has only the solution

$$y = 0.$$

If $\lambda = \frac{1}{3}$, the equation has an infinite number of solutions of the form $y = mx$, where m is an arbitrary constant.

B. We now consider the nonhomogeneous equation

$$\lambda y - E[y] = g, \tag{3}$$

where $g(x)$ is a given function. First, let us suppose that $\lambda \neq 0$.

If we multiply both members of (3) by x and then integrate from 0 to 1, we obtain

$$\lambda \int_0^1 xy(x)\, dx - \int_0^1 \int_0^1 x^2 sy(s)\, dx\, ds = \int_0^1 xg(x)\, dx$$

or

$$\lambda \int_0^1 xy(x)\, dx - \tfrac{1}{3} \int_0^1 sy(s)\, ds = \int_0^1 xg(x)\, dx,$$

so that

$$(\lambda - \tfrac{1}{3}) \int_0^1 sy(s)\, ds = \int_0^1 xg(x)\, dx. \tag{4}$$

Thus

(a) if $\lambda \neq \frac{1}{3}$, we have

$$\int_0^1 sy(s)\, ds = \frac{1}{\lambda - \frac{1}{3}} \int_0^1 xg(x)\, dx$$

and hence, by virtue of (3),

$$y(x) = \frac{1}{\lambda} g(x) + \frac{x}{\lambda(\lambda - \frac{1}{3})} \int_0^1 sg(s)\, ds. \tag{5}$$

The integral equation has one and only one solution, which is given explicitly by (5).

(b) If $\lambda = \frac{1}{3}$, the equation is impossible, except when

$$\int_0^1 xg(x)\, dx = 0. \tag{6}$$

In this case, there are an infinite number of solutions depending on the parameter m. These solutions are of the form

$$y(x) = 3g(x) + mx, \tag{7}$$

where m is an arbitrary constant.

C. As an example we choose $g(x) = x + \alpha$. The results of the preceding problem show that

(a) if $\lambda \neq 0$ and $\lambda \neq \frac{1}{3}$, there exists the unique solution

$$y(x) = \frac{3}{2}\frac{2\lambda + \alpha}{\lambda(3\lambda - 1)} x + \frac{\alpha}{\lambda};$$

(b) if $\lambda = \frac{1}{3}$, the equation has no solution unless $\alpha = -\frac{2}{3}$, in which case

$$y(x) = mx - 2.$$

D. Let us return to the original equation, assuming that $\lambda = 0$. The homogeneous equation

$$\int_0^1 xsy(s)\, ds = 0$$

is satisfied by all functions $y(s)$ that are orthogonal to s on the interval $[0, 1]$. These solutions are much more general than linear and homogeneous functions.

For example,

$$y(x) = x - \tfrac{2}{3} \qquad \text{verifies} \qquad E[y] = 0$$

and this function is not of the form $y = mx$.

Similarly, the equation $-E[y] = g$ may be written in the form

$$\int_0^1 sy(s)\, ds + \frac{g(x)}{x} = 0.$$

Thus if $g(x)$ is a linear function Ax, there are infinitely many solutions such that

$$\int_0^1 sy(s)\, ds = -A. \tag{8}$$

If $g(x)$ is not of the form Ax, the above equation has no solution.

It is easy to find functions $y(s)$ satisfying condition (8). Let $y_0(x)$ be any function such that

$$\int_0^1 sy_0(s)\, ds \neq 0.$$

The function $y(x) = Ky_0(x)$ is a solution of this problem if

$$K = -A\Big/\int_0^1 sy_0(s)\, ds.$$

Summary.

$$\lambda y(x) - \int_0^1 xsy(s)\,ds = 0$$

$$\begin{cases} \lambda \neq 0 & \begin{cases} \lambda \neq \frac{1}{3} & y = 0 \quad \text{unique solution} \\ \lambda = \frac{1}{3} & y = mx \quad \text{infinitely many solutions} \end{cases} \\ \lambda = 0 & \text{infinitely many solutions} \end{cases}$$

$$\lambda y(x) - \int_0^1 xsy(s)\,ds = g(x)$$

$$\begin{cases} \lambda \neq 0 & \begin{cases} \lambda \neq \frac{1}{3} & y = \dfrac{1}{\lambda} g(x) + \dfrac{x}{\lambda(\lambda - \frac{1}{3})} \displaystyle\int_0^1 sg(s)\,ds \quad \text{unique solution} \\ \lambda = \frac{1}{3} & \begin{cases} \displaystyle\int_0^1 sg(s)\,ds \neq 0 & \text{no solution} \\ \displaystyle\int_0^1 sg(s)\,ds = 0 & y = 3g(x) + mx \quad \text{infinitely many solutions} \end{cases} \end{cases} \\ \lambda = 0 & \begin{cases} g(x) \neq Ax & \text{no solution} \\ g(x) = Ax & \text{infinitely many solutions} \end{cases} \end{cases}$$

*7. Summation in the sense of Cesaro. Uniform distributions modulo 1

This exercise deals with the simplest of the *summation methods.*

Let $\{x_n\}$ be a sequence of real numbers. Consider an "infinite matrix" (a_{nk}) such that the series

$$u_a = \sum_k a_{nk} x_k$$

converges for all n. The following questions arise: Can it happen that for every convergent sequence $\{x_n\}$, the sequence $\{u_n\}$ tends to the same limit as the sequence $\{x_n\}$ as $n \to \infty$? Can the sequence $\{u_n\}$ have a limit even when the sequence $\{x_n\}$ does not converge?

Ordinary convergence corresponds to the case in which

$$a_{nk} = \delta_{nk}, \qquad u_n = x_n .$$

The Cesaro process consists in choosing

$$a_{n1} = a_{n2} = \cdots = a_{nn} = \frac{1}{n}, \qquad a_{nk} = 0 \quad \text{if } k > n,$$

so that

$$u_n = \frac{x_1 + \cdots + x_n}{n} .$$

Let us first apply it to sequences that are equally distributed on [0, 1]. It will consist of characterizing divergent sequences $\{x_n\}$ whose terms lie between 0 and 1 and which have the property that, for every integrable function $f(x)$, the sequence $\{f(x_n)\}$ converges in the Cesaro sense to the mean value $\int_0^1 f(x)\,dx$.

The application of this result to the Fourier series of continuous functions constitutes *Fejér's theorem.*

Problems. A. Show that if a sequence $\{x_n\}$ converges, the sequence $\{u_n\}$ defined by

$$u_n = \frac{x_1 + \cdots + x_n}{n}$$

converges also and to the same limit. Can it happen that the sequence $\{u_n\}$ converges when the sequence $\{x_n\}$ diverges?

B. Let $\{v_n\}$ be a sequence of positive numbers. Show that if v_{n+1}/v_n tends to a limit, then $\sqrt[n]{v_n}$ tends to the same limit.

C. Let $\{x_n\}$ be a sequence of numbers between 0 and 1. Let L denote an arbitrary interval of length l contained in the open interval (0, 1). Consider the first N terms of the sequence $\{x_n\}$. Let N' denote the number of these terms belonging to the interval L. A sequence $\{x_n\}$ is said to be **equally distributed** on (0, 1) if, for every such L, the ratio N'/N tends to 1 as $N \to \infty$. Show that, for a sequence to be equally distributed on (0, 1), it is necessary and sufficient that, for every integrable function on (0, 1), the sequence $\{f(x_n)\}$ have a limit in the sense of Cesaro equal to $\int_0^1 (f_2 - f_1)\,dx$.

[Hint: Minorize and majorize $f(x)$ by two functions $f_1 < f$ and $f_2 > f$ such that $\int_0^1 (f_2 - f_1)\,dx < \epsilon$.]

Solutions. **A.** By hypothesis, the sequence $\{x_n\}$ has a finite limit x as n increases without bound. We wish to show that the sequence $\{u_n\}$ defined by

$$u_n = \frac{x_1 + x_2 + \cdots + x_n}{n} \tag{1}$$

converges to this same limit x. We write

$$u_n - x = \frac{(x_1 - x) + \cdots + (x_n - x)}{n} \tag{2}$$

and then decompose the right-hand member into two parts:

$$u_n - x = \frac{(x_1 - x) + \cdots + (x_p - x)}{n} + \frac{(x_{p+1} - x) + \cdots + (x_n - x)}{n}. \tag{3}$$

By hypothesis, for any $\epsilon > 0$, there exists a positive integer N_1 such that the inequality $p > N_1$ implies that $| x_p - x | < \epsilon/2$. Let us assign to p in (3) a fixed value satisfying this condition. We then have the following majorization:

$$| u_n - x | < \frac{| (x_1 - x) + \cdots + (x_p - x) |}{n} + \frac{\epsilon}{2} \frac{n - p}{n}$$

$$< \frac{| (x_1 - x) + \cdots + (x_p - x) |}{n} + \frac{\epsilon}{2}. \tag{4}$$

Keeping p fixed, we choose n sufficiently large that

$$\frac{| (x_1 - x) + \cdots + (x_p - x) |}{n} < \frac{\epsilon}{2}.$$

This will be the case if n exceeds

$$N_2 = \frac{2}{\epsilon} [| (x_1 - x) + \cdots + (x_p - x) |] .$$

Then, from inequality (4), the inequality

$$| u_n - x | < \epsilon$$

will be satisfied for all n such that

$$n > \max (N_1 , N_2).$$

This proves that $\lim_{n\to\infty} u_n = x$.

Remark 1. It may happen that the sequence $\{u_n\}$ converges even though the sequence $\{x_n\}$ diverges. We need only look at the example in which $x_n = (-1)^n$ or, more generally, the case in which $x_n = e^{2i\pi\theta n}$, where θ is not an integer.

Note that in this example, $x_1 + \cdots + x_n$ is the sum of the first n terms of a geometric progression:

$$x_1 + \cdots + x_n = e^{2\pi i\theta} + e^{4i\pi\theta} + \cdots + e^{2i\pi\theta n}$$

$$= \frac{e^{2i\pi\theta}(1 - e^{2i\pi n\theta})}{1 - e^{2i\pi\theta}} = \frac{e^{i\pi\theta}(1 - e^{2i\pi n\theta})}{-2i \sin \pi\theta},$$

where θ is not an integer and where $| nu_n |$ is less than $1/| \sin \pi\theta |$, which is independent of n. Therefore, u_n tends to 0.

Remark 2. The terminology of sequences can be adapted for *series*. We need only consider x_n as the sum of the first n terms of a series whose general term is $y_n = x_n - x_{n-1}$. If $(x_1 + \cdots + x_n)/n$ tends to a limit, we say that the series is summable in the sense of Cesaro or that it is *Cesaro-summable*. It may happen that the series is not convergent in the ordinary sense.

Cesaro summability is useful in the theory of Fourier series (Féjer's theorem).

B. *An application.* Let $v_0, v_1, ..., v_n ...$ be a sequence of positive numbers. We set $x_n = \log v_n - \log v_{n-1}$. Then

$$\frac{\nu_n}{\nu_{n-1}} = \exp(x_n),$$

$$\nu_n = \nu_0 \exp(x_1 + x_2 + \cdots + x_n),$$

$$\sqrt[n]{\nu_n} = \sqrt[n]{\nu_0} \exp\left(\frac{x_1 + x_2 + \cdots + x_n}{n}\right).$$

If v_n/v_{n-1} tends to a finite nonzero limit L as $n \to \infty$, then x_n tends to $\log L$. Therefore, $(x_1 + \cdots + x_n)/n$ also tends to $\log L$. Since $\sqrt[n]{\nu_0}$ tends to 1, $\sqrt[n]{\nu_n}$ tends to L.

This reasoning enables us to state the following theorem: *If $\{v_n\}$ is a sequence of positive terms such that the ratio v_{n+1}/v_n tends to a limit L as $n \to \infty$, then $\sqrt[n]{v_n}$ also tends to the limit L as $n \to \infty$.*

Remark. In the above calculations we assumed that $L \neq 0$ and that $L \neq +\infty$. A slight modification enables us to include these two cases in the statement of the preceding theorem.

C. Consider a sequence of numbers $x_1, ..., x_n, ...$ between 0 and 1. Let L be *any* interval of length l contained in (0, 1). Consider the first N points $x_1, ..., x_N$ of the sequence. By definition, N' of these points belong to L (it is immaterial whether L is open or closed). It is assumed that the ratio N'/N tends to a limit exactly equal to l as $N \to \infty$. We then say that the sequence $\{x_n\}$ is uniformly distributed on the interval [0, 1].

If this is the case, the sequence $\{x_n\}$ is *dense* in [0, 1]: Every interval contained in [0, 1] holds at least one of the x_n. In fact, it is "uniformly" dense. From this it follows in particular that it is not convergent.

We shall now show that *for a sequence $\{x_n\}$ to be uniformly distributed, it is necessary and sufficient that the equation*

$$\int_0^1 f(x)\,dx = \lim_{N\to\infty} \frac{f(x_1) + \cdots + f(x_N)}{N} \tag{5}$$

hold for every function $f(x)$ that is integrable on $[0, 1]$. The expression on the left is the mean value of $f(x)$ on $[0, 1]$. The expression on the right is the arithmetic mean of the $f(x_n)$, which is also the *limit of $f(x_n)$ in the sense of Cesaro.*

If (5) holds, we may choose for f the function whose value is 1 on the interval L and 0 elsewhere. Then the integral is equal to the length l of L. Only those $f(x_n)$ such that $f \neq 0$ appear on the right. These are the ones for which $f = 1$. The sum $f(x_1) + \cdots + f(x_N)$ is equal to the number N' of points $x_1, \ldots, x_N$ belonging to L. Equation (5) expresses the fact that $N'/N \to l$, so that the sequence $\{x_n\}$ is evenly distributed.

Conversely, let us suppose that the sequence $\{x_n\}$ is evenly distributed. Then (5) will hold if $f(x)$ is equal to 1 on the interval L but 0 elsewhere. It immediately follows from this that if $f(x)$ is a step function, (5) will hold. Now, since $f(x)$ is integrable, there exist two step functions $f_1(x)$ and $f_2(x)$ such that

$$f_1 \leqslant f \leqslant f_2, \qquad \int_a^b (f_2 - f_1)\, dx \leqslant \epsilon,$$

where ϵ is any positive number. We then have

$$\int_0^1 f_1(x)\, dx \leqslant \int_0^1 f(x)\, dx \leqslant \int_0^1 f_2(x)\, dx, \tag{6}$$

$$\frac{f_1(x_1) + \cdots + f_1(x_N)}{N} \leqslant \frac{f(x_1) + \cdots + f(x_N)}{N} \leqslant \frac{f_2(x_1) + \cdots + f_2(x_N)}{N}, \tag{7}$$

$$\int_a^b f_2(x)\, dx - \int_a^b f_1(x)\, dx \leqslant \epsilon. \tag{8}$$

In inequality (7), the first sum tends to $\int_0^1 f_1(x)\,dx$ and the third to $\int_0^1 f_2(x)\,dx$. We may choose N sufficiently great that these sums are within an ϵ distance of their limits.

Let us set

$$\int_0^1 f_1\, dx = A, \qquad \int_0^1 f_2\, dx = B, \qquad \int_0^1 f\, dx = I, \qquad \frac{f(x_1) + \cdots + f(x_N)}{N} = S.$$

Under these conditions,

$$A - \epsilon \leqslant S \leqslant B + \epsilon,$$

$$A \leqslant I \leqslant B,$$

$$B - A < \epsilon.$$

From this it follows that, for the value N chosen, $|I - S|$ is at most equal to 2ϵ. The sum S does indeed tend to the integral I.

*8. Summation in the sense of Euler

This exercise is devoted to a study of a summation process (see Exercise 7) that is somewhat more complicated than the Cesaro process. Here

$$a_{n0} = \frac{1}{2^n}, \quad a_{n1} = \frac{C_n^{\,1}}{2^n}, \quad \ldots, \quad a_{nn} = \frac{C_n^{\,n}}{2^n}, \quad a_{nk} = 0 \qquad \text{for} \quad k < n.$$

Again, we shall apply the method to a numerical example which will show the significance of the method.

Problems. A. Suppose that a sequence $\{x_n\}$ converges. Show that the sequence $\{u_n\}$ defined by

$$u_n = \frac{1}{2^n}(x_0 + C_n^{\,1}x_1 + \cdots + C_n^{\,n}x_n),$$

where

$$C_n^{\,k} = \frac{n!}{k!(n-k)!},$$

converges to the same limit. Give an example of a sequence $\{u_n\}$ that converges even though the sequence $\{x_n\}$ does not.

B. Consider an infinite sequence $u_0, u_1, \ldots, u_n, \ldots$ of real numbers. Define a linear operation on that sequence by

$$Au_k = u_k + u_{k+1}, \qquad A^p u_k = A^{p-1}u_k + A^{p-1}u_{k+1}.$$

Show that $A^p u_0$ is given by the formula

$$A^p u_0 = (1 + u)^p$$

provided we replace any exponent k as applied to u with a subscript k when the binomial expansion is performed.

C. Suppose that the series

$$u_0 + u_1 + \cdots + u_n + \cdots$$

converges. We set

$$s_n = u_0 + \cdots + u_n, \qquad S_n = \frac{u_0}{2} + \cdots + \frac{A^n u_0}{2^{n+1}}.$$

Show by induction that

$$S_n = \frac{1}{2^{n+1}}(C_{n+1}s_0 + \cdots + C_{n+1}^{n+1}s_n).$$

From this, show that the series whose general term is $A^n u_0/2^{n+1}$ converges to the same value as does the series whose general term is u_n.

D. *Example.* Suppose that $u_n = (-1)^n/(n+1)$. The sum of the series $\{u_n\}$ is then equal to log 2. Find the general term of the series $A^n u_0/2^{n+1}$. If we use the first series to calculate log 2, how many terms must we evaluate to ensure an error less than 10^{-3}? How many if we use the second series?

Solutions. **A.** The hypothesis is that the sequence $\{x_n\}$ tends to a limit x as $n \to +\infty$. We wish to show that the sequence $\{u_n\}$ defined by

$$u_n = \frac{1}{2^n}(x_0 + C_n{}^1 x_1 + C_n{}^2 x_2 + \cdots + C_n{}^n x_n) \tag{1}$$

converges to the same limit x. The reasoning is analogous to that used in connection with the Cesaro mean (see Exercise 7). The equation

$$2^n = 1 + C^1{}_n + C^2{}_n + \cdots + C_n{}^n$$

enables us to write

$$u_n - x = \frac{1}{2^n}[(x_0 - x) + C_n{}^1(x_1 - x) + \cdots + C_n{}^p(x_p - x)$$

$$+ \cdots + C_n(x_n - x)]. \tag{2}$$

By hypothesis, there exists an integer p such that, for $n > p$,

$$| x_n - x | < \frac{\epsilon}{2}.$$

We now set

$$S_1(n) = \frac{1}{2^n}[(x_0 - x) + C_n(x_1 - x) + \cdots + C_n{}^p(x_p - x)],$$

$$S_2(n) = \frac{1}{2^n}[C_n{}^{p+1}(x_{p+1} - x) + \cdots + C_n{}^n(x_n - x)],$$

so that

$$u_n - x = S_1(n) + S_2(n). \tag{3}$$

We see that

$$| S_2(n) | \leqslant \frac{1}{2^n}\frac{\epsilon}{2}[C_n^{p+1} + C_n^{p+2} + \cdots + C_n{}^n]$$

$$< \frac{1}{2^n}\frac{\epsilon}{2}[1 + C_n{}^1 + \cdots + C_n^{p+1} + \cdots + C_n{}^n] = \frac{\epsilon}{2}. \tag{4}$$

On the other hand, if we let A denote the maximum of $| x_j - x |$ as j assumes the values 0, 1, 2, ..., p, we have

$$| S_1(n) | \leqslant \frac{1}{2^n} A[1 + C_n^1 + C_n^2 + \cdots + C_n^p].$$

Now,

$$C_n^p = \frac{n(n-1)(n-2)\cdots(n-p+1)}{p!} \leqslant \frac{n^p}{p!},$$

so that we may write

$$| S_1(n) | \leqslant \frac{A}{2^n}\left[1 + n + \frac{n^2}{2!} + \cdots + \frac{n^p}{p!}\right],$$

where p is fixed and the bracketed expression in the right-hand member of the inequality is polynomial $P(n)$ in the variable n.

As n increases without bound,

$$\frac{P(n)}{2^n} \to 0.$$

Therefore, there exists a number N such that, for $n > N$,

$$| S_1(n) | < \frac{\epsilon}{2}. \tag{5}$$

By adding inequalities (4) and (5), we see that the inequality

$$n > \max(p, N)$$

implies the inequality $| u_n - x | < \epsilon$.

Remark. There exist divergent sequences $\{x_n\}$ such that the sequence $\{u_n\}$ defined by

$$u_n = \frac{1}{2^n}[x_0 + C_n^1 x_1 + \cdots + C_n^n x_n]$$

converges. For example, if θ is not an integer, the sequence $\{x_n\}$ defined by $x_n = e^{2i\pi n\theta}$ diverges. Now

$$u_n = \frac{1}{2^n}[1 + C_n^1 e^{2i\pi\theta} + \cdots + C_n^n e^{2i\pi n\theta}] = \left(\frac{1 + e^{2\pi i\theta}}{2}\right)^n,$$

$$| u_n | = | \cos \pi\theta |^n.$$

Since $| \cos \pi\theta |$ is assumed different from 1, u_n tends to 0 as $n \to +\infty$.

B. Consider the sequence $u_0, u_1, ..., u_k, ...$. Let us map this sequence into the sequence whose general term is

$$u_k + u_{k+1}.$$

This mapping is a linear operation on the set of numerical sequences. We need only define a vector space by the following laws:

(1) The sequence $\{u_k + v_k\}$ is the sum of the sequences $\{u_k\}$ and $\{v_k\}$.

(2) The sequence $\{mu_k\}$ is the product of the sequence $\{u_k\}$ and the real number m.

We set

$$Au_k = u_k + u_{k+1}, \qquad A^p u_k = A^{p-1}u_k + A^{p-1}u_{k+1}.$$

Let us prove the formula

$$A^p u_0 = (1+u)^p = u_0 + C_p{}^1 u_1 + C_p{}^2 u_2 + \cdots + u_p, \tag{6}$$

where $(1+u)^p$ is defined as the expression on the right, which follows from the (symbolic) binomial expansion of $1+u$. We do this by induction, noting that the binomial coefficients verify the relation

$$C_p^q + C_p^{q+1} = C_{p+1}^{q+1}. \tag{7}$$

C. By hypothesis, the series $\sum_{n=0}^{\infty} u_n$ converges. Let us show that the series

$$\sum_{n=0}^{\infty} \frac{A^n u_0}{2^{n+1}}$$

converges to the same sum. To do this, we set

$$s_n = u_0 + u_1 + \cdots + u_n, \qquad S_n = \frac{u_0}{2} + \cdots + \frac{A^n u_0}{2^{n+1}},$$

which represents the partial sums of the two series. We shall first show that

$$S_n = \frac{1}{2^{n+1}} [C_{n+1}^1 s_0 + \cdots + C_{n+1}^{n+1} s_n]. \tag{8}$$

This formula holds for $n = 0$. Let us suppose it true for some unspecified n. Let us calculate S_{n+1}:

$$\begin{aligned} S_{n+1} &= \frac{u_0}{2} + \cdots + \frac{A^{n+1}u_0}{2^{n+2}} \\ &= S_n + \frac{A^{n+1}u_0}{2^{n+2}} \end{aligned}$$

$$= \frac{1}{2^{n+1}} [C_{n+1}^1 s_0 + \cdots + C_{n+1}^{n+1} s_n] + \frac{1}{2^{n+2}} [u_0 + C_{n+1}^1 u_1 + \cdots + C_{n+1}^{n+1} u_{n+1}]$$

$$= \frac{1}{2^{n+2}} [C_{n+1}^1 s_0 + C_{n+1}^2 s_1 + C_{n+1}^3 s_2 + \cdots + C_{n+1}^{n+1} s_n$$
$$+ u_0 + C_{n+1}^1 u_1 + C_{n+1}^2 u_2 + C_{n+1}^3 u_3 + \cdots + C_{n+1}^{n+1} u_{n+1}$$
$$+ C_{n+1}^1 s_0 + C_{n+1}^2 s_1 + \cdots + C_{n+1}^{n+1} s_n].$$

By means of the two equations

$$s_j + u_{j+1} = s_{j+1}, \qquad C_p^{\,q} + C_p^{q+1} = C_{p+1}^{q+1},$$

we arrive at the formula

$$S_{n+1} = \frac{1}{2^{n+2}} (C_{n+2}^1 s_0 + C_{n+2}^2 s_1 + \cdots + C_{n+2}^{n+2} s_{n+1}),$$

which shows that (8) holds for $n + 1$. This completes the proof of (8).

The result of Problem A shows now that

$$\lim_{n\to\infty} S_n = \lim_{n\to\infty} s_n .$$

D. *An application.* Let us apply this result to the series whose general term is $u_n = (-1)^n/(n+1)$, the sum of which is log 2. From the formula of the preceding problem [Eq. (8)], we have

$$\frac{A^n u_0}{2^{n+1}} = \frac{1}{2^{n+1}} \left[1 - \frac{1}{2} C_n^{\,1} + \frac{1}{3} C_n^{\,2} - \frac{1}{4} C_n^{\,3} + \cdots + \frac{(-1)^{n-1}}{n+1} C_n^{\,n}\right].$$

To evaluate the sum in the right-hand member, we first apply the binomial formula to $(1 - x)^n$:

$$1 - C_n^{\,1} x + C_n^{\,2} x^2 - \cdots + (-1)^n x^n = (1 - x)^n.$$

Let us integrate this identity from 0 to 1:

$$1 - \frac{1}{2} C_n^{\,1} + \frac{1}{3} C_n^{\,2} - \cdots + \frac{(-1)^n}{n+1} = \int_0^1 (1 - t)^n \, dt = \frac{1}{n+1}.$$

Therefore,

$$\frac{A^n u_0}{2^{n+1}} = \frac{1}{(n+1)2^{n+1}}.$$

Thus log 2 is also the sum of the series

$$\log 2 = \frac{1}{2} + \frac{1}{2 \cdot 2^2} + \frac{1}{3 \cdot 2^3} + \cdots + \frac{1}{n \cdot 2^n} + \cdots.$$

The series $\Sigma_0^\infty (-1)^n/(n+1)$ converges much more slowly than does the series $\Sigma_1^\infty 1/n \cdot 2^n$. Let us see how these two series behave in a numerical calculation. Suppose that we wish to calculate log 2 with an error less than 10^{-3} by using these two series.

We know from a classical theorem that the absolute value of any remainder of an alternating series is less than the absolute value of the first term discarded. Thus we would need 1000 terms to be sure (on this basis) of getting the value of log 2 with an error not exceeding 10^{-3} by use of the series $\Sigma(-1)^n/(n+1)$.

Let us now majorize the remainder of the series $\Sigma_{n=1}^{\infty} 1/n \cdot 2^n$. We have

$$R_N = \sum_{n=N+1}^{\infty} \frac{1}{2^n \cdot n} < \frac{1}{N+1} \sum_{n=N+1}^{\infty} \frac{1}{2^n} = \frac{1}{(N+1)2^N}.$$

The inequality $R_N < 1/10^3$ may be written

$$2^N(N+1) > 10^3.$$

This inequality is satisfied from $N = 7$ on. Thus we have

$$R_7 < \frac{1}{8 \cdot 2^7} = \frac{1}{2^{10}} = \frac{1}{1024}.$$

Therefore, we need only seven terms instead of 1000.

Let us complete the calculation by this procedure. It will be more convenient to take eight terms so as to calculate each term with less precision. We have

$$R_8 < \frac{1}{9 \cdot 2^8} < 5 \cdot 10^{-4}.$$

Since we know five of these terms exactly, we need only assure ourselves of making a total error less than $5 \cdot 10^{-4}$ in the other three. If we calculate each of these to four decimal places (discarding the fifth whether it be less or greater than 5), we obtain a value for log 2 which the precise value exceeds by an amount less than $9 \cdot 10^{-4}$. In practice, we make these calculations taking the first five decimals.

n	$1/n \cdot 2^n$
1	0.5
2	0.125
3	0.04166 ...
4	0.01562 ...
5	0.00625
6	0.00260 ...
7	0.00111 ...
8	0.00048 ...
	0.69272

The finite sum $\Sigma_{n=1}^{\infty} 1/n \cdot 2^n$ lies between 0.69272 and

$$0.69272 + 0.00005 = 0.69277.$$

Since the error is less than 0.0005, we know that

$$0.6927 < \log 2 < 0.6933.$$

The value to be found in tables is 0.69315.

Remark. This example shows the practical power of this method of summation. In the present case, of course, the series representing log 2 can be obtained without resort to Euler's method. We need only expand $-\log(1 + x)$ in a power series and set $x = -\frac{1}{2}$.

9. The metric tensor

Let $\mathscr{E}$ be a real Euclidean space relative to a nonorthonormal basis $E_1, ..., E_n$. Let g_{ij} denote the scalar products of the vectors E_i and E_j. The square of the length of the vector X whose components are $x_1, ..., x_n$ (that is, the square of the distance from the point represented by the vector X to the origin) is equal to the quadratic form

$$\sum g_{ij}x_ix_j .$$

These g_{ij} are real numbers and $g_{ii} > 0$. An arbitrary symmetric matrix with positive diagonal elements is not in general the matrix of the g_{ij} of a Euclidean space. In other words, certain inequalities must hold among the g_{ij} for these to be the components of a metric tensor. The purpose of this exercise is to display these inequalities in two equivalent forms, one symmetric, one not. We shall confine ourselves to the three-dimensional case.

Problems. A. Consider nine real numbers g_{ij} constituting a 3×3 matrix. For $\Sigma g_{ij}x_ix_j$ to represent the square of the length of the real vector X relative to a nonorthonormal Cartesian system, it is necessary and sufficient that one or the other of the following sets of conditions be satisfied:

I. $g_{11} > 0, \quad g_{11}g_{22} - g_{12}^2 > 0, \quad g > 0$ (where g is the determinant of the g_{ij}).

II. $g_{11} + g_{22} + g_{33} > 0, \quad g_{22}g_{33} + g_{33}g_{11} + g_{11}g_{22} - g_{23}^2 - g_{31}^2 - g_{12}^2 > 0$ $(g > 0)$.

B. Verify the equivalence of conditions I and II. Show that, for every i and j with $j \neq i$, each of them implies that

$$g_{ii} > 0, \quad g_{ii}g_{jj} - g_{ij}^2 > 0.$$

C. Choose

$$g_{11} = g_{22} = g_{33} = 1$$

and set

$$g_{23} = \cos\alpha, \qquad g_{31} = \cos\beta, \qquad g_{12} = \cos\gamma,$$

where the angles α, β, and γ are all between 0 and π.

Show that the condition $g > 0$ is equivalent to the following conditions:

$$\alpha + \beta + \gamma < 2\pi, \qquad \alpha < \beta + \gamma, \qquad \beta < \gamma + \alpha, \qquad \gamma < \alpha + \beta.$$

D. Still assuming $g_{11} = g_{22} = g_{33} = 1$, take the number g_{23}, between -1 and 1. Set $g_{31} = \xi$ and $g_{12} = \eta$. Show that the points (ξ, η) must necessarily lie in the interior of a certain ellipse E. Place this ellipse in the square $\xi = \pm 1$, $\eta = \pm 1$.

Solutions. **A.** For the quadratic form $\varphi = \sum g_{ij}x_ix_j$ to be the square of the length of a vector, it is necessary and sufficient that it be non-negative for all the x_i. The condition for this, in turn, is that it be representable in the form of the sum of three squares of independent linear forms. This decomposition can be made in two different ways.

First method. We use an "arbitrary" linear transformation. The meaning of the word "arbitrary" will be made precise upon comparison with the second method.

We know that g_{11} is positive, since g_{11} is the square of the length of the first basic vector. We write

$$\begin{aligned} g_{11}\varphi &= \sum g_{11}g_{ij}x_ix_j \\ &= \Big(\sum_i g_{1j}x_j\Big)^2 + (g_{11}g_{22} - g_{12}^2){x_2}^2 + (g_{11}g_{33} - g_{13}^2){x_3}^2 \\ &\quad + 2(g_{11}g_{23} - g_{12}g_{13})x_2x_3 . \end{aligned}$$

This is the first step of the decomposition in squares of φ. The quantity $g_{11}\varphi$ is the sum of the square of a linear form in x_1, x_2, and x_3 and a quadratic form in x_2 and x_3. We can proceed in the same way with the second form and represent it as a linear combination of the squares of two linear forms, one in x_2 and x_3, the other in x_3. Thus we obtain three linear forms, which are independent, since the first contains three variables, the second only two, and the third only one. For φ (or $g_{11}\varphi$) to be nonnegative, it is sufficient that the quadratic form in x_2 and x_3 be nonnegative. Now, for a form in two variables, the conditions for positiveness are elementary. The coefficient of $x_2{}^2$ must be positive and the discriminant negative:

$$g_{11}g_{22} - g_{12}^2 > 0$$

$$(g_{11}g_{23} - g_{12}g_{13})^2 - (g_{11}g_{22} - g_{12}^2)(g_{11}g_{33} - g_{13}^2) < 0.$$

The first member of the last inequality is simply the product of g_{11} by the determinant g of the g_{ij}. This is true because

$$g = \begin{vmatrix} g_{11} & g_{12} & g_{13} \\ g_{21} & g_{22} & g_{23} \\ g_{31} & g_{32} & g_{33} \end{vmatrix} = g_{11}g_{22}g_{33} + 2g_{23}g_{31}g_{12} - g_{11}g_{23}^2 - g_{22}g_{31}^2 - g_{33}g_{12}^2,$$

and this multiplied by g_{11} is equal to the first member of the above inequality.

The condition for the quadratic form φ to be positive is thus provided by the following set of inequalities:

$$g_{11} > 0, \qquad g_{11}g_{22} - g_{12}^2 > 0, \qquad g > 0.$$

These inequalities are shown in a nonsymmetric form.

Second method. In contrast with our initial point of view, let us *interpret* the x_i as the Cartesian coordinates of a point relative to an orthonormal basis. Then $\varphi = \Sigma g_{ij}x_ix_j$ is a quadratic form to which no particular metric significance is attached. But this form is positive, and we wish to write the condition for this to be the case.

Let us make a change of orthonormal basis in the space. (The word "orthonormal" is in contrast with the word "arbitrary" of the first method.)

This amounts to setting

$$x_1 = X_1(y_i), \qquad x_2 = X_2(y_i), \qquad x_3 = X_3(y_i),$$

where X_1, X_2, and X_3 are three linear forms in the three new coordinates y_1, y_2, and y_3. We know that φ is then

$$\varphi = \lambda_1 y_1{}^2 + \lambda_2 y_2{}^2 + \lambda_3 y_3{}^2,$$

where λ_1, λ_2, and λ_3 are the *eigenvalues* of the matrix g_{ij}.

Thus the condition for φ to be positive is that these three eigenvalues be positive, that is, that the roots of the equation

$$D(\lambda) = \begin{vmatrix} g_{11} - \lambda & g_{12} & g_{13} \\ g_{12} & g_{22} - \lambda & g_{23} \\ g_{13} & g_{23} & g_{33} - \lambda \end{vmatrix}$$

be positive, When we expand this determinant, we have

$$D(\lambda) = -\lambda^3 + a\lambda^2 - b\lambda + g,$$

where the coefficients a, b, and g are three functions of the g_{ij} and, of course, g is the determinant of the g_{ij} .

Now, from the relations between the coefficients and the roots of the polynomial $D(\lambda)$, we have

$$a = \lambda_1 + \lambda_2 + \lambda_3 ,$$
$$b = \lambda_2\lambda_3 + \lambda_3\lambda_1 + \lambda_1\lambda_2 ,$$
$$g = \lambda_1\lambda_2\lambda_3 .$$

If the λ_i are positive, so are a, b, and g.

Conversely, the inequality $g > 0$ implies that λ_1 , λ_2 , and λ_3 are positive or that λ_1 is positive and λ_2 and λ_3 negative. Let us show that this last combination is impossible. We set

$$\lambda_2 = -\lambda_2', \qquad \lambda_3 = -\lambda_3'.$$

Then

$$\lambda_1 - \lambda_2' - \lambda_3' > 0, \qquad \lambda_2'\lambda_3' - \lambda_1(\lambda_2' + \lambda_3') > 0,$$

so that

$$\lambda_2' + \lambda_3' < \lambda_1 < \frac{\lambda_2'\lambda_3'}{\lambda_2' + \lambda_3'}$$

which implies

$$(\lambda_2' + \lambda_3')^2 < \lambda_2'\lambda_3', \qquad \lambda_2'^2 + \lambda_3'^2 + \lambda_2'\lambda_3' < 0,$$

and this inequality contradicts the positiveness of λ_2' and λ_3'.

Therefore, the conditions $a > 0$, $b > 0$, and $g > 0$ are equivalent to the conditions $\lambda_1 > 0$, $\lambda_2 > 0$, and $\lambda_3 > 0$. These conditions may be written

$$g_{11} + g_{22} + g_{33} > 0,$$
$$g_{22}g_{33} + g_{33}g_{11} + g_{11}g_{22} - g_{23}^2 - g_{31}^2 - g_{12}^2 > 0,$$
$$g > 0.$$

B. *Equivalence of conditions I and II.* This equivalence follows from the fact that I and II both express the positiveness of φ. Condition II also implies that

$$g_{ii} > 0, \qquad g_{ii}g_{jj} - g_{ij}^2 > 0,$$

whatever the values of the indices i and j.

However, we can proceed to a direct algebraic verification without appealing to any comprehension of the problem.

Suppose that conditions I are satisfied. The first two imply that

$$g_{22} > 0.$$

The third inequality, which we may write

$$(g_{11}g_{22} - g_{12}^2)(g_{11}g_{33} - g_{13}^2) > (g_{11}g_{23} - g_{12}g_{13})^2,$$

implies similarly that

$$g_{11}g_{33} - g_{13}^2 > 0$$

and hence

$$g_{33} > 0.$$

It follows that

$$g_{11} + g_{22} + g_{33} > 0 \qquad \text{(the first of conditions II)}.$$

But we can also write g in the form

$$gg_{33} = (g_{11}g_{33} - g_{13}^2)(g_{22}g_{33} - g_{23}^2) - (g_{33}g_{12} - g_{31}g_{32})^2,$$

which implies that $g_{22}g_{33} - g_{23}^2 > 0$. Thus the second of inequalities II is verified.

Let us assume now that conditions II are verified. Since

$$g_{11} + g_{22} + g_{33} > 0,$$

at least one of the g_{ii} is positive. For definiteness, let us suppose that $g_{11} > 0$. Let us set

$$A = g_{22}g_{33} - g_{23}^2, \qquad B = g_{33}g_{11} - g_{13}^2, \qquad C = g_{11}g_{22} - g_{12}^2 .$$

Conditions II in these notations become

$$A + B + C > 0$$

and

$$g_{11}[BC - (\ )^2] > 0, \qquad g_{22}[CA - (\ )^2] > 0, \qquad g_{33}[AB - (\ )^2] > 0.$$

Since g_{11} is positive, BC cannot be negative. If B and C are both positive, we conclude that $g_{22} > 0$ and $g_{33} > 0$. This returns us to conditions I.

If B and C could be negative, the inequality $A + B + C > 0$ would imply that $A > 0$ and hence that G_{22} and G_{33} would be negative. Let us set $g'_{22} = -g_{22}$ and $g'_{33} = -g_{33}$. The condition $A + B + C > 0$ becomes

$$\underline{g_{11}(g'_{22} + g'_{33}) - g'_{22}g'_{33}} + g_{23}^2 + g_{31}^2 + g_{12}^2 < 0.$$

Since

$$g_{11} > g'_{22} + g'_{33},$$

the underlined part of the preceding inequality must be positive, which leads to a contradiction.

The only possible consequence of conditions II is therefore

$$g_{11} > 0, \qquad g_{22} > 0, \qquad g_{33} > 0, \qquad B > 0, \qquad C > 0,$$

which, with $g > 0$, is equivalent to I.

C. *Geometric interpretation.* We can always arrange to have

$$g_{11} = g_{22} = g_{33} = 1.$$

The numbers g_{23}, g_{31}, and g_{12}, which lie between -1 and 1, are then the cosines of the angles α, β, and γ between the pairs of basis vectors. It remains for us to interpret the condition $g > 0$. It may be written

$$(\cos\alpha - \cos\beta\cos\gamma)^2 - \sin^2\beta\sin^2\gamma < 0,$$

or

$$[\cos\alpha - \cos(\beta - \gamma)][\cos\alpha - \cos(\beta + \gamma)] < 0,$$

or

$$\sin\frac{\alpha+\beta+\gamma}{2}\sin\frac{\beta+\gamma-\alpha}{2}\sin\frac{\gamma+\alpha-\beta}{2}\sin\frac{\alpha+\beta-\gamma}{2} > 0.$$

The angles α, β, and γ lie between 0 and π. Let us assume for definiteness that $\alpha < \beta < \gamma$. From this we conclude that $(\beta + \gamma - \alpha)/2$ and $(\gamma + \alpha - \beta)/2$ assume values only between 0 and π. Therefore, their sines are positive, so that we have

$$\sin\frac{\alpha+\beta+\gamma}{2}\sin\frac{\alpha+\beta-\gamma}{2} > 0.$$

If both these sines are negative, we find that

$$2\pi < \alpha + \beta + \gamma < 3\pi, \qquad -\pi < \alpha + \beta - \gamma < 0.$$

It follows from these inequalities that $\alpha + \beta + \gamma$ is both greater than 2π and less than 2γ. This conclusion contradicts the fact that γ is less than π.

Therefore, the only possibility is that both sines are positive. This implies that

$$0 < \alpha + \beta + \gamma < 2\pi, \qquad 0 < \alpha + \beta - \gamma < 2\pi.$$

Since $\alpha + \beta - \gamma$ is less than $\alpha + \beta + \gamma$, the inequality

$$\alpha + \beta - \gamma < 2\pi$$

is superfluous. Therefore, we have

$$\gamma < \beta + \alpha, \qquad \alpha + \beta + \gamma < 2\pi,$$

where, by convention,

$$\alpha < \beta < \gamma.$$

In sum, the condition $g > 0$ expresses the fact that each face angle in a trihedral angle is less than the sum of the other two and that the sum of all three is less than four right angles. These properties of a trihedral angle are familiar from elementary solid geometry.

D. *An example.* We still take

$$g_{11} = g_{22} = g_{33} = 1.$$

Suppose that g_{23} is given. We set

$$g_{31} = \xi, \qquad g_{12} = \eta.$$

If $|g_{23}| < 1$, conditions I or II reduce to

$$2g_{23}\xi\eta + 1 - g_{23}^2 - \xi^2 - \eta^2 > 0.$$

These conditions express the fact that the point (ξ, η) is in the interior of a certain *ellipse* E. This ellipse is inscribed in the square whose sides are segments of the straight lines $\xi = \pm 1$, $\eta = \pm 1$. It touches the side $\xi = 1$ at the point whose ordinate is $\eta = g_{23}$ and its axes are the diagonals of the square.

Thus if g_{23} is given, the point (ξ, η) must fall in the interior of the square $|\xi| < 1$, $|\eta| < 1$, but it cannot fall just anywhere. It must necessarily be in the interior of the ellipse E.

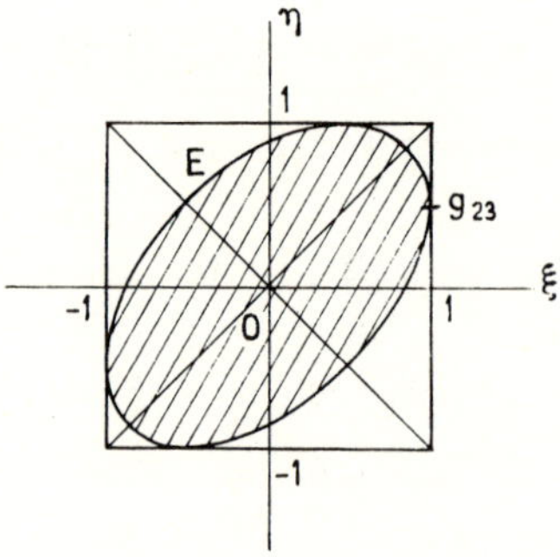

FIGURE 7

Analogous reasoning can be applied in three-dimensional space. The condition $g > 0$ expresses the fact that the point whose coordinates are g_{23}, g_{31}, and g_{12} is in the interior of a certain cubical surface (more difficult to construct than an ellipse). The equation of this surface is

$$1 + 2\xi\eta\zeta - \xi^2 - \eta^2 - \zeta^2 = 0.$$

Its intersection with the plane $\zeta = g_{23}$ (constant) is the ellipse E. For $\zeta = 0$, E becomes the unit circle. For $\zeta = 1$, the ellipse E degenerates to a line segment included in a diagonal of the square.

10. The correlation tensor of *n* reduced random variables

This exercise is worded in probability-theory language. However, it is easy to reword it in purely algebraic language. It will be sufficient to *define* Γ_{ij} as the metric tensor of a real Euclidean space relative to a nonorthogonal basis of normed vectors. More precisely, we denote by $X_1, ..., X_n$ a set of n linearly independent normed vectors in an n-dimensional real Euclidean space. Let Γ_{ij} be the scalar product of the vectors X_i and X_j. We introduce the determinant Δ_n of the Γ_{ij} and we wish to show that

$$0 \leqslant \Delta_n \leqslant \Delta_{n-1}$$

Problems. A. Suppose that $X_1, ..., X_n$ are n real random variables. We assume that $EX_i = 0$ and $EX_i^2 = 1$, where E is the symbol of mathematical expectation. We denote by $\Gamma_{ij} = EX_iX_j$ the correlation coefficient between X_i and X_j (where $\Gamma_{ii} = 1$) and we introduce the determinant

$$\Delta_n = \begin{vmatrix} 1 & \Gamma_{12} & \cdots & \Gamma_{1n} \\ \cdots & \cdots & \cdots & \cdots \\ \Gamma_{n1} & \Gamma_{n2} & \cdots & 1 \end{vmatrix}.$$

We denote by Γ^{ij} the elements of the inverse matrix of Γ_{ij}. Show that $\Gamma^{nn} = \Delta_{n-1}/\Delta_n$ and that $0 \leqslant \Delta_n \leqslant \Delta_{n-1}$.

B. *Examples.* (a) Calculate Δ_n when the random variables X_n are uncorrelated with the exception of X_1 and X_2.

(b) Calculate Δ_n when the Γ_{ij} all have the same value Γ.

In the first question, one may think of the quadratic form $\sum \Gamma_{ij}x_ix_j$ as representing the square of the distance from the point $(x_1, ..., x_n)$ to the origin in a Euclidean space with metric tensor Γ_{ij}. The ratio Δ_n/Δ_{n-1} will then represent the square of the cosine of the angle between the two vectors.

Solutions. **A.** $X_1, ..., X_i, ..., X_n$ are reduced variables ($EX_i = 0$, $EX_i^2 = 1$) that are not independent. We set

$$\Gamma_{ij} = EX_iX_j, \qquad \Gamma_{ii} = EX_i^2 = 1.$$

$$\Delta_n = \begin{vmatrix} 1 & \Gamma_{12} & \cdots & \Gamma_{1n} \\ \cdots & \cdots & \cdots & \cdots \\ \Gamma_{n1} & \Gamma_{n2} & \cdots & 1 \end{vmatrix}.$$

Note that Δ_{n-1} is the minor of the lower right term $\Gamma_{nn} = 1$ in Δ_n. Clearly,

$$\Delta_1 = 1, \qquad \Delta_2 = 1 - \Gamma_{12}^2 \leqslant 1.$$

Consider a real Euclidean space relative to a basis consisting of normed but not orthogonal vectors $E_1, ..., E_n$. Let Γ_{ij} be the cosine of the angle between E_i and E_j. The square of the distance from the point $M(x_i)$ to the origin is equal to

$$OM^2 = \sum_j \Gamma_{ij}x_ix_j = \sum_j x_i^2 + 2\sum_{i<i} \Gamma_{ij}x_ix_j.$$

Let us denote by Γ^{ij} the elements of the inverse matrix of the Γ_{ij}. We have

$$\sum_i \Gamma_{ij}\Gamma^{ik} = \delta_{jk} \qquad \text{and} \qquad \Gamma^{nn} = \frac{\Delta_{n-1}}{\Delta_n},$$

since Δ_{n-1} is the minor of Γ_{nn} in the determinant Δ_n.

In particular,

$$\sum_i \Gamma_{ij}\Gamma^{in} = 0 \qquad \text{for} \qquad j \neq n.$$

Now, the condition for orthogonality of two vectors x_i and y_i is

$$\sum_{ij} \Gamma_{ij}x_i y^j = 0.$$

In particular, the condition for the vector $(x_1, ..., x_n)$ to be orthogonal to the basis vector E_j (0, ..., 0, 1, 0, ..., 0) is

$$\sum_i \Gamma_{ij}x_i = 0 \qquad (j \text{ fixed}).$$

From this it follows that the vector whose components are

$$u_1 = \Gamma^{1n}, ..., u_n = \Gamma^{nn}$$

is orthogonal to all the basis vectors except E_n.

Let us evaluate the angle θ that this vector makes with E_n (0, 0, ..., 0, 1). The scalar product of these two vectors is equal to

$$\sum \Gamma_{in}u_i = \sum \Gamma_{in}\Gamma^{in} = 1.$$

The squares of their lengths are respectively Γ_{nn} and

$$\sum_{ij} \Gamma_{ij}\Gamma^{in}\Gamma^{jn} = \sum_i \Gamma^{in}\Big(\sum_j \Gamma_{ij}\Gamma^{jn}\Big) = \sum_i \delta_{in}\Gamma^{in} = \Gamma^{nn}$$

Therefore,

$$\cos^2\theta = \frac{1}{\Gamma_{nn}\Gamma^{nn}} = \frac{1}{\Gamma^{nn}} = \frac{\Delta_n}{\Delta_{n-1}},$$

from which it follows that

$$0 \leqslant \Delta_n \leqslant \Delta_{n-1}.$$

Since $\Delta_2 \leqslant 1$, the sequence of the Δ_n is nonincreasing and all its terms lie between 0 and 1.

B. *Example* 1. The random variables are uncorrelated except for X_1 and X_2:

$$\Delta_2 = 1 - \Gamma_{12}^2,$$

$$\Delta_n = \begin{vmatrix} 1 & \Gamma_{12} & 0 & \cdots & 0 \\ \Gamma_{12} & 1 & 0 & \cdots & 0 \\ \cdots & \cdots & \cdots & \cdots & \cdots \\ 0 & 0 & \cdots & \cdots & 1 \end{vmatrix} = 1 - \Gamma_{12}^2 = \Delta_2.$$

The terms of the sequence of the Δ_n become invariant from $n = 2$.

Example 2. All the Γ_{ij} have the same value Γ. Then

$$\Delta_n = \begin{vmatrix} 1 & \Gamma & \Gamma & \cdots & \Gamma \\ \Gamma & 1 & \Gamma & \cdots & \Gamma \\ \cdots & \cdots & \cdots & \cdots & \cdots \\ \Gamma & \Gamma & \Gamma & \cdots & 1 \end{vmatrix}.$$

To calculate Δ_n, we replace each row of the determinant with the row obtained by subtracting from it the following row:

$$\Delta_n = \begin{vmatrix} 1-\Gamma & \Gamma-1 & 0 & \cdots & 0 & 0 \\ 0 & 1-\Gamma & \Gamma-1 & \cdots & 0 & 0 \\ \cdots & \cdots & \cdots & \cdots & \cdots & \cdots \\ 0 & 0 & 0 & \cdots & 1-\Gamma & \Gamma-1 \\ \Gamma & \Gamma & \Gamma & \cdots & \Gamma & 1 \end{vmatrix}$$

$$= (1-\Gamma)^{n-1} \begin{vmatrix} 1 & -1 & 0 & \cdots & 0 & 0 \\ 0 & 1 & -1 & \cdots & 0 & 0 \\ \cdots & \cdots & \cdots & \cdots & \cdots & \cdots \\ 0 & 0 & 0 & \cdots & 1 & -1 \\ \Gamma & \Gamma & \Gamma & \cdots & \Gamma & 1 \end{vmatrix}.$$

If we add all the columns to the first column, we obtain

$$\Delta_n = (1-\Gamma)^{n-1} \begin{vmatrix} 0 & -1 & 0 & \cdots & 0 & 0 \\ 0 & 1 & -1 & \cdots & 0 & 0 \\ \cdots & \cdots & \cdots & \cdots & \cdots & \cdots \\ 0 & 0 & 0 & \cdots & 1 & -1 \\ 1+(n-1)\Gamma & \Gamma & \Gamma & \cdots & \Gamma & 1 \end{vmatrix}.$$

If we now expand this determinant according to the first column, we see that its value is $(-1)^{n-1}[1+(n-1)\Gamma]$ multiplied again by $(-1)^{n-1}$. Therefore,

$$\Delta_n = (1-\Gamma)^{n-1}[(n-1)\Gamma + 1].$$

Here we have

$$\frac{\Delta_{n+1}}{\Delta_n} = (1-\Gamma)\frac{n\Gamma + 1}{(n-1)\Gamma + 1}.$$

This ratio is a decreasing homographic function that tends to $1-\Gamma$ as $n \to \infty$.

Remark. This calculation can be made more simply. Consider the symmetric matrix

$$\begin{pmatrix} \Gamma & \Gamma & \cdots & \Gamma \\ \cdots & \cdots & \cdots & \cdots \\ \Gamma & \Gamma & \cdots & \Gamma \end{pmatrix}.$$

Its eigenvectors are solutions of the linear system

$$\Gamma s = \lambda x_i \qquad (i = 1, 2, ..., n)$$
$$s = x_1 + \cdots + x_n .$$

The quantity $\lambda = n\Gamma$ is the simple eigenvalue corresponding to the eigenvector $x_1 = x_2 = \cdots = x_n$; $\lambda = 0$ is a multiple eigenvalue of order $n - 1$ for the eigen subspace $x_1 + \cdots + x_n = 0$. Therefore, the characteristic polynomial is equal to $(-1)^n \lambda^{n-1}(\lambda - n\Gamma)$.

When we set $\lambda = \Gamma - 1$, the characteristic determinant reduces to Δ_n, from which we again find its value.

CHAPTER 2

Uniform Convergence: Fourier Series and Integrals

This chapter is devoted primarily to functions defined by series and integrals, including Fourier series and integrals, treated from the point of view of either pointwise or uniform convergence. Most of the exercises on mean-square convergence, orthogonal functions, and orthogonal polynomials are deferred until Chapter 5 in connection with second-order differential equations.

Problems on the uniform convergence of integrals sometimes lead to preliminary questions on the convergence and the conditions for the existence of improper numerical integrals.

We assume that the reader is familiar with the elementary theorems on uniform convergence. *Suppose that* $\{f_n(x)\}$ *is a sequence of continuous functions that converges uniformly to a limit function* $f(x)$ *on a closed interval* Δ. *Under these conditions,*

(a) $f(x)$ *is continuous;*

(b) $\int_a^b f_n(x)\,dx$ *tends to a limit, which is equal to* $\int_a^b f(x)\,dx$.

If the sequence $\{f_n(x)\}$ *converges to* $f(x)$, *if* $f_n'(x)$ *exists and is continuous for every* n, *and if the sequence* $\{f_n'(x)\}$ *converges uniformly to a limit* $g(x)$ *in a closed interval* Δ, then the sequence $\{f_n(x)\}$ *converges uniformly on* Δ *to the function* $f(x)$ *and, at every interior point of* Δ, $f(x)$ *has a derivative equal to* $g(x)$.

Analogous theorems hold for improper integrals depending on a parameter which assert the validity, on the basis of uniform convergence, of differentiation and integration under the integral sign.

In several exercises, we shall encounter sequences $\{f_n(x)\}$ that do not converge uniformly on an interval $[a, b]$. With regard to these sequences, we shall pose the question as to whether

$$\lim \int_a^b f_n(x)\,dx = \int_a^b f(x)\,dx.$$

To answer this question, we may make use of the concept of uniform convergence several times in a row. It is possible to integrate over a subinterval in which the series $\{f_n(x)\}$ converges uniformly and then take the limit of the resulting integral as the end points of this subinterval approach a and b. It is also possible to show directly, by a special process, that the integral

$$\int_a^b [f(x) - f_n(x)]\,dx$$

tends to 0 as $n \to \infty$.

Finally, one may apply a celebrated theorem of Lebesgue. In its general form (which is suited for its proof), this theorem applies to functions that are summable with respect to a measure. However, it can be applied in the following form to functions that are integrable in the sense of Riemann: *Suppose that* $\{f_n(x)\}$ *is a sequence of functions defined on an interval* Δ *that converges pointwise to a limit function* $f(x)$. *Suppose that a nonnegative function* $\mu(x)$ *exists such that the integrals*

$$\int_\Delta |f_n(x)|\,\mu(x)\,dx, \qquad \int_\Delta |f(x)|\,\mu(x)\,dx$$

exist (possibly as improper integrals). Suppose also that there exists a nonnegative function $g(x)$ *such that*

$$\int_\Delta g(x)\mu(x)\,dx$$

exists and such that the inequality

$$|f_n(x)| \leqslant g(x)$$

holds for every n *and every* x *(in* Δ*). Under these conditions,*

$$\lim_{n\to\infty} \int_\Delta f_n(x)\mu(x)\,dx = \int_\Delta f(x)\mu(x)\,dx.$$

The Δ of this theorem is not necessarily a finite interval. In contrast, the theorem on the integration of uniformly convergent sequences is valid if the interval of integration is finite. However, it is possible to extend it in certain cases to an infinite interval, as well be shown in Exercise 7.

We assume that the reader is familiar with the basic properties of power series. Nonetheless, we call his attention to the following theorem of Abel: *Suppose that a series* $\sum a_n x^n$ *has a finite radius of convergence* R *and that it converges at the point* $x = R$. *Then the convergence is uniform not only in every closed interval contained in the open interval* $(-R, R)$ *but also in every interval of the form* $[a, R]$, *where* a *is any number in the open interval* $(-R, R)$. *It follows that the function representing the limit of the series is continuous from the left at* $x = R$.

With regard to Fourier series, we shall treat them from the point of view of pointwise and uniform convergence. We shall apply Jordan's theorem: *If* $f(x)$ *is periodic and of bounded variation, its Fourier series converges for every* x *and the convergence is uniform in every closed interval throughout which* $f(x)$ *is continuous. At a point at which* $f(x)$ *is discontinuous, the sum of the series is*

$$\tfrac{1}{2}[f(x+0)+f(x-0)].$$

The series can be integrated term by term.

An important special case is that in which f and f' are continuous except at points, finitely many in each period, at which f or f' has a finite discontinuity. The theory rests on a theorem (Riemann-Lebesgue) that asserts, in the case of integrable functions, that

$$\lim_{n\to\infty}\int_a^b f(x)\sin nx\,dx = 0.$$

This result holds for an infinite interval if $|f|$ is integrable over that interval.

From this theorem we can derive, by use of Dirichlet's integral, the theorem on the Fourier integral in the following form: *If* f *is of bounded variation and is absolutely integrable from* $-\infty$ *to* $+\infty$, *we have the following reciprocal (nonsymmetric) formulas:*

$$g(x) = \int_{-\infty}^{\infty} f(t)e^{2\pi ixt}\,dt, \qquad f(x) = \lim_{\lambda\to\infty}\int_{-\lambda}^{\lambda} g(t)e^{-2\pi ixt}\,dt.$$

We assume that the reader is familiar with Riemann's zeta function:

$$\zeta(s) = 1 + \frac{1}{2^s} + \cdots + \frac{1}{n^s} + \cdots$$

and that he knows that $\zeta(2) = \pi^2/6$. From time to time in this chapter we shall use the properties of Euler's beta and gamma functions, the formula on complements, and Stirling's formula for the approximation of $n!$.

1. Uniform convergence

Problems. Suppose that we have

(a) a numerical sequence $\{u_n\}$ that converges to a number u,

(b) a sequence of continuous functions $S_n(x)$ that converges to a function $S(x)$ uniformly on a closed interval Δ to which u and u_n (for every n) belong.

A. Show that $S_n(u_n)$ tends to $S(u)$.

B. Verify this result for the case in which

$$S_n(x) = \frac{nx}{1+nx}, \qquad u_n = \frac{n}{n-1}.$$

What happens if we apply this same sequence of functions $S_n(x)$ to the numerical sequence $\{u_n\}$, where $u_n = 1/n$?

Solutions. **A.** First of all, we have the inequality

$$| S_n(u_n) - S(u) | \leqslant | S_n(u_n) - S(u_n) | + | S(u_n) - S(u) |.$$

By hypothesis, the sequence $\{S_n(x)\}$ converges uniformly on Δ to $S(x)$. This means that, for arbitrary ϵ, there exists an N_1 such that, for $n > N_1$,

$$\sup_x | S_n(x) - S(x) | < \epsilon,$$

so that

$$| S_n(u_n) - S(u_n) | < \epsilon.$$

On the other hand, since the functions $S_n(x)$ are continuous, their uniform limit $S(x)$ is also continuous. In other words, for arbitrary ϵ, there exists a number N_2 such that

$$| S(u_n) - S(u) | < \epsilon$$

for all $n > N_2$. Therefore, for $n \geqslant \sup\{N_1, N_2\}$, we have

$$| S_n(u_n) - s(u) | < 2\epsilon.$$

B. *An example of the application of uniform convergence.*

$$S_n(x) = \frac{nx}{1+nx}, \quad u_n = \frac{n}{n-1} \qquad \text{for} \quad n \geqslant 2.$$

Obviously u_n belongs to the interval [1, 2]. On this interval, the sequence of functions $S_n(x)$ converges uniformly to 1, since

$$|\,1 - S_n(x)\,| = \frac{1}{1+nx},$$

$$\sup_x |\,1 - S_n(x)\,| = \frac{1}{1+n} \to 0 \qquad \text{as} \quad n \to \infty.$$

Therefore, the theorem is applicable. One can verify that

$$S_n(u_n) = \frac{n[n/(n-1)]}{1 + n[n/(n-1)]}$$

tends to a limit equal to $S(u)$, that is, 1. Direct verification is immediate:

$$S_n(u_n) = \frac{n^2}{n^2+n-1} \to 1 \qquad \text{as} \quad n \to \infty.$$

Counterexample. If the sequence $\{S_n(x)\}$ does not converge uniformly, this result does not necessarily follow. Let us take

$$u_n = \frac{1}{n} \quad \text{and} \quad S_n(x) = \frac{nx}{1+nx} \qquad (0 \leqslant x \leqslant 1),$$

$$u_n \to 0 \qquad \text{as } n \to \infty.$$

Since $\{S_n(x)\}$ does not converge uniformly on any interval containing 0, the theorem is not applicable. And, in fact, we can see that

$$S_n(u_n) = \frac{n(1/n)}{1 + n(1/n)} = \frac{1}{2}, \qquad S(0) = 1.$$

2. Termwise integration of series in various cases of uniform and nonuniform convergence

In this exercise we propose to evaluate an integral by integrating termwise a series expansion of a function under the integral sign.

With one procedure we use a power series that converges everywhere on the interval of integration, including the limits. Therefore, the convergence is uniform and termwise integration is permissible.

By a second process we use a series that converges *nonuniformly* on the interval of integration. In this case, we may justify termwise integration by various methods: Direct study of the remainder, application of Lebesgue's convergence theorem to the remainder, and application of the theorems on uniform convergence in two stages.

We finally obtain the value of the integral as the sum of a numerical series which leads to the series $\zeta(2)$, which is well known.

Problems. A. Consider the three integrals

$$A = \int_0^1 \frac{\log x}{1 - x^2}\, dx, \qquad B = \int_0^1 \frac{\log x}{1 - x}\, dx, \qquad C = \int_0^1 \frac{\log x}{1 + x}\, dx.$$

Show that they converge and that

$$B + C = 2A, \qquad A - C = \frac{B}{4}.$$

B. Show that

$$C = -\int_0^1 \frac{\log(1 + x)}{x}\, dx.$$

Expand the function $x^{-1} \log(1 + x)$ in powers of x. Show that, for $0 \leqslant x \leqslant 1$, this expansion is a uniformly convergent alternating series. Show from this that we may integrate termwise from 0 to 1 to find the value of C. Recall that

$$1 + \frac{1}{2^2} + \cdots + \frac{1}{n^2} + \cdots = \frac{\pi^2}{6}.$$

C. Study the convergence of the series

$$x^2 \log x + \cdots + x^{2n} \log x + \cdots$$

for $0 \leqslant x \leqslant 1$. Calculate the sum of the first n terms, the remainder $R_n(x)$, and the sum $S(x)$ of the series. Represent $R_n(x)$ and $S(x)$ graphically.

Is the function $S(x)$ continuous? Is the convergence of the series uniform for $0 \leqslant x \leqslant 1$? for $0 \leqslant x \leqslant a$, where $0 < a < 1$?

Show by direct calculation that $\int_0^1 R_n(x)\, dx$ tends to 0 as $n \to \infty$. Show from this that the series may be integrated termwise from 0 to 1. Then find the value of the integral A and finally that of the integral C.

D. Show on the basis of uniform convergence that the series

$$\sum_{n=1}^{\infty} x^{2n} \log x,$$

may be integrated termwise from 0 to $a < 1$. The general term of the series obtained is a function of a. Show that this series converges uniformly for $0 \leqslant a \leqslant 1$. Use this result to show that the function defined by this series is continuous from the right at $a = 0$ and use this result to get the answer to Question C in a different manner.

E. Show that $\int_0^1 R_n(x)\, dx$ tends to 0 by using Lebesgue's convergence theorem (see the introduction to this chapter).

Solutions. **A.** Let us first show that the three integrals

$$A = \int_0^1 \frac{\log x}{1 - x^2}\, dx, \qquad B = \int_0^1 \frac{\log x}{1 - x}\, dx, \qquad C = \int_0^1 \frac{\log x}{1 + x}\, dx$$

converge. Note first of all that the three integrands are bounded on any interval of the form $0 < a \leqslant x \leqslant 1$. As $x \to 0$, these integrands become infinite.

Now, if, for example, $0 < x \leqslant \frac{1}{2}$,

$$\left| \frac{\log x}{1 + x} \right| < | \log x |,$$

$$\left| \frac{\log x}{1 - x} \right| \leqslant 2 \, | \log x |,$$

$$\left| \frac{\log x}{1 - x^2} \right| \leqslant \tfrac{4}{3} \, | \log x |.$$

Since the integral $\int_0^a | \log x |\, dx$, where $a > 0$, converges, we may conclude that the three integrals A, B, and C converge. Also,

$$B + C = \int_0^1 \left(\frac{1}{1 - x} + \frac{1}{1 + x} \right) \log x \, dx = 2 \int_0^1 \frac{\log x}{1 - x^2}\, dx = 2A,$$

$$A - C = \int_0^1 \frac{x \log x}{1 - x^2}\, dx = \int_0^1 \frac{\log \sqrt{t}}{1 - t} \frac{dt}{2} = \frac{1}{4} \int_0^1 \frac{\log t}{1 - t}\, dt = \frac{B}{4}.$$

B. Integrating by parts we obtain

$$-C = \int_0^1 \frac{\log(1 + x)}{x}\, dx.$$

The function $x^{-1}\log(1+x)$ can be expanded in a power series whose radius of convergence is 1:

$$\frac{\log(1+x)}{x} = 1 - \frac{x}{2} + \cdots + (-1)^{n-1}\frac{x^{n-1}}{n} + \cdots \qquad \text{for} \quad |x| < 1.$$

Now, for $x = 1$, the series

$$1 - \frac{1}{2} + \cdots + \frac{(-1)^{n-1}}{n} + \cdots$$

converges. We know from Abel's theorem that if a power series converges at one end R of the interval of convergence, the function defined by it is continuous on the entire closed interval $a \leqslant x \leqslant R$, where a is any interior point of the interval of convergence. Therefore, the series representing $x^{-1}\log(1+x)$ converges uniformly on the interval $0 \leqslant x \leqslant 1$, so that we may integrate it term by term.

As it happens, this result is easy to verify directly. For $x > 0$, the series $(-1)^{n-1}x^{n-1}/n$ is an alternating series. The ratio x^{n-1}/n decreases with increasing n for $x < 1$, since the inequality $x^{n-1}/n > x^n/(n+1)$ is equivalent to $x < (n+1)/n$. The first member of this inequality is less than or equal to 1, whereas the second member is greater than 1.

According to a familiar property of the remainder of an alternating series, the remainder of the series in question is less in absolute value than $x^n/(n+1) \leqslant 1/(n+1)$. Thus it is majorized by an expression that is *independent of* x provided x does not exceed 1. The series converges uniformly for $0 \leqslant x \leqslant 1$.

Integrating term by term, we obtain

$$\int_0^1 \frac{\log(1+x)}{x}\,dx = 1 - \frac{1}{2^2} + \cdots + \frac{(-1)^{n-1}}{n^2} + \cdots.$$

This series is related to the series $\zeta(s) = 1 + \cdots + (1/n^s) + \cdots$ for $s = 2$. Its sum is $\zeta(2)/2$, as can be seen by observing the expression for $\zeta(2) - 2[\zeta(2)/4]$. Now we know that $\zeta(2) = \pi^2/6$. Therefore,

$$C = -\frac{\pi^2}{12}.$$

Since

$$2A = B + C, \qquad B = 4(A - C),$$

we obtain

$$A = -\frac{\pi^2}{8}, \qquad B = -\frac{\pi^2}{6}.$$

C. To evaluate the integral

$$A = \int_0^1 \frac{\log x}{1 - x^2}\, dx,$$

we have found it convenient to convert it by a simple algebraic process to the integral C, which we evaluated by integrating by parts with the aid of an *alternating series* that was easy to handle.

Let us now try to calculate A directly by an analogous procedure. We have

$$\frac{\log x}{1 - x^2} - \log x = x^2 \log x + \cdots + x^{2n} \log x + \cdots .$$

Let us first study the series whose general term is

$$u_n(x) = x^{2n} \log x,$$

starting with the term $x^2 \log x$ to avoid the discontinuity at the origin. For $x = 1$, $u_n(x)$ is 0. For $x = 0$, $u_n(x)$ is not defined. Let us *agree* to set

$$u_n(0) = \lim_{x \to 0} u_n(x) = 0.$$

Then $u_n(x)$ is continuous at $x = 0$. The curve for $u_n(x) = x^{2n} \log x$ is shown in Figure 1.

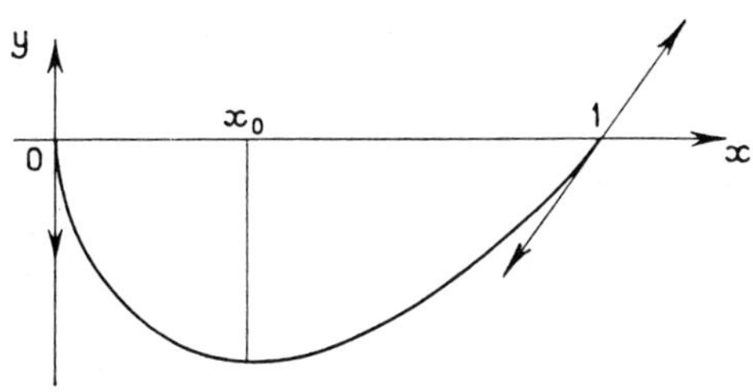

FIGURE 1

We have

$$u_n'(x) = x^{2n-1}(2n \log x + 1).$$

The derivative vanishes for $x_0 = e^{-1/2n}$. For this value of x_0, we have

$$u_n(x_0) = -\frac{1}{2ne}.$$

If we fix x, the series $\Sigma\, u_n(x)$ becomes a numerical series. For $x = 0$ or 1, $u_n(x) = 0$, so that the series converges and its sum is 0.

For $0 < x < 1$, $\Sigma\, u_n(x)$ is a geometric series whose first term is $x^2 \log x$. Since $x^2 < 1$, the series converges. The sum of the series is

$$S(x) = \begin{cases} \log x \dfrac{x^2}{1 - x^2} & \text{for} \quad x \neq \begin{cases} 0, \\ 1, \end{cases} \\ 0 & \text{for} \quad x = 0 \quad \text{or} \quad x = 1. \end{cases}$$

Since $S(x) \to 0$ as $x \to 0$, the function $S(x)$ is continuous at $x = 0$. Since $S(x) \to -\frac{1}{2}$ as $x \to 1$, $S(x)$ is not continuous at $x = 1$.

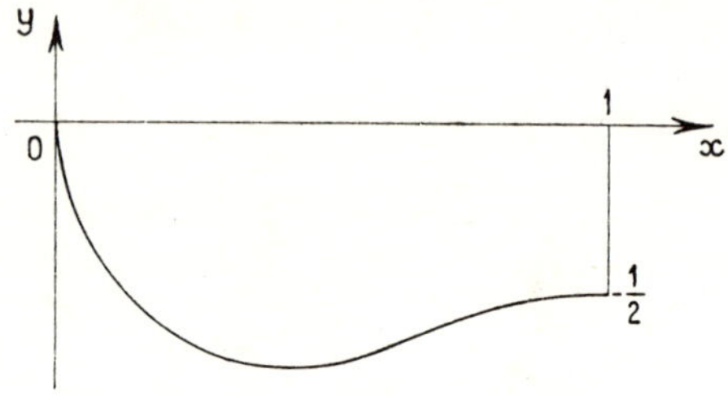

FIGURE 2

The *curve representing the remainder* $R_n(x)$ is shown in Figure 2. For $0 \leqslant x < 1$, we have

$$R_n(x) = x^{2n+2} \frac{\log x}{1 - x^2}.$$

Note that for fixed n, $R_n(x) \to -\frac{1}{2}$ as $x \to 1$.

For every fixed x in the interval $0 \leqslant x < 1$, $R_n(x)$ tends to 0 as $n \to \infty$. At the point $x = 1$, $R_n(x)$ is the remainder of a series whose general term is 0. Therefore, $R_n(1) = 0$. This does not contradict the fact that

$$\lim_{x-1} R_n(x) = -\tfrac{1}{2}.$$

It simply means that the distance between $R_n(x)$ and the zero function [this distance meaning, in this case, $|R_n(x)|$] does not tend to 0 as $n \to \infty$. Thus the series $\Sigma\, u_n(x)$ does not converge uniformly to the zero function on the closed interval $0 \leqslant x \leqslant 1$. It does converge uniformly, however, on any interval of the form $0 \leqslant x \leqslant a < 1$. To show this, let us majorize the remainder directly. Let M be the least upper bound of the continuous function $x^2|\log x|/(1 - x^2)$ in the closed interval [0, 1]. Then

$$|R_n(x)| \leqslant Mx^{2n}.$$

Therefore, if $|x| \leqslant a < 1$, we have

$$|R_n(x)| \leqslant Ma^{2n}.$$

The expression on the right is independent of x and it tends to 0 as $n \to \infty$. The uniform convergence follows from this by definition.

Now, we wish to integrate the series $u_n(x)$ term by term from 0 to 1. Since this series does not converge uniformly on [0, 1], we need to show directly that

$$\lim_{n \to \infty} \int_0^1 R_n(x)\,dx = 0.$$

For this we use the majorization $|R_n(x)| < Mx^{2n}$. When we integrate, we obtain

$$\left|\int_0^1 R_n(x)\,dx\right| \leqslant \int_0^1 |R_n(x)|\,dx \leqslant M\int_0^1 x^{2n}\,dx = \frac{M}{2n+1}.$$

The value of $\int_0^1 R_n(x)\,dx$ tends to 0 as $n \to \infty$. Since the integral of the remainder tends to 0, this series can be integrated term by term.

Consequently,

$$\int_0^1 \frac{x^2 \log x}{1 - x^2}\,dx = \sum_{n=1}^{\infty} \int_0^1 u_n(x)\,dx = -\sum_{n=1}^{\infty} \frac{1}{(2n+1)^2}$$

or

$$\int_0^1 \frac{\log x}{1 - x^2}\,dx = -\sum_{n=0}^{\infty} \frac{1}{(2n+1)^2}.$$

The series $\Sigma\,(2n + 1)^{-2}$ leads to the series $\zeta(2)$. Its sum is equal to

$$\zeta(2) - \frac{\zeta(2)}{2^2} = \frac{\pi^2}{8}.$$

Therefore, once again we find

$$A = -\frac{\pi^2}{8}.$$

D. *Another proof.* Since the convergence is uniform for $0 \leqslant x \leqslant a < 1$, we can integrate the series $x^{2n} \log x$ from 0 to a:

$$\int_0^a \frac{x^2 \log x \, dx}{1 - x^2} = \sum_{n=1}^{\infty} \int_0^a x^{2n} \log x \, dx$$

$$= \sum_{n=1}^{\infty} \frac{a^{2n+1} \log a}{2n + 1} - \sum_{n=1}^{\infty} \frac{a^{2n+1}}{(2n + 1)^2} .$$

The expression on the right is the difference of two series. The second series is a power series majorized by the series $(2n + 1)^{-2}$. It is uniformly convergent for $| a | \leqslant 1$. Its sum tends to

$$- \sum_{n=1}^{\infty} \frac{1}{(2n + 1)^2}$$

as $a \rightarrow 1$.

We wish to show that the sum of the series

$$\sum_{n=1}^{\infty} \frac{a^{2n+1} \log a}{2n + 1}$$

tends to 0 as $a \rightarrow 1$. If this actually is the case, we shall have in the limit

$$\int_0^1 \frac{x^2 \log x}{1 - x^2} \, dx = - \sum_{n=1}^{\infty} \frac{1}{(2n + 1)^2} ,$$

since the integral $\int_0^a x^2 \log x \, dx/(1 - x^2)$ is a continuous function of a.

The series $\Sigma \, [a^{2n+1}/(2n + 1)] \log a$ converges faster than the initial series $x^{2n} \log x$. The remainder of this series is less than

$$\frac{a^{2n+1}}{2n + 1} | \log a | (1 + a^2 + \cdots) \leqslant \frac{a \, | \log a \, |}{1 - a^2} \frac{1}{2n + 1} .$$

Since the function $a | \log a |/(1 - a^2)$ is continuous, it has an upper bound M. The remainder is less than $M/(2n + 1)$, which is independent of a and tends to 0 as $n \rightarrow \infty$. Therefore, the series is uniformly convergent, which enables us to complete the proof.

We see that brutal utilization of the theorems on uniform convergence leads in such a case to unnecessary complications and that the methods used in Problems A and C are faster.

E. The remainder

$$R_n(x) = x^{2n+2} \frac{\log x}{1 - x^2} = x^{2n} \frac{x^2 \log x}{1 - x^2}$$

possesses the following properties:

(1) It is integrable on [0, 1].

(2) It is bounded by a constant M.

(3) It tends to 0 (although not uniformly) for every x such that $0 \leqslant x \leqslant 1$.

A constant is "summable" with respect to Lebesgue measure on [0, 1]. Therefore, the conditions under which Lebesgue's convergence theorem (see the introduction to this chapter) are satisfied. The integral of $R_n(x)$ tends to the integral of $\lim R_n(x)$, that is, to the zero function. This amounts to saying that the series $\sum x^{2n} \log x$ can be integrated term by term from 0 to 1 despite the lack of uniform convergence in a neighborhood of $x = 1$.

These two problems amount to proving or applying Lebesgue's theorem in this particular case.

3. Evaluation of an integral by differentiation under the integral sign. Generalization of the result

Problems. Consider the integral

$$I(x) = \int_0^\infty \frac{e^{-t} - e^{-tx}}{t} \, dt.$$

A. Examine the convergence of this integral.

B. Examine the continuity of this integral as a function of x.

C. Show that we can differentiate under the integral sign for all positive values of x.

D. Evaluate the derivative $I'(x)$ and from it evaluate $I(x)$.

E. Let $f(t)$ be a function defined for $t > 0$ possessing the following properties:

(1) $f(t)$ is integrable over every finite interval.

(2) $f(t)$ tends to a limit L as $t \to 0$.

(3) $\int_\epsilon^\infty t^{-1} f(t) \, dt$ exists for every positive number ϵ.

Show that the integral

$$\int_0^\infty \frac{f(t) - f(tx)}{t}\, dt$$

exists for all $x > 0$ and that its value if $L \log x$. (This integral may be thought of as the limit of

$$\int_\epsilon^A \frac{f(t) - f(tx)}{t}\, dt,$$

as $\epsilon \to 0$ and $A \to \infty$.)

(N.B.: The solution of Problem E is independent of the other four.)

Solutions. **A.** *Convergence of the integral.* We set

$$K(x, t) = \frac{e^{-t} - e^{-tx}}{t}.$$

In a neighborhood of infinity, we have $e^{-t} < 1$ and, for $x < 0$,

$$| K(x, t) | > \frac{e^{|tx|} - 1}{| t |},$$

so that $| K(x, t)|$ increases without bound as t increases. Therefore, the integral diverges.

If $x = 0$,

$$K(x, t) = \frac{e^{-t} - 1}{t} = \frac{e^{-t}}{t} - \frac{1}{t}.$$

Now, the inequality

$$\int_1^\infty \frac{e^{-i}}{t}\, dt < \int_1^\infty e^{-i}\, dt$$

shows that the first integral converges. Since the integral of $1/t$ diverges, the integral $\int_0^\infty K(x, t)\, dt$ also diverges.

If $x > 0$, the integral is, in a neighborhood of infinity (for example, for $t > 1$), the difference between the two integrals

$$\int_1^\infty \frac{e^{-t}}{t}\, dt \qquad \text{and} \qquad \int_1^\infty \frac{e^{-tx}}{t}\, dt,$$

both of which converge. Therefore, the integral

$$\int_1^\infty \frac{e^{-t} - e^{-tx}}{t}\, dt$$

converges.

In a neighborhood of the origin, we have

$$K(x, t) = \frac{1 - t - 1 + tx + \epsilon t}{t} = x - 1 + \epsilon,$$

where ϵ is a function of t that tends to 0 with t.

Thus the function $K(x, t)$ is continuous at the origin. There is no problem of convergence. Finally, the integral $I(x)$ exists for $x > 0$.

B. *Continuity of* $I(x)$. In the xt-plane, $K(x, t)$ is a continuous function for $t \geqslant 0$ and for $x \geqslant 0$.

The integral $\int_0^\infty K(x, t)\, dt$ converges uniformly with respect to x for $x \geqslant x_0$, where x_0 is a given positive number. This is true because, for $x \geqslant x_0 > 0$, we have

$$\frac{e^{-tx}}{t} < \frac{e^{-tx_0}}{t},$$

and the integral

$$\int_0^\infty \frac{e^{-tx_0}}{t}\, dt,$$

which is independent of x, converges.

Thus $I(x)$ is continuous for $x > 0$ (that is, it is continuous at every positive value of x).

C. *Differentiation.* The partial derivative $\partial K(x, t)/\partial x = e^{-tx}$ is a continuous function of t and x throughout the xt-plane.

The integral

$$J(x) = \int_0^\infty e^{-tx}\, dt$$

exists and converges uniformly with respect to x over the entire interval $[x_0, x_1]$, where $0 < x_0 < x_1$, because, in that interval, $J(x)$ is majorized by

$$\int_0^\infty \exp(-tx_0)\, dt = \left[-\frac{1}{x_0}\exp(-tx_0)\right]_0^\infty = \frac{1}{x_0}.$$

The integral $\int_0^\infty K(x, t)\, dt$ exists for $x = 1$ because $I(1) = 0$ (and the same applies for all $x > 0$). Thus $J(x)$ is the derivative of $I(x)$ for all $x > 0$.

D. Now

$$J(x) = \int_0^\infty e^{-tx}\, dt = \frac{1}{x}.$$

Therefore,

$$I(x) = \log x + C,$$

where C is a constant. Since $I(1) = 0$, we see that $C = 0$. Finally,

$$\boxed{I(x) = \log x \qquad \text{for } x > 0.}$$

E. This result is a particular case of a more general formula, which can be proved by a special procedure not involving uniform convergence.

Suppose that $f(t)$ is integrable for $t > 0$. Suppose also that $f(t)$ tends to a limit L as $t \to 0$ and that the integral

$$\int_{\epsilon}^{\infty} \frac{f(t)}{t}\, dt$$

exists for all $\epsilon > 0$. Examples are the functions

$$\left.\begin{aligned} f(t) &= e^{-t} \\ f(t) &= \cos t \end{aligned}\right\} (L = 1).$$

Consider the integral

$$\int_0^{\infty} \frac{f(t) - f(tx)}{t}\, dt \qquad (x > 0).$$

This integral is the limit of

$$Z = \int_{\epsilon}^{A} \frac{f(t) - f(tx)}{t}\, dt$$

as $\epsilon \to 0$ and $A \to \infty$. Now

$$Z = \int_{\epsilon}^{A} \frac{f(t)}{t}\, dt - \int_{\epsilon}^{A} \frac{f(t'x)}{t'}\, dt'.$$

Let us set $t'x = t$. The second integral becomes

$$\int_{\epsilon x}^{Ax} \frac{f(t)\, dt}{t}.$$

Therefore,

$$Z = \int_{\epsilon}^{\epsilon x} \frac{f(t)}{t}\, dt - \int_{A}^{Ax} \frac{f(t)}{t}\, dt.$$

Since the integral $\int t^{-1}f(t)\,dt$ converges in a neighborhood of infinity, the integral

$$\int_A^{Ax} \frac{f(t)}{t}\,dt = \int_A^{\infty} \frac{f(t)}{t}\,dt - \int_{Ax}^{\infty} \frac{f(t)}{t}\,dt$$

tends to 0 as $A \to \infty$. Therefore,

$$\int_0^{\infty} \frac{f(t) - f(tx)}{t}\,dt = \lim_{\epsilon \to 0} \int_{\epsilon}^{\epsilon x} \frac{f(t)}{t}\,dt.$$

Since the sign of $1/t$ does not change, we may apply the mean-value theorem. In the second member, we have

$$f(\xi) \int_{\epsilon}^{\epsilon x} \frac{dt}{t} = f(\xi) \log x,$$

where ξ lies between ϵx and ϵ. Also, $\xi \to 0$ as $\epsilon \to 0$, and $f(\xi)$, by hypothesis, approaches a limit L. Finally, by virtue of the hypotheses made on $f(t)$, we have

$$\int_0^{\infty} \frac{f(t) - f(tx)}{t}\,dt = L \log x \qquad (x > 0).$$

For $f(t) = e^{-t}$ we again obtain the result of Problem D. For $f(t) = \cos t$, we have, similarly,

$$\int_0^{\infty} \frac{\cos t - \cos tx}{t}\,dt = \log x.$$

Remark. We have not assumed f to be differentiable. Therefore, it is not in general permissible to differentiate under the integral sign. The method used for solving Questions A–D is not applicable in the general case.

4. Integration of a sequence that does not converge to a function

We saw in Exercise 3 that, under certain hypotheses made on the function $f(t)$, the integral

$$\int_0^{\infty} \frac{f(t) - f(tx)}{t}\,dt$$

has a value independent of the form of $f(t)$. It is equal to $L \log x$, where $L = \lim_{t\to 0} f(t)$ and does not depend on any local property of f. The reader may be familiar with the following result (Dirichlet's integral):

$$\lim_{\lambda\to\infty} \int_0^a f(t) \frac{\sin \lambda t}{t} \, dt = \frac{\pi}{2} f(+0) \qquad (a > 0).$$

Here it is assumed that $f(t)$ is locally integrable and possesses a right-hand derivative at the origin. Therefore, it has a definite limit, which does not depend on the local structure of $f(t)$. We should note also that this integral has a limit even though the integrand does not.

The present exercise will exhibit another example of a limit that does not depend on the local structure of a function. Again, it is a question of the integral of a sequence of functions that does not converge to any function.

All these examples deal with the Dirac distribution δ.

Problem. Suppose that $f(x)$ is a continuous function for $0 \leqslant x \leqslant a$, where a is given positive number. Show that

$$\lim_{h\to 0} \int_0^a \frac{h}{h^2 + x^2} f(x) \, dx = \frac{\pi}{2} f(0).$$

Solution. We wish to calculate the limit of the integral

$$\int_0^a \frac{h}{h^2 + x^2} f(x) \, dx \qquad (a > 0)$$

as $h \to 0$, where $f(x)$ is a continuous function.

Note that the function

$$F(x, h) = \frac{h}{h^2 + x^2} f(x),$$

is not continuous over the region $0 \leqslant x \leqslant a$, h real. To see this observe that

$$F(x, 0) = 0 \qquad [\text{therefore, } \lim_{x\to 0} F(x, 0) = 0],$$

$$F(0, h) = \frac{1}{h} f(0) \qquad [\text{therefore, } \lim_{h\to 0} F(0, h) = \infty]$$

[except of course if $f(0) = 0$].

We set

$$I(h) = \int_0^a \frac{h}{h^2 + x^2} [f(x) - f(0)] \, dx.$$

Since the function $f(x)$ is continuous, we may assign to every $\epsilon > 0$ a number $\eta(\epsilon)$ such that the inequality $|x| < \eta(\epsilon)$ implies

$$|f(x) - f(0)| < \epsilon.$$

If we fix ϵ, and consequently η, we may write

$$\begin{aligned} I(h) &= \int_0^\eta \frac{h}{h^2 + x^2} [f(x) - f(0)]\, dx + \int_\eta^a \frac{h}{h^2 + x^2} [f(x) - f(0)]\, dx \\ &= J(h) + K(h). \end{aligned}$$

(a) $\lim_{h\to 0} K(h) = 0$. This is true because $F(x, h)$ is continuous over the region $\eta \leqslant x \leqslant a$, h arbitrary, so that $K(h)$ is a continuous function and

$$\lim_{h\to 0} K(h) = K(0) = 0.$$

(b)

$$|J(h)| \leqslant \epsilon \int_0^\eta \frac{|h|}{h^2 + x^2}\, dx = \epsilon \arctan \frac{\eta}{|h|} \leqslant \epsilon \frac{\pi}{2}.$$

This inequality implies that

$$\lim_{h\to 0} I(h) = 0.$$

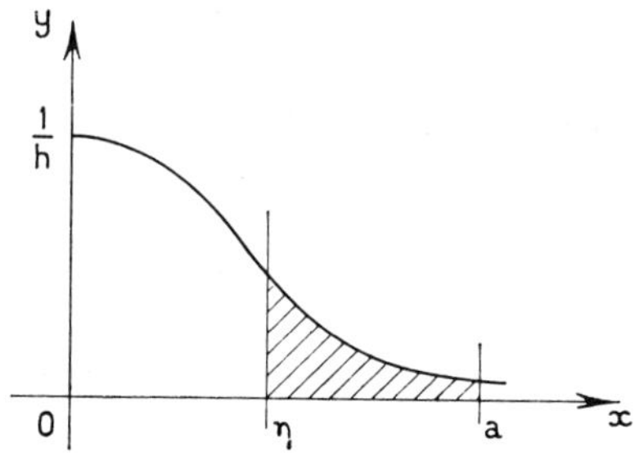

FIGURE 3

Therefore,

$$\begin{aligned} \lim_{h\to 0} \int_0^a \frac{h}{h^2 + x^2} f(x)\, dx &= \lim_{h\to 0} f(0) \int_0^a \frac{h}{h^2 + x^2}\, dx \\ &= f(0) \lim_{h\to 0} \left[\arctan \frac{x}{h}\right]_0^a \\ &= f(0) \lim_{h\to 0} \arctan \frac{a}{h} = \frac{\pi}{2} f(0). \end{aligned}$$

Finally,

$$\lim_{h\to 0} \int_0^a \frac{h}{h^2 + x^2} f(x)\, dx = \frac{\pi}{2} f(0).$$

5. Differential equations involving elliptic integrals

The integral

$$I = \int_a^b \frac{dx}{[(x-a)(b-x)(c-x)]^{1/2}}$$

is an elliptic integral. It is studied most naturally in the complex plane. However, if a, b, and c are real numbers such that $a < b < c$, I is real. It is interesting to treat I as a function of a, b, c. In this case, by certain modifications in the notation, I appears as a particular hypergeometric function.

As a function of a, b, or c, I verifies a second-order linear differential equation with variable coefficients.

It is easy to calculate $\partial I/\partial c$ and $\partial^2 I/\partial c^2$ by differentiation under the integral sign and making use of the principle of uniform convergence. The calculation of $\partial I/\partial b$ and $\partial^2 I/\partial b^2$, on the other hand, requires certain precautions.

The relation between I and its derivatives is obtained by application of a general procedure for reducing elliptic (or, more generally, Abelian) integrals. It can be shown, for example, that, if $R(x)$ is a rational function, all the indefinite integrals of the form

$$\int \frac{R(x)\, dx}{[(x-a)(b-x)(c-x)]^{1/2}}$$

are linear combinations (nonhomogeneous) of integrals of the form

$$\int \frac{dx}{[(x-a)(b-x)(c-x)]^{1/2}}, \qquad \int \frac{x\, dx}{[(x-a)(b-x)(c-x)]^{1/2}},$$

$$\int \frac{dx}{(x-h)\,[(x-a)(b-x)(c-x)]^{1/2}}.$$

Problems. Consider the integral

$$I = \int_a^b \frac{dx}{[(x-a)(b-x)(c-x)]^{1/2}},$$

where a, b, and c are three real numbers such that $a < b < c$.

A. Show that we may calculate $\partial I/\partial c$ and $\partial^2 I/\partial c^2$ by differentiating under the integral sign.

B. Show that I, treated as a function of c, verifies a homogeneous second-order linear differential equation. Use the method of reducing elliptic integrals that consists in differentiating the function

$$\frac{[(x-a)(b-x)(c-x)]^{1/2}}{(c-x)^2}$$

and integrating the formula obtained (after arranging the numerators in decreasing powers of $c-x$).

C. It is not possible to differentiate I under the integral sign with respect to b. By a homographic change of variable, show that

$$I = \int_c^\infty \frac{dx}{[(x-a)(b-x)(c-x)]^{1/2}}$$

From this, show that I, treated as a function of b, verifies a homogeneous second-order linear differential equation.

D. Transform I by the change of variable

$$x = a + (b-a)u.$$

Write the differential equation that the function $J = I(b-a)^{1/2}$ satisfies in the variable $\lambda = (c-a)/(b-a)$. Then again find the result of Problem C.

Solutions. **A.** It can be shown immediately that the integral

$$I = \int_a^b \frac{dx}{[(x-a)(b-x)(c-x)]^{1/2}},$$

where a, b, and c are real parameters such that $a < b < c$, converges.

I is a function of a, b, and c. As a function of c, I can be differentiated under the integral sign. Specifically,

$$\frac{\partial I}{\partial c} = -\frac{1}{2}\int_a^b \frac{dx}{[(x-a)(b-x)(c-x)]^{1/2}}\,\frac{1}{c-x}.$$

This integral converges uniformly in the interval

$$c_1 \leqslant c \leqslant c_2,$$

where c_1 is a number exceeding b, since the integrand is majorized by

$$\frac{1}{[(x-b)(b-x)(c_1-x)]^{1/2}}\frac{1}{c_1-x},$$

which is an integrable function between a and b. Similarly,

$$\frac{\partial^2 I}{\partial c^2}=\frac{3}{4}\int_a^b\frac{dx}{[(x-a)(b-x)(c-x)]^{1/2}}\frac{1}{(c-x)^2}.$$

B. The integral I and its derivatives with respect to c are integrals of the form

$$\int_a^b\frac{1}{(c-x)^n}\frac{dx}{[(x-a)(b-x)(c-x)]^{1/2}}.$$

Consider the function

$$\frac{[(x-a)(b-x)(c-x)]^{1/2}}{(c-x)^2}.$$

Its derivative is

$$\frac{d}{dx}\frac{[(x-a)(b-x)(c-x)]^{1/2}}{(c-x)^2}=\frac{2(x-a)(b-x)(c-x)}{(c-x)^3\,[(x-a)(b-x)(c-x)]^{1/2}}$$
$$+\frac{(c-x)(a+b-2x)-(x-a)(b-x)}{2(c-x)^2\,[(x-a)(b-x)(c-x)]^{1/2}}.$$

We arrange the numerators in decreasing powers of $c-x$:

$$(x-a)(b-x)=-(c-x)^2+(2c-a-b)(c-x)+(c-a)(b-c)$$
$$a+b-2x=a+b-2c+2(c-x).$$

After regrouping the terms, we get

$$\frac{d}{dx}\frac{[(x-a)(b-x)(c-x)]^{1/2}}{(c-x)^2}=-\frac{3}{2\sqrt{}}+\frac{3}{2}\frac{2c-a-b}{(c-x)\sqrt{}}+\frac{3}{2}\frac{(c-a)(b-c)}{(c-x)^2\sqrt{}}$$
$$+\frac{a+b-2c}{2}\frac{1}{(c-x)\sqrt{}}+\frac{1}{\sqrt{}}$$
$$=\frac{3}{2}\frac{(c-a)(b-c)}{(c-x)^2\sqrt{}}+\frac{(2c-a-b)}{(c-x)\sqrt{}}-\frac{1}{2\sqrt{}},$$

where the symbol $\sqrt{}$ denotes $[(x-a)(b-x)(c-x)]^{1/2}$.

Let us integrate from a to b, keeping the expressions for I, $\partial I/\partial c$, and $\partial^2 I/\partial c^2$ in mind. The first member vanishes. We have finally,

$$4(a-c)(c-b)\frac{\partial^2 I}{\partial c^2}+4(a+b-2c)\frac{\partial I}{\partial c}-I=0.$$

Thus I verifies a homogeneous linear second-order differential equation. For example, the integral

$$I=\int_0^1\frac{dx}{[x(1-x)(c-x)]^{1/2}},$$

where $c>1$, verifies the equation

$$4c(1-c)I''+4(1-2c)I'-I=0.$$

C. It is not possible to make an analogous calculation to obtain $\partial I/\partial b$. The expression that we would obtain by performing a differentiation of an integral of the form $I=\int_a^b K(x,b)\,dx$ with respect to b would be

$$\frac{\partial I}{\partial b}=\int_a^b\frac{\partial K}{\partial b}\,dx+K(b,b).$$

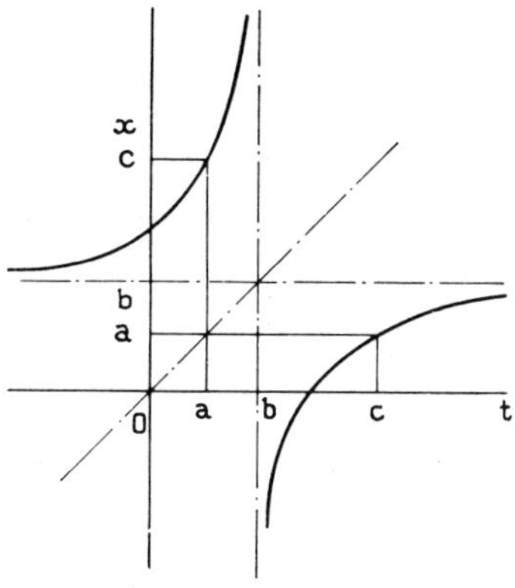

FIGURE 4

Now, the integral in this expression diverges, and we must add to it the term $K(b, b)$, which in this case is infinite.

To calculate $\partial I/\partial b$, we must therefore, first make a transformation of I. This can be done by making a homographic change of variable.

We wish a transformation that will satisfy the following correspondence:

x	a	b	c	∞
t	c	∞	a	b

The transformation

$$x = \frac{bt + h}{t - b}$$

is compatible with this correspondence if

$$h = ac - ab - bc.$$

We then have

$$x - a = \frac{(b - a)(t - c)}{t - b},$$

$$b - x = \frac{(b - a)(c - b)}{t - b},$$

$$c - x = \frac{(c - b)(t - a)}{t - b},$$

$$(x - a)(b - x)(c - x) = \frac{(b - a)^2(c - b)^2(t - c)(t - a)}{(t - b)^3},$$

$$\frac{dx}{dt} = \frac{(c - b)(b - a)}{(t - b)^2}.$$

As x increases from a to b, t increases from c to ∞. Thus our new expression for I is

$$I = \int_c^\infty \frac{dt}{[(t - a)(b - t)(c - t)]^{1/2}}.$$

In this form, b no longer appears as a limit of integration and we can show, as in Problem A, that $\partial I/\partial b$ and $\partial^2 I/\partial b^2$ can be evaluated by differentiation under the integral sign. We have

$$\frac{\partial I}{\partial b} = -\frac{1}{2}\int_c^\infty \frac{dt}{[(t - a)(b - t)(c - t)]^{1/2}}\,\frac{1}{b - t},$$

$$\frac{\partial^2 I}{\partial b^2} = \frac{3}{4}\int_c^\infty \frac{dt}{[(t - a)(b - t)(c - t)]^{1/2}}\,\frac{1}{(b - t)^2}.$$

The calculation in Problem B remains valid if we permute b and c. In this way, we obtain the differential equation

$$4(c - b)(b - a)\frac{\partial^2 I}{\partial b^2} + 4(a + c - 2b)\frac{\partial I}{\partial b} - I = 0.$$

D. To differentiate I with respect to b, we may also make a linear change of variable that will yield constant limits of integration. For example, we may set

$$x = a + (b - a)u.$$

We then obtain

$$I = \int_0^1 \frac{du}{\{u(1-u)[c-a-(b-a)u]\}^{1/2}}.$$

If we now set

$$\lambda = \frac{c-a}{b-a}, \qquad J = I(b-a)^{1/2},$$

we obtain

$$J = \int_0^1 \frac{du}{[u(1-u)(\lambda-u)]^{1/2}},$$

where λ is a parameter exceeding 1. The integral J is of the type studied in Problem B. Thus it satisfies the differential equation

$$4\lambda(1-\lambda)\frac{d^2J}{d\lambda^2} + 4(1-2\lambda)\frac{dJ}{d\lambda} - J = 0.$$

Now we only need to return to the variables b and I. We have

$$\frac{db}{d\lambda} = -\frac{(b-a)^2}{c-a}, \ \lambda(1-\lambda) = -\frac{(c-a)(c-b)}{(b-a)^2}, \ 1-2\lambda = \frac{b+a-2c}{b-a}.$$

$$\frac{dJ}{d\lambda} = \frac{dJ}{db}\frac{db}{d\lambda} = -\frac{(b-a)^{5/2}}{c-a}\frac{dI}{db} - \frac{1}{2}\frac{(b-a)^{3/2}}{c-a}I,$$

$$\frac{d^2J}{d\lambda^2} = \frac{(b-a)^{9/2}}{(c-a)^2}\frac{d^2I}{db^2} + \frac{3(b-a)^{7/2}}{(c-a)^2}\frac{dI}{db} + \frac{3}{4}\frac{(b-a)^{5/2}}{(c-a)^2}I.$$

Finally, as in Problem C, we obtain

$$4(c-b)(b-a)\frac{\partial^2 I}{\partial b^2} + 4(a+c-2b)\frac{\partial I}{\partial b} - I = 0.$$

Remark. The general solution of the differential equation

$$4(c-b)(b-a)y'' + 4(a+c-2b)y' - y = 0$$

can be solved in quadrature. If we set

$$y = Iu,$$

we obtain

$$\frac{u''}{u'} + 2\frac{I'}{I} + \frac{a+c-2b}{(c-b)(b-a)} = 0.$$

This can be integrated once to yield

$$\log\frac{u'I^2}{K} + \log(b-a)(b-c) = 0, \qquad u' = \frac{K}{I^2(b-a)(b-c)},$$

where K is an arbitrary constant. Since I is known, we can get u by evaluating an indefinite integral (not, of course, in terms of elementary functions).

6. Evaluation of Gauss' integral

There are many classical procedures for evaluating the integrals

$$\int_0^\infty \exp(-t^2)\,dt, \qquad \int_0^\infty \exp(-t^2)\cos xt\,dt,$$

for example, by use of multiple integrals, by differentiation under the integral sign, by integration in the complex plane. In each particular case, it is instructive to examine whichever of the properties of the exponential functions come into play in connection with that procedure. Let us recall some of these:

(a) $e^{x+y} = e^x e^y$ (functional property).

(b) $de^t/dt = e^t$ (property of the derivative).

(c) $e^t = \lim_{n\to\infty} [1 + (t/n)]^n$.

(d) $e^t = 1 + t + (t^2/2) + \cdots + (t^n/n!) + \cdots$.

In the present exercise, we shall use property (c) to evaluate Gauss' integral. In the following exercise, we shall evaluate the Fourier transform of $\exp(-x^2)$ by using property (d).

Problems. A. Evaluate the integral

$$I_n = \int_0^1 x^n \frac{dx}{(1-x^2)^{1/2}},$$

by the method of reduction to Abelian integrals.

B. Calculate I_n by using the gamma and beta functions.

C. Transform I_{2n+1} by making the change of variable

$$x = \left(1 - \frac{t^2}{n}\right)^{1/2}.$$

The integral $n^{1/2}I_{2n+1}$ takes the form

$$J_n = \int_0^{\sqrt{n}} \left(1 - \frac{t^2}{n}\right)^n dt.$$

Show that J_n tends to Gauss' integral,

$$\int_0^\infty \exp(-t^2)\,dt,$$

as $n \to \infty$.

D. Show that $2nI_{2n}I_{2n+1}$ tends to $\pi/2$ and that I_n/I_{n+1} tends to 1 as $n \to \infty$. From this, find the limit J_n and evaluate Gauss' integral.

Solutions. **A.** I_n is an Abelian integral associated with the conic (circle)

$$y^2 = 1 - x^2.$$

The method of reduction consists in writing

$$\begin{aligned}\frac{d}{dx}\,[x^n\,(1-x^2)] &= nx^{n-1}\,(1-x^2)^{1/2} - \frac{x^{n+1}}{(1-x^2)^{1/2}}\\ &= \frac{nx^{n-1}-(n+1)x^{n+1}}{(1-x^2)^{1/2}},\end{aligned}$$

and integrating term by term. In the general case, it is a matter of indefinite integrals. Here, we find the definite integral from 0 to 1. We obtain

$$0 = nI_{n-1} - (n+1)I_{n+1}, \qquad (n \geqslant 1),$$

with

$$I_0 = \int_0^1 \frac{dx}{(1-x^2)^{1/2}} = \frac{\pi}{2},$$

$$I_1 = \int_0^1 \frac{x\,dx}{(1-x^2)^{1/2}} = -\,[(1-x^2)^{1/2}]_0 = 1.$$

Consequently,

$$I_{2n} = \frac{1\cdot 3 \cdots (2n-1)}{2\cdot 4 \cdots 2n}\,\frac{\pi}{2} = \frac{(2n)!}{2^{2n}(n!)^2}\,\frac{\pi}{2},$$

$$I_{2n+1} = \frac{2\cdot 4 \cdots 2n}{3\cdot 5 \cdots (2n+1)} = \frac{2^{2n}(n!)^2}{(2n+1)!}.$$

B. If we set $x^2 = t$, we obtain

$$I_n = \frac{1}{2}\int_0^1 t^{(n-1)/2}\,(1-t)^{-1/2}\,dt = \frac{1}{2}\,B\left(\frac{n+1}{2},\frac{1}{2}\right),$$

where

$$\mathrm{B}(p,q) = \int_0^1 t^{p-1}(1-t)^{q-1}\,dt$$

is Euler's beta function.

We know that

$$\mathrm{B}(p, q) = \frac{\Gamma(p)\Gamma(q)}{\Gamma(p+q)},$$

where $\Gamma(p) = \int_0^\infty t^{p-1}e^{-t}\,dt$ is Euler's gamma function.

Thus

$$I_n = \frac{1}{2}\left[\Gamma\left(\frac{n+1}{2}\right)\Gamma\left(\frac{1}{2}\right)\Big/\,\Gamma\left(\frac{n}{2}+1\right)\right] = \frac{\sqrt{\pi}}{2}\left[\Gamma\left(\frac{n}{2}+\frac{1}{2}\right)\Big/\,\Gamma\left(\frac{n}{2}+1\right)\right].$$

Now if m is an integer,

$$\Gamma(m) = (m-1)!, \qquad \Gamma\left(m+\frac{1}{2}\right) = \frac{(2m)!}{2^{2m}m!}\sqrt{\pi}.$$

Thus we again find the results of the preceding problem.

C. In the integral I_{2n+1}, let us make the change of variable

$$x = \left(1 - \frac{t^2}{n}\right)^{1/2}.$$

We obtain

$$I_{2n+1} = \frac{1}{\sqrt{n}}\,J_n \qquad \text{where} \qquad J_n = \int_0^{\sqrt{n}} \left(1 - \frac{t^2}{n}\right)^n dt.$$

For every point t, the expression $(1 - t^2/n)^n$ tends to $\exp(-t^2)$ as $n \to \infty$. Thus J_n tends *formally* to $\int_0^\infty \exp(-t^2)\,dt$.

However, one difficulty presents itself. Although the sequence $\{S_n(t)\}$ defined by $S_n(t) = (1 - t^2/n)^n$ tends uniformly to $S(t) = \exp(-t^2)$ (as we shall see below) in the interval $0 \leqslant t \leqslant \infty$, we still are not justified in asserting that

$$\lim_{n\to\infty} \int_0^{\sqrt{n}} S_n(t)\,dt = \int_0^\infty \exp(-t^2)\,dt,$$

because the region of integration depends on n and becomes infinite in the limit.

We proceed in the following fashion:

(a) First, let us prove that, for $0 \leqslant \alpha \leqslant n$,

$$1 - \frac{\alpha}{n} \leqslant e^{-\alpha/n}, \qquad \left(1 + \frac{\alpha}{n}\right) \leqslant e^{\alpha/n}, \qquad 1 - \frac{\alpha^2}{n} \leqslant \left(1 - \frac{\alpha^2}{n^2}\right)^n.$$

The first of these inequalities is established by considering the function

$$f(\alpha) = e^{-\alpha/n} - 1 + \frac{\alpha}{n},$$

which vanishes at $\alpha = 0$ and increases with increasing positive values of α.

The second and third inequalities are proved in an analogous fashion.

(b) From these inequalities it is easy to show that

$$\left(1 - \frac{\alpha^2}{n}\right) e^{-\alpha} \leqslant \left(1 - \frac{\alpha}{n}\right)^n \leqslant e^{-\alpha}$$

or

$$0 \leqslant e^{-\alpha} - \left(1 - \frac{\alpha}{n}\right)^n \leqslant \frac{\alpha^2 e^{-\alpha}}{n}. \tag{1}$$

Next, we show that the sequence whose general term is $(1 - \alpha/n)^n$ for $0 \leqslant \alpha \leqslant n$ and equal to 0 for $\alpha \geqslant n$ tends uniformly to $e^{-\alpha}$ on the interval $0 \leqslant \alpha < \infty$ as $n \to \infty$. This is true because

$$\sup_\alpha \frac{\alpha^2 e^{-\alpha}}{n} = \frac{4e^{-2}}{n} \to 0 \qquad \text{as} \qquad n \to \infty.$$

(c) Let us set $\alpha = t^2$ and integrate the double inequality (1) from $t = 0$ to $t = \sqrt{n}$:

$$0 \leqslant \int_0^{\sqrt{n}} e^{-t^2}\, dt - \int_0^{\sqrt{n}} \left(1 - \frac{t^2}{n}\right)^n dt \leqslant \int_0^{\sqrt{n}} \frac{t^4 \exp(-t^2)}{n}\, dt.$$

The integral $\int_0^{\sqrt{n}} [t^4 \exp(-t^2)/n]\, dt$ tends to 0 as $n \to \infty$, since it is positive and less than the quotient of $\int_0^\infty t^4 \exp(-t^2)\, dt$ (which is a positive constant) divided by n. Therefore,

$$\lim_{n\to\infty} \int_0^{\sqrt{n}} \left(1 - \frac{t^2}{n}\right)^n dt = \int_0 \exp(-t^2)\, dt.$$

D. We have

$$2nI_{2n}I_{2n+1} = \frac{(2n)!2n}{(2n+1)!} \frac{\pi}{2} = \frac{2n}{2n+1} \frac{\pi}{2}.$$

The limit of this expression as $n \to \infty$ is $\pi/2$.

On the other hand, I_n is a decreasing function of n. Therefore,

$$1 < \frac{I_{2n}}{I_{2n+1}} < \frac{I_{2n-1}}{I_{2n+1}} = \frac{2n+1}{2n} \qquad \text{and} \qquad \lim_{n\to\infty} \frac{I_n}{I_{n+1}} = 1.$$

From these two limiting expressions, we obtain

$$\lim_{n\to\infty} nI_n{}^2 = \frac{\pi}{2},$$

whether n assumes even or odd values.

However,

$$J_n = I_{2n+1}\sqrt{n} = I_{2n+1}(2n+1)^{1/2}\left(\frac{n}{2n+1}\right)^{1/2}.$$

Therefore, J_n tends to

$$\left(\frac{\pi}{2}\right)^{1/2} \times \frac{1}{\sqrt{2}} = \frac{1}{2}\sqrt{\pi}.$$

Finally,

$$\int_0^\infty \exp(-t^2)\,dt = \frac{1}{2}\sqrt{\pi}.$$

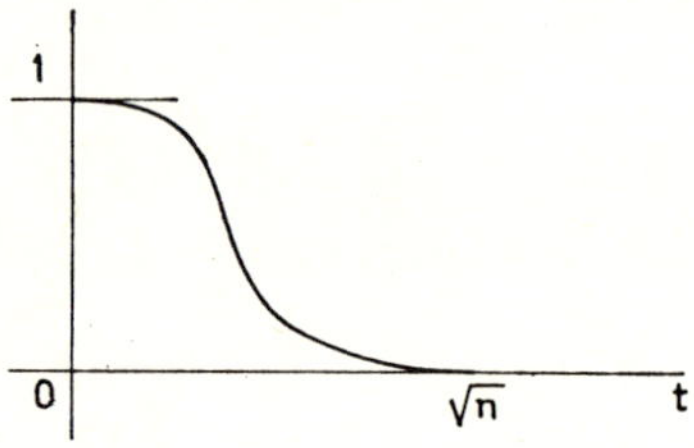

FIGURE 5

E. Let us set

$$f_n(t) = \left(1 - \frac{t^2}{n}\right)^n \qquad \text{for} \qquad 0 \leqslant t \leqslant \sqrt{n},$$

$$f_n(t) = 0 \qquad \text{for} \qquad t \geqslant \sqrt{n}.$$

Then

$$\int_0^{\sqrt{n}} \left(1 - \frac{t^2}{n}\right)^n dt = \int_0^\infty f_n(t)\,dt,$$

and $f_n(t)$ tends uniformly to $\exp(-t^2)$. Now this interval of integration is independent of n. However, we still may not assert that the limit of the integral is the integral of the limit, since the theorem on the integration of uniformly convergent sequences applies only when the interval of integration is finite.

The general theorem that justifies passage to the limit in this case is that of Lebesgue. It can be applied if the following three conditions hold with regard to the measure in question, which in the present case is ordinary Lebesgue measure:

(a) $f_n(t)$ is summable. This is satisfied, since $f_n(t)$ is Riemann-integrable and vanishes outside a finite interval.

(b) $f_n(t)$ approaches a limit for each fixed t. In the present case, this limit is $\exp(-t^2)$.

(c) $f_n(t)$ is majorized by a summable function. Let us prove that this condition is satisfied in the present case. It follows from the inequality

$$\left(1 - \frac{t^2}{n}\right)^n \leqslant e^{-t^2},$$

proved above, and it applies also for the function $f_n(t)$. The function $\exp(-t^2)$ is summable since the integral $\int_0^\infty \exp(-t^2)\, dt$ exists.

Thus passage to the limit is permissible.

7. The Fourier transform of exp($-t^2$)

This exercise will enable us to evaluate the integral

$$\int_0^\infty \exp(-t^2/2) \cos xt\, dt$$

by term-by-term integration of a uniformly convergent series. However, the interval of integration in this case is infinite, so that the elementary integration theorems cannot be applied directly. Therefore, we begin with a simple extension of the theorem on integrating uniformly convergent series (by implicit introduction of a bounded measure). The rest of the exercise is of an elementary nature.

Let us note the difference between uniform convergence on a semi-infinite interval of the form $[a, \infty)$ (i.e., for $x \geqslant a$) and uniform convergence over an entire closed interval $[a, A]$, where A is an arbitrarily large number. To illustrate, note that the sequence $\{f_n(x)\}$ defined by $f_n(x) = x/n$ converges uniformly on the interval $[a, A]$, since

$$\sup_{a \leqslant x \leqslant A} \left| \frac{x}{n} \right| = \frac{A}{n},$$

which tends to 0 as $n \to \infty$. But this sequence does not converge uniformly "for all x," that is, for the entire interval $a \leqslant x$ ($< \infty$), since

$$\sup_{x \geqslant a} \left| \frac{x}{n} \right| = \infty.$$

Up to the factor $1/(2\pi)^{1/2}$, the integrals evaluated in Problem B are the even moments of the normal law (or the Laplace-Gauss law) in probability theory. The last integral is the characteristic function of the normal law. The problem consists in calculating the characteristic function by starting with the moments. Conversely, we know that the characteristic function $\varphi(x)$ of the random variable has the expansion

$$\varphi(x) = 1 + m_1 ix - \frac{m_2}{2} x^2 + \cdots + \frac{m_k}{k!} (ix)^k + \cdots$$

provided the moments m_k exist and the series has a nonzero radius of convergence. Here the radius of convergence is infinite.

Problems. A. Suppose that $\{f_n(x)\}$ is a sequence of continuous functions defined on an interval Δ, finite or infinite. Suppose that $f_n(x)$ converges uniformly on Δ to a function $f(x)$.

If $\mu(x)$ is a function such that the integral $\int_\Delta |\mu(x)| \, dx$ converges, then

$$\lim_{n\to\infty} \int_\Delta f_n(x)\mu(x) \, dx = \int_\Delta f(x)\mu(x) \, dx.$$

B. Evaluate the integral

$$J_n = \int_0^\infty t^{2n} \exp(-t^2/2) \, dt,$$

either by integration by parts, by differentiation of $I = \int_0^\infty \exp(-\lambda t^2/2) \, dt$ under the integral sign, or by reducing the above integral to a gamma function.

C. Show that, for fixed x, the series

$$\exp(-t^2/2) - \frac{x^2}{2} \exp(-t^2/2) + \cdots + (-1)^n \frac{x^{2n} t^{2n}}{(2n)!} \exp(-t^2/2) + \cdots$$

can be integrated term by term from $t = 0$ to $t = \infty$.

D. From these results, show that

$$\int_0^\infty \exp(-t^2/2) \cos xt \, dt = (\pi/2)^{1/2} \exp(-x^2/2).$$

Solutions. **A.** We reason exactly as in the proof of the elementary theorem on the integration of uniformly convergent series over a finite interval.

The problem amounts to showing that

$$\lim_{n\to\infty} \int_{\Delta} [f_n(x) - f(x)]\, \mu(x)\, dx = 0.$$

Now the absolute value of this integral is less than

$$\int_{\Delta} | f_n(x) - f(x) | \, | \mu(x) | \, dx.$$

By hypothesis, there exists a number $n_0(\epsilon)$, independent of x, such that

$$| f_n(x) - f(x) | < \epsilon$$

everywhere on Δ whenever $n > n_0(\epsilon)$. Therefore, the above integral is less than

$$\epsilon \int_{\Delta} | \mu(x) | \, dx.$$

Since $\mu(x)$ is absolutely integrable on Δ, this quantity exists and can be made arbitrarily small. The conclusion then follows.

B. (a) By successive integrations by parts, we obtain the recursion formula

$$J_n = (2n - 1) J_{n-1} ,$$

where

$$J_0 = \int_0^{\infty} \exp(-t^2/2)\, dt = \left(\frac{\pi}{2}\right)^{1/2}.$$

Consequently,

$$\boxed{J_n = 1 \cdot 3 \cdot 5 \cdots (2n - 1)\, J_0 = \frac{(2n)!}{2^n n!} \left(\frac{\pi}{2}\right)^{1/2}.}$$

(b) Let us set $K(\lambda, t) = \exp(-\lambda t^2/2)$. The kernels

$$\frac{\partial^p K}{\partial \lambda^p}(\lambda, t) = (-1)^p \frac{t^{2p}}{2^p} \exp(-\lambda t^2/2)$$

are all continuous for all $t \geqslant 0$ and $\lambda \geqslant 1$. The integrals

$$\int_0^{\infty} \frac{\partial^p K}{\partial \lambda^p}(\lambda, t)\, dt \qquad (p = 0, 1, ..., n)$$

all converge uniformly with respect to λ in the interval Δ (for $\lambda \geqslant 1$) since, for $\lambda \in \Delta$, we have

$$\left| \frac{\partial^p K}{\partial \lambda^p}(\lambda, t) \right| < \frac{t^{2p}}{2^p} \exp(-t^2/2).$$

(This is the test for uniform convergence by means of a majorizing function.)

Therefore, we may differentiate under the integral sign. Consequently, for $\lambda \geqslant 1$, we have

$$\begin{aligned} I^{(n)}(\lambda) = \int_0^\infty (-1)^n \frac{t^{2n}}{2^n} \exp(-\lambda t^2/2)\, dt &= \left(\frac{\pi}{2}\right)^{1/2} \frac{d^n}{d\lambda^n}(\lambda^{-1/2}) \\ &= \left(\frac{\pi}{2}\right)^{1/2} \left(-\frac{1}{2}\right) \cdot \left(-\frac{3}{2}\right) \cdots \left(-\frac{2n-1}{2}\right) \lambda^{-(2+n1)/2}. \end{aligned}$$

Because of the uniform convergence, the function $I^{(n)}(\lambda)$ is continuous for $\lambda = 1$. By taking the limit as $\lambda \to 1$, we obtain

$$I^{(n)}(1) = \int_0^\infty (-1)^n \frac{t^{2n}}{2^n} \exp(-t^2/2)\, dt = \left(\frac{\pi}{2}\right)^{1/2} (-1)^n \frac{1 \cdot 3 \cdot (2n-1)}{2^n}.$$

Since $I^{(n)}(1) = (-1)^n J_n/2^n$, we again obtain the value of J_n.

(c) If we set $t^2 = 2s$, we obtain

$$\begin{aligned} J_n &= 2^{n-(1/2)} \int_0^\infty s^{n-(1/2)} e^{-s}\, ds = 2^{n-(1/2)} \Gamma(n + \tfrac{1}{2}) \\ &= 2^{n-(1/2)} \frac{(2n)!}{2^{2n} n!} \sqrt{\pi} \\ &= \frac{(2n)!}{2^n n!} \left(\frac{\pi}{2}\right)^{1/2} \end{aligned}$$

C. The series whose general term is

$$u_n(t) = (-1)^n \frac{x^{2n}}{(2n)!} t^{2n} \exp(-t^2/2),$$

being an exponential series (multiplied by a term independent of n) converges. To show that this series converges uniformly, we majorize the general term:

$$\sup_t |u_n(t)| = \frac{x^{2n}}{(2n)!} \sup_t [t^{2n} \exp(-t^2/2)] = \left(2 \frac{x^2}{e}\right)^n \frac{n^n}{(2n)!} = v_n.$$

It can be verified, for example, by use of d'Alembert's rule, that v_n is the general term of a convergent series.

But this property is not sufficient for integrating from 0 to ∞ with respect to t.

Therefore, let us apply the result of the preceding problem. We set

$$w_n(t) = (-1)^n \frac{x^{2n}}{(2n)!} t^{2n} \exp(-t^2/4)$$

and

$$\mu(t) = \exp(-t^2/4).$$

This $\mu(t)$ is a positive function that is integrable from 0 to ∞.

As before, we find

$$\sup_t | w_n(t) | = \left(4 \frac{x^2}{e}\right)^n \frac{n^n}{(2n)!} \cdot$$

The series $w_n(t)$ converges uniformly for $t \geqslant 0$. Therefore,

$$\lim_{n\to\infty} \sum_{p=0}^{n} \int_0^\infty w_p(t)\mu(t)\, dt = \int_0^\infty \lim_{n\to\infty} \sum_{p=0}^{n} w_p(t)\mu(t)\, dt$$

or

$$\lim_{n\to\infty} \sum_{p=0}^{n} \int_0^\infty u_p(t)\, dt = \int_0^\infty \lim_{n\to\infty} \sum_{p=0}^{n} u_p(t)\, dt.$$

D. The second member is equal to

$$\int_0^\infty \cos xt \exp(-t^2/2).$$

The first member is equal to

$$\lim_{n\to\infty} \sum_{p=0}^{n} (-1)^p \frac{x^{2p}}{(2p)!} \int_0^\infty t^{2p} \exp(-t^2/2)\, dt = \sum_{p=0}^{\infty} (-1)^p \frac{x^{2p}}{2^p p!} \left(\frac{\pi}{2}\right)^{1/2}.$$

Now,

$$\sum_0^\infty \frac{1}{p!} \left(-\frac{x^2}{2}\right)^p = \exp(-x^2/2).$$

Therefore,

$$\boxed{\int_0^\infty \cos xt \exp(-t^2/2)\, dt = (\pi/2)^{1/2} \exp(-x^2/2).}$$

8. The Fourier transform of $(1 + t^2)^{-1}$

This is an exercise on uniform convergence and differentiation under the integral sign. It yields a Fourier integral and, consequently, the characteristic function in the sense of probability theory, of a random variable that obeys Cauchy's law or, alternatively, Laplace's first law.

Problems. A. Consider the integrals

$$I(x) = \int_0^\infty \frac{\sin xt}{t(1+t^2)}\,dt, \qquad J(x) = \int_0^\infty \frac{\cos xt}{1+t^2}\,dt.$$

Show that these integrals converge uniformly with respect to x for every closed interval of the real axis.

B. Show that the integral

$$K(x) = \int_0^\infty \frac{t}{1+t^2}\sin xt\,dt$$

converges for all x and that the convergence is uniform in every interval of the form $0 < x_0 \leqslant x \leqslant x_1$.

C. Show that $J = dI/dx$ and $K = -d^2I/dx^2$. Then show that $I(x)$ verifies a second-order linear differential equation with constant coefficients. Solve this equation and calculate first $J(x)$, then $K(x)$ and $I(x)$.

D. Apply the inverse Fourier integral formulas to the integral $J(x)$.

Solutions. **A.** In any given interval $-x_0 \leqslant x \leqslant x_0$, the integrals $I(x)$ and $J(x)$ are respectively majorized by the convergent integrals

$$x_0 \int_0^\infty \frac{dt}{1+t^2} \qquad \text{and} \qquad \int_0^\infty \frac{dt}{1+t^2}.$$

These integrals therefore converge uniformly (and absolutely) with respect to x in every closed interval of the real axis.

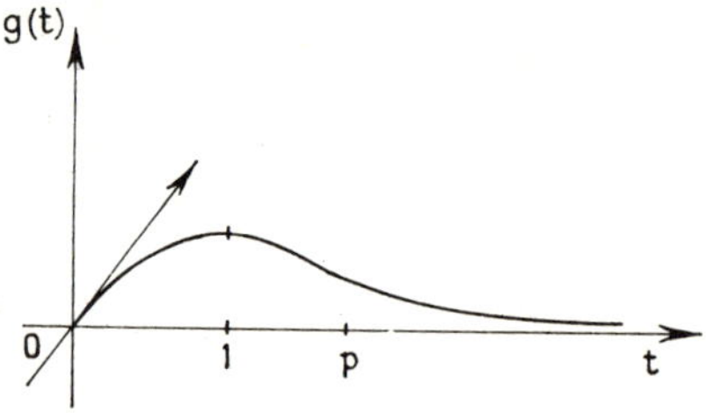

FIGURE 6

B. Let us apply the second mean-value theorem. The function

$$g(t) = \frac{1}{1 + t^2}$$

is nonnegative and decreasing in the interval $(1, \infty)$.

The function $\sin xt$ is integrable. For $x \geqslant x_0$, we have

$$\int_p^q \frac{t}{1 + t^2} \sin xt \, dt = \frac{p}{1 + p^2} \int_p^\xi \sin xt \, dt = \frac{p}{(1 + p^2)x} \int_{px}^{\xi x} \sin u \, du$$

(where ξ lies between p and q) whenever p is sufficiently great. Therefore,

$$\left| \int_p^q \frac{t}{1 + t^2} \sin xt \, dt \right| \leqslant \frac{2p}{(1 + p^2)x} \leqslant \frac{2p}{(1 + p^2)x_0}.$$

Let us suppose $q > p$. We may choose p sufficiently great that

$$\frac{2p}{(1 + p^2)x_0} < \epsilon.$$

The integral

$$\left| \int_p^q \frac{t}{1 + t^2} \sin xt \, dt \right|$$

is therefore itself less than ϵ. From this it follows that the integral

$$K(x) = \int_0^\infty \frac{t}{1 + t^2} \sin xt \, dt$$

converges. With p fixed, we let q approach infinity. We obtain

$$\left| \int_p^\infty \frac{t}{1 + t^2} \sin xt \, dt \right| \leqslant \frac{2p}{(1 + p^2)x_0}.$$

The remainder of the integral is majorized by a function of p that is independent of x and that tends to zero as $p \to \infty$. By definition, this integral is uniformly convergent in every interval such that $x \geqslant x_0 > 0$.

C. Consequently, if x is positive, we have

$$\frac{dI}{dx} = J, \qquad \frac{dJ}{dx} = -K = \frac{d^2I}{dx^2}.$$

Now,

$$I(x) + K(x) = \int_0^\infty \frac{\sin xt}{t} \, dt = \begin{cases} +\dfrac{\pi}{2} & \text{for} \quad x > 0, \\ 0 & \text{for} \quad x = 0, \\ -\dfrac{\pi}{2} & \text{for} \quad x < 0. \end{cases}$$

Therefore, for $x > 0$,

$$I - I'' = \frac{\pi}{2}.$$

The general solution of this differential equation is

$$I = \frac{\pi}{2} + Ae^{x} + Be^{-x},$$

where A and B are arbitrary constants. From this, we conclude that

$$J(x) = Ae^{x} - Be^{-x}.$$

Since

$$|J(x)| < \int_0^\infty \frac{dt}{1+t^2} = \frac{\pi}{2},$$

$J(x)$ is bounded and $A = 0$.

Since $J(0) = \pi/2$, we have $B = -\pi/2$. Therefore,

$$J(x) = \frac{\pi}{2} e^{-x} \qquad \text{for} \qquad x > 0.$$

Since $J(x)$ is an even function, we have

$$J(x) = \frac{\pi}{2} e^{+x} \qquad \text{for} \qquad x < 0.$$

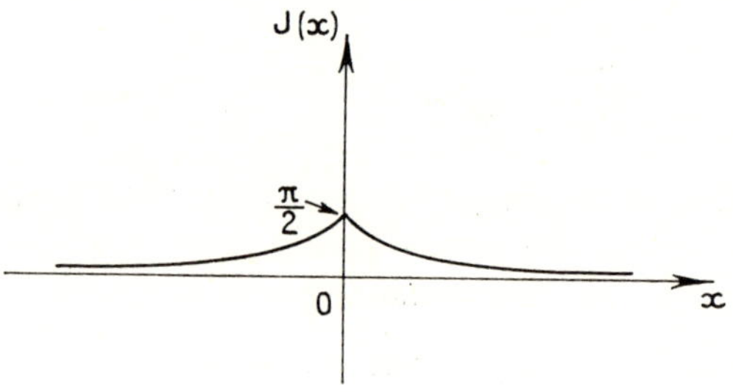

FIGURE 7

For $x = 0$, $J(x) = \pi/2$, and we already know that $J(x)$ is continuous. In sum,

$$\boxed{J(x) = \frac{\pi}{2} e^{-|x|}.}$$

We have

$$K(x) = -\frac{dJ}{dx} = \begin{cases} +\dfrac{\pi}{2}e^{-x} & \text{if} \quad x > 0, \\ -\dfrac{\pi}{2}e^{x} & \text{if} \quad x < 0. \end{cases}$$

Thus $K(x)$ is an odd function.

$K(x)$ is discontinuous at the origin. Therefore, it is not surprising that the integral representing it does not converge uniformly in an interval containing the origin.

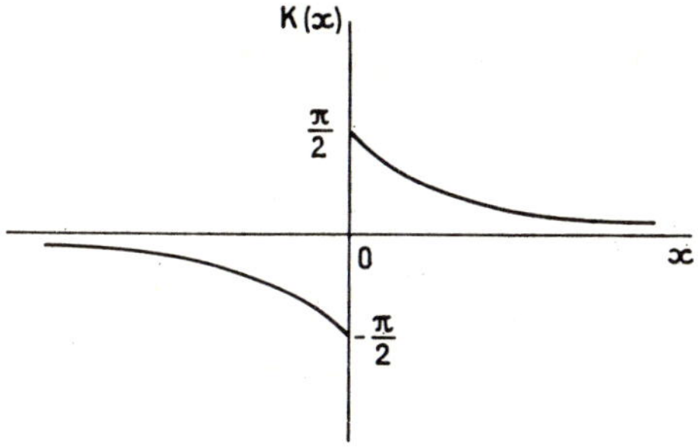

FIGURE 8

Finally,

$$I(x) = \begin{cases} \dfrac{\pi}{2}(1 - e^{-x}) & \text{if} \quad x > 0, \\ \dfrac{\pi}{2}(-1 + e^{x}) & \text{if} \quad x < 0. \end{cases}$$

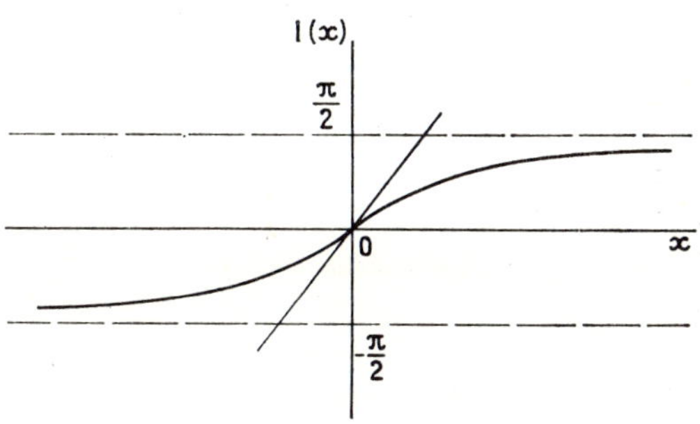

FIGURE 9

D. The integral

$$J(x) = \int_0^\infty \frac{\cos xt}{1+t^2}\,dt = \frac{\pi}{2}\,e^{-|x|}$$

is a Fourier integral. We may write

$$J(x) = \frac{1}{2}\int_{-\infty}^{\infty} \frac{\cos xt}{1+t^2}\,dt = \frac{1}{2}\int_{-\infty}^{\infty} \frac{e^{ixt}}{1+t^2}\,dt.$$

The function $(1+t^2)^{-1}$ is integrable from $-\infty$ to $+\infty$ and it is everywhere differentiable. Therefore, we may apply the inverse Fourier formulas:

$$\frac{1}{2}\frac{1}{1+t^2} = \lim_{\lambda\to\infty} \frac{1}{2\pi}\int_{-\lambda}^{\lambda} J(x)e^{-ixt}\,dx$$

$$= \lim_{\lambda\to\infty} \frac{1}{2\pi}\int_{-\lambda}^{\lambda} \frac{\pi}{2}\,e^{|x|}e^{-ixt}\,dx.$$

Here we may identify

$$\lim_{\lambda\to\infty}\int_{-\lambda}^{\lambda} \qquad \text{with} \qquad \int_{-\infty}^{\infty}$$

and write

$$\frac{1}{1+t^2} = \frac{1}{2}\int_{-\infty}^{\infty} e^{-|x|}e^{ixt}\,dx$$

$$= \frac{1}{2}\int_{-\infty}^{\infty} e^{-|x|}\cos xt\,dx$$

$$= \int_0^\infty e^{-|x|}\cos xt\,dx.$$

Therefore,

$$\int_0^\infty e^{-x}\cos xt\,dx = \frac{1}{1+t^2}\,.$$

This formula can be verified by elementary calculations. It can be used as the starting point for calculating the integral $J(x)$ by using the inverse Fourier formulas in the opposite direction.

9. The Fourier-series expansions of $|\cos x|$ and $|\sin x|$. Applications

In this section we shall first study from an elementary point of view the Fourier-series expansions of $|\cos x|$ and $|\sin x|$ and that of $(\sin x + |\sin x|)/2$, which is equal to $\sin x$ if x is positive and equal to 0 if $\sin x$ is negative.

We can use this exercise to give an example of a ring of functions with zero divisors: a product of two nonidentically zero functions that vanishes for all values of the argument.

We shall then use it to evaluate an interesting limit. We know that, under quite broad hypotheses,

$$\lim_{\lambda\to\infty} \int_a^b f(t) \sin \lambda t \, dt = 0.$$

For example, it is sufficient that $f(t)$ be integrable.

The result is due to the alternations in the sign of $\sin \lambda x$ and to cancellation of the positive and negative portions of the integral. What will happen if we give up these alternations by replacing $\sin \lambda x$ with its absolute value? What is the limit of

$$\int_a^b f(t) \mid \sin \lambda t \mid dt\,?$$

The result is easily obtained by integrating the Fourier-series expansion of $\mid \sin \lambda t \mid$, which is absolutely and uniformly convergent. We shall then prove the same result directly by use of the properties of integrable functions.

Problems. **A.** Define

$$y_1(x) = \begin{cases} \sin x & \text{if} \quad \sin x \geqslant 0, \\ 0 & \text{if} \quad \sin x \leqslant 0. \end{cases}$$

Expand this function, which is periodic with period 2π, in a Fourier series.

B. Expand the function $y_2 = y_1(x - \pi)$ in a Fourier series. Show that the product $y_1 y_2$ is identically 0, although neither of the factors is identically 0.

C. Expand the functions $y_3(x) = \mid \sin x \mid$ and $y_4(x) = \mid \cos x \mid$.

D. Use the result of Problem C to show that, if $f(t)$ is integrable,

$$\lim_{\lambda\to\infty} \int_a^b f(t) \mid \sin \lambda t \mid dt = \frac{2}{\pi} \int_a^b f(t)\, dt.$$

E. Prove the preceding result without using Fourier series. First prove it for $f(t) = 1$, then for $f(t)$ equal to any step function, and finally for $f(t)$ any integrable function.

(N.B.: Problem E is independent of the other four.)

Solutions. **A.** The Fourier coefficients a_n and b_n of the function

$$y_1(x) = \begin{cases} \sin x & \text{if} \quad \sin x \geqslant 0, \\ 0 & \text{if} \quad \sin x \leqslant 0, \end{cases}$$

are given by

$$\begin{aligned} a_n &= \frac{1}{\pi} \int_0^{2\pi} y_1(x) \cos nx \, dx \\ &= \frac{1}{\pi} \int_0^{\pi} \sin x \cos nx \, dx = \frac{1}{\pi} \frac{1 + (-1)^n}{1 - n^2} \qquad \text{if} \quad n \neq 1 \\ &\hphantom{= \frac{1}{\pi} \int_0^{\pi} \sin x \cos nx \, dx} = 0 \qquad \text{if} \quad n = 1. \end{aligned}$$

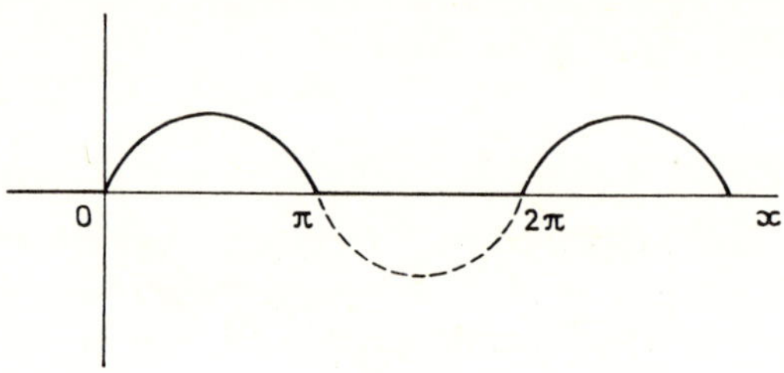

FIGURE 10

Therefore,

$$a_n = \begin{cases} 0 & \text{if } n \text{ is odd (result valid for } n = 1), \\ \dfrac{2}{\pi(1 - n^2)} & \text{if } n \text{ is even (including } n = 0). \end{cases}$$

On the other hand,

$$b_n = \frac{1}{\pi} \int_0^{\pi} \sin x \sin nx \, dx = 0 \qquad \text{if} \quad n \neq 1,$$

$$b_1 = \frac{1}{\pi} \int_0^{\pi} \sin^2 x \, dx = \tfrac{1}{2}.$$

The function $y_1(x)$ is continuous. It has a continuous derivative except at the points $x = k\pi$, at which it has a left-hand and a right-hand derivative. Therefore, the Fourier series with the Fourier coefficients of $y_1(x)$ converges for all x. Its sum is $y_1(x)$ and the convergence is uniform. Therefore,

$$y_1(x) = \frac{1}{\pi} - \frac{2}{\pi} \left[\frac{\cos 2x}{2^2 - 1} + \frac{\cos 4x}{4^2 - 1} + \cdots + \frac{\cos 2px}{(2p)^2 - 1} + \cdots \right] + \frac{1}{2} \sin x.$$

B. We have

$$y_2(x) = \frac{1}{\pi} + \frac{1}{2}\sin(x-\pi) - \frac{2}{\pi}\left[\frac{\cos 2(x-\pi)}{2^2-1} + \cdots + \frac{\cos 2p(x-\pi)}{(2p)^2-1} + \cdots\right]$$

$$= \frac{1}{\pi} - \frac{1}{2}\sin x - \frac{2}{\pi}\left[\frac{\cos 2x}{2^2-1} + \cdots + \frac{\cos 2px}{(2p)^2-1} + \cdots\right].$$

Note that the product $y_1 y_2 = 0$, although neither y_1 nor y_2 is identically zero. This property of the ring of functions of period 2π distinguishes it from the "usual" rings for which the equation $y_1 y_2 = 0$ and the inequality $y_1 \neq 0$ together imply that $y_2 = 0$.

Rings for which the equation $y_1 y_2 = 0$ implies that either y_1 or y_2 is zero are called rings without proper zero-divisors.† A second example of a ring that is not an integral domain is that of square matrices. The two matrices

$$\begin{pmatrix} 1 & 0 \\ 0 & 0 \end{pmatrix} \quad \text{and} \quad \begin{pmatrix} 0 & 0 \\ 0 & 1 \end{pmatrix}$$

are not zero matrices, but their product is the zero matrix.

Direct verification that $y_1 y_2 = 0$ can be made as follows: We have

$$y_1(x) = \varphi(x) + \tfrac{1}{2}\sin x, \qquad y_2(x) = \varphi(x) - \tfrac{1}{2}\sin x,$$

where

$$\varphi(x) = \frac{1}{\pi} - \frac{2}{\pi}\left[\frac{\cos 2x}{2^2-1} + \cdots + \frac{\cos 2p^2}{(2p)^2-1} + \cdots\right].$$

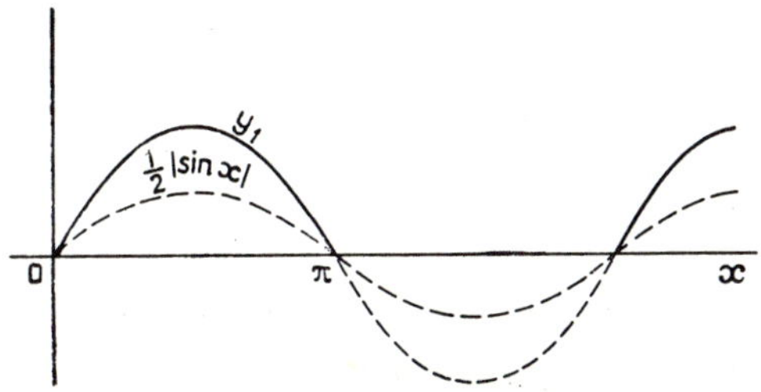

FIGURE 11

† A commutative ring (with identity element) having no proper zero-divisors is called an integral domain.

Therefore, we need to verify that

$$\varphi^2(x) = \tfrac{1}{4}\sin^2 x.$$

This means simply that $y_1(x) - (\tfrac{1}{2}) \sin x$ is equal to $(\tfrac{1}{2})|\sin x|$, which is obvious. The result obtained thus yields at the same time the Fourier-series expansion of $|\sin x|$.

C. The function $|\cos x|$ is even and periodic with period π. Its Fourier series is of the form

$$\frac{a_0}{2} + a_1 \cos 2x + \cdots + a_p \cos 2px + \cdots,$$

where

$$a_p = \frac{2}{\pi}\int_{-\pi/2}^{\pi/2} \cos 2px \cos x \, dx$$

$$= \frac{2}{\pi}\left(\frac{\cos p\pi}{2p+1} - \frac{\cos p\pi}{2p-1}\right) = -\frac{4}{\pi}\frac{(-1)^p}{(2p)^2-1}.$$

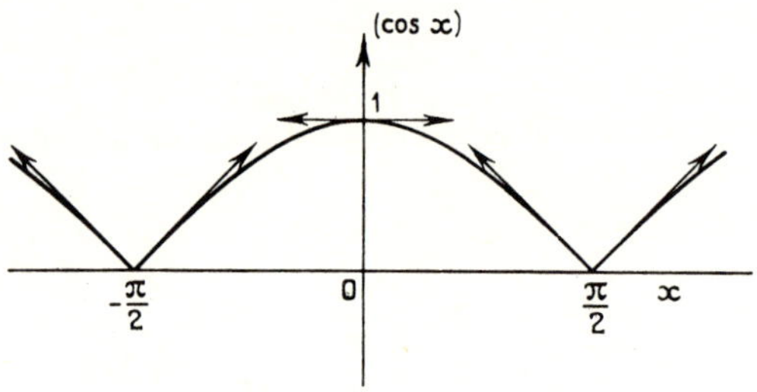

FIGURE 12

Thus, for every value of x,

$$|\cos x| = \frac{2}{\pi} - \frac{4}{\pi}\left[-\frac{\cos 2x}{2^2-1} + \frac{\cos 4x}{4^2-1} + \cdots + \frac{(-1)^p \cos 2px}{(2p)^2-1} + \cdots\right],$$

the convergence of this series being uniform for the entire interval.

For the function $|\sin x|$, we write

$$|\sin x| = \left|\cos\left(\frac{\pi}{2} - x\right)\right|.$$

Since

$$\cos 2p\left(\frac{\pi}{2} - x\right) = \cos(p\pi - 2px) = (-1)^p \cos 2px,$$

we have

$$| \sin x | = \frac{2}{\pi} - \frac{4}{\pi}\left(\frac{\cos 2x}{4-1} + \frac{\cos 4x}{16-1} + \cdots + \frac{\cos 2px}{4p^2-1} + \cdots\right).$$

This gives us again the result of Problem B.

D. Let us show that if $f(t)$ is integrable, then

$$\lim_{\lambda\to\infty} \int_a^b f(t) \, | \sin \lambda t | \, dt = \frac{2}{\pi} \int_a^b f(t) \, dt.$$

We write

$$f(t) \, | \sin \lambda t | = \frac{2}{\pi} f(t) - \frac{4}{\pi}\left[\frac{\cos 2\lambda t}{2^2-1} f(t) + \cdots + \frac{\cos 2p\lambda t}{(2p)^2-1} f(t) + \cdots\right].$$

We know that the sequence $\{S_n(t)\}$ defined by

$$S_n(t) = \frac{2}{\pi} - \frac{4}{\pi} \sum_{p=1}^{n} \frac{\cos 2p\lambda t}{(2p)^2-1}$$

converges uniformly to the function $S(t) = | \sin \lambda t |$. On the other hand, the function $f(t)$ is integrable. Therefore, $|f(t)|$ is bounded, so that the sequence $\{S_n(t) f(t)\}$ converges uniformly with the sequence $\{S_n(t)\}$.

Therefore, we may integrate termwise the series whose general term is $\{\cos 2p\lambda t/[(2p)^2 - 1]\} f(t)$:

$$\int_a^b f(t) \, | \sin \lambda t | \, dt = \frac{2}{\pi} \int_a^b f(t) \, dt - \frac{4}{\pi} \sum_{p=1}^{\infty} \frac{1}{4p^2-1} \int_a^b \cos 2p\lambda t \, f(t) \, dt.$$

In the expression on the right, we have a series of the form

$$\sum_{p=1}^{\infty} \frac{1}{4p^2-1} u_p(\lambda),$$

the general term of which depends on λ. According to the Riemann-Lebesgue theorem (which is proved in an elementary fashion for an integrable function),

$$\lim_{\lambda\to\infty} u_p(\lambda) = \lim_{\lambda\to\infty} \int_a^b \cos 2p\lambda t f(t) \, dt = 0.$$

Every term of the series tends to 0. Is its limit the zero function?

To see that it is, we need only show that the series $\Sigma\,[u_p(\lambda)/(4p^2-1)]$ converges uniformly. Now $f(t)\cos 2p\lambda t$ is less in absolute value than the least upper bound M of $|f|$. Therefore, the series is majorized by the convergent series $M/(4p^2-1)$, which is independent of λ. The uniform convergence of the series then follows.

Finally,

$$\lim_{\lambda\to\infty}\int_a^b f(t)\,|\sin\lambda t|\,dt = \frac{2}{\pi}\int_a^b f(t)\,dt.$$

E. Without using the properties of Fourier series, we shall show that, if $f(t)$ is integrable, then

$$\lim_{\lambda\to\infty}\int_a^b f(t)\,|\sin\lambda t|\,dt = \frac{2}{\pi}\int_a^b f(t)\,dt.$$

We proceed in steps:

Case 1. $f(t) = 1$. Note that $\sin\lambda t$ vanishes for $t = k\pi/\lambda$.

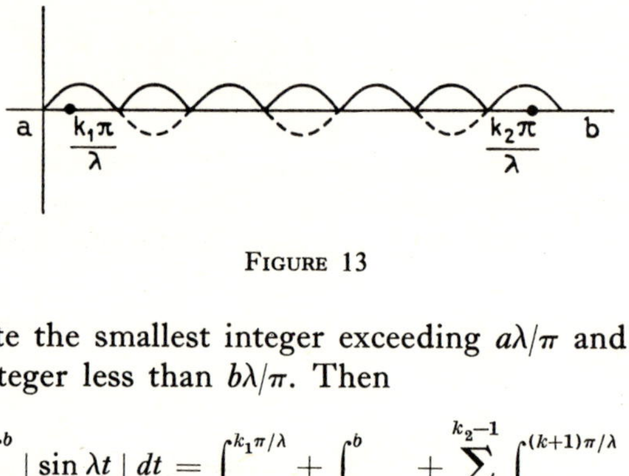

FIGURE 13

Let k_1 denote the smallest integer exceeding $a\lambda/\pi$ and let k_2 denote the greatest integer less than $b\lambda/\pi$. Then

$$\int_a^b |\sin\lambda t|\,dt = \int^{k_1\pi/\lambda} + \int_{k_2\pi/\lambda}^b + \sum_{k=k_1}^{k_2-1}\int_{k\pi/\lambda}^{(k+1)\pi/\lambda}$$

The first and the second integral in the right-hand member tend to zero as $\lambda\to\infty$. These are, in fact, integrals of bounded functions over an interval whose length, which is less than π/λ, tends to zero.

On the other hand, on every interval of the form $[k\pi/\lambda, (k+1)\pi/\lambda]$, the sign of $\sin\lambda t$ is constant. Therefore,

$$\int_{k\pi/\lambda}^{(k+1)\pi/\lambda} |\sin\lambda t|\,dt = \left|\int_{k\pi/\lambda}^{(k+1)\pi/\lambda}\sin\lambda t\,dt\right| = \frac{2}{\lambda}.$$

It follows that

$$\int_{k_1\pi/\lambda}^{k_2\pi/\lambda} |\sin\lambda t|\,dt = \frac{2n}{\lambda},$$

where n is the number of intervals of the form indicated between $k_1\pi/\lambda$ and $k_2\pi/\lambda$ and is equal to $k_2 - k_1$. Now

$$\frac{(k_2 - k_1)\pi}{\lambda} \leqslant b - a < \frac{(k_2 - k_1)\pi}{\lambda} + \frac{2\pi}{\lambda}.$$

Therefore, $n/\lambda = (k_2 - k_1)/\lambda$ tends to $(b - a)/\pi$ as $\lambda \to \infty$. Finally,

$$\lim_{\lambda\to\infty} \int_a^b | \sin \lambda t | \, dt = \frac{2(b - a)}{\pi}.$$

Case 2. *$f(t)$ is a step function.* Suppose that f has a constant value f_i in each of the intervals t_i, t_{i+1}. Then

$$\int_a^b f(t) | \sin \lambda t | \, dt = \sum f_i \int_{t_i}^{t_{i+1}} | \sin \lambda t | \, dt.$$

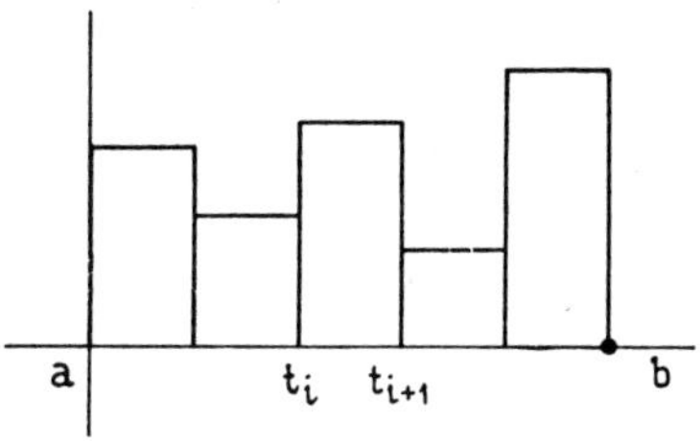

FIGURE 14

The desired limit as $\lambda \to \infty$ is therefore

$$\sum f_i \times \frac{2}{\pi}(t_{i+1} - t_i) = \frac{2}{\pi} \int_a^b f(t) \, dt.$$

Case 3. *$f(t)$ is any integrable function.* For any $\epsilon > 0$, there exist two step functions $f_1(t)$ and $f_2(t)$ such that

$$0 < \int_a^b f_2(t) \, dt - \int_a^b f_1(t) \, dt < \epsilon, \qquad f_1(t) \leqslant f(t) \leqslant f_2(t).$$

If we multiply the first of these double inequalities by the positive quantity $| \sin \lambda t |$ and integrate, we get

$$\int_a^b f_1(t) | \sin \lambda t | \, dt - \int_a^b f(t) | \sin \lambda t | \, dt - \int_a^b f_2(t) | \sin \lambda t | \, dt.$$

To simplify our notation, we set

$$u_1 = \frac{2}{\pi}\int_a^b f_1(t)\,dt, \qquad u_2 = \frac{2}{\pi}\int_a^b f_2(t)\,dt, \qquad u = \frac{2}{\pi}\int_a^b f(t)\,dt,$$

$$\nu_1(\lambda) = \int_a^b f_1(t)\,|\sin\lambda t|\,dt, \qquad \nu_2(\lambda) = \int_a^b f_2(t)\,|\sin\lambda t|\,dt,$$

$$\nu(\lambda) = \int_a^b f(t)\,|\sin\lambda t|\,dt.$$

Our hypotheses are summarized in the following equations and inequalities:

$$0 < u_2 - u_1 < \frac{2}{\pi}\epsilon,$$

$$u_1 < u < u_2\,, \qquad \nu_1(\lambda) < \nu(\lambda) < \nu_2(\lambda),$$

$$\lim_{\lambda\to\infty} \nu_1(\lambda) = u_1\,, \qquad \lim_{\lambda\to\infty} \nu_2(\lambda) = u_2\,.$$

There exists a number $\lambda_0(\epsilon)$ such that the inequality $\lambda > \lambda_0(\epsilon)$ implies that

$$\nu_1(\lambda) > u_1 - \epsilon, \qquad \nu_2(\lambda) < u_2 + \epsilon,$$

so that

$$u_1 - \epsilon < \nu_1(\lambda) < \nu(\lambda) < \nu_2(\lambda) < u_2 + \epsilon,$$

where

$$u_1 < u < u_2 \qquad \text{and} \qquad 0 < u_2 - u_1 < \frac{2}{\pi}\epsilon < \epsilon.$$

If we mark points A, B, C, and D with abscissas, respectively, equal to $u_1 - \epsilon$, u_1, u_2, $u_2 + \epsilon$ on a straight line, we see that, if $\lambda > \lambda_0(\epsilon)$, the point whose abscissa is $v(\lambda)$ lies between A and D and the point whose abscissa is u lies between B and C. The distance between these two points is therefore less than 2ϵ:

$$|\, u - v(\lambda)\,| < 2\epsilon$$

and hence

$$\left| \frac{2}{\pi}\int_a^b f(t)\,dt - \int_a^b f(t)\,|\sin\lambda t|\,dt \right| < 2\epsilon \qquad \text{if} \qquad \lambda > \lambda_0(\epsilon).$$

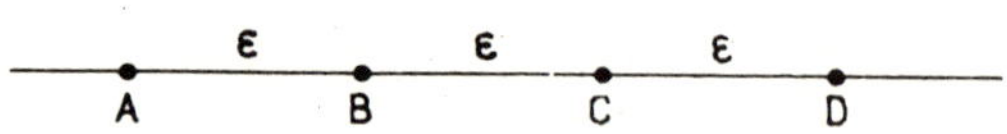

FIGURE 15

Since ϵ may be chosen arbitrarily small, this inequality means that $\int_a^b f(t) \mid \sin \lambda t \mid dt$ tends to a limit equal to

$$\frac{2}{\pi} \int_a^b f(t)\, dt$$

as $\lambda \to \infty$.

10. Evaluation of two definite integrals by different processes

The purpose of the present exercise is to indicate different methods of calculating the two integrals

$$\int_0^\pi \log(1 - 2x \cos \theta + x^2)\, d\theta \qquad \text{and} \qquad \int_0^{\pi/2} \log \sin t\, dt.$$

These processes consist of utilization of the definition of the integral and of Riemann sums, functional properties, and the application of uniform convergence and integration under the integral sign.

In the following exercise, we shall show that we can also use certain Fourier expansions to evaluate these integrals.

We note that the second integral can be evaluated by integrating the function $(1 - z)^{-1} \log z$ of the complex variable z around a circle of radius 1 with center at the point $z = 1$ with a suitable indentation made to avoid the point $z = 0$.

We note also that the first integral is related to potential theory and the theory of harmonic functions of two variables. In fact, $(1-2x \cos \theta + x^2)^{1/2}$ is the distance r from the fixed point $(x, 0)$ to a variable point $(\cos \theta, \sin \theta)$ on the unit circle, and $\log r$ is the logarithmic potential at the point $(x, 0)$ of a uniform distribution of masses on the circle with density 1.

Problems. A. Show that the integral

$$I(x) = \int_0^\pi \log(1 - 2x \cos \theta + x^2)\, d\theta$$

exists for all values of the variable x including ± 1 and that it is a continuous function of x. [*Hint* : Set $(1 + x^2)/2x = z$ and integrate by parts.]

B. Prove the identity

$$\left(1 - 2x\cos\frac{\pi}{n} + x^2\right)\left(1 - 2x\cos\frac{2\pi}{n} + x^2\right)\cdots$$
$$\left(1 - 2x\cos\frac{n-1}{n}\pi + x^2\right) = \frac{x^{2n} - 1}{x^2 - 1}.$$

Use the results to evaluate the integral I for $|x| > 1$, for $|x| < 1$, and for $|x| = 1$.

C. Reevaluate $I(x)$ by the following method: Show that

$$I(x) = \tfrac{1}{2}I(x^2) = \tfrac{1}{4}I(x^4) = \cdots.$$

Use the result to show that $I(x) = 0$ for $0 < x < 1$. Evaluate $I(x)$ for $x > 1$ and $x < 0$ with the aid of the identities

$$I(-x) = I(x), \qquad I\left(\frac{1}{x}\right) = I(x) - \pi \log x^2.$$

Finally, use the continuity of I at the origin.

D. Evaluate $I(1)$ by integrating the function

$$f(x) = \int_0^{\pi/2} \frac{dt}{1 + x^2 \tan^2 t}.$$

under the integral sign from 0 to 1.

(N.B.: Problems B, C, and D are independent.)

Solutions. **A.** Consider the integral

$$I(x) = \int_0^{\pi} \log(1 - 2x\cos\theta + x^2)\, d\theta, \tag{1}$$

where f is a real parameter.

The trinomial $1 - 2x\cos\theta + x^2$ is positive when $|x| \neq 1$ and $0 \leqslant \theta \leqslant \pi$. Therefore, the integral $I(x)$ exists if $|x| \neq 1$.

Suppose that $x = 1$. Then the function $1 - 2x\cos\theta + x^2$ reduces to

$$2(1 - \cos\theta) = 4\sin^2\frac{\theta}{2}.$$

As $\theta \to 0$, the function $4\sin^2(\theta/2)$ is an infinitesimal of the same order as θ^2. Therefore, the convergence of $I(1)$ is equivalent to the convergence of the integral

$$\int^{\pi} \log\theta^2\, d\theta \qquad \text{or} \qquad \int^{\pi} \log\theta\, d\theta.$$

Since $\log \theta$ is integrable in a neighborhood of the origin, we may conclude that $I(x)$ exists for $x = 1$. Analogous reasoning shows that $I(-1)$ also exists. Thus the integral $I(x)$ exists for all values of x.

Let us show now that $I(x)$ is a continuous function of x.

In a neighborhood of any value of x other than ± 1, the expression $\log(1 - 2x \cos \theta + x^2)$ is a continuous function of both x and θ. Therefore, the integral is continuous in the neighborhood of any value of x other than ± 1.

For $x = 1$, $I(x)$ is an improper integral, which we shall find it convenient to transform. We set $(1 + x^2)/2x = z$. This z is greater than 1 in a deleted neighborhood of 1, and it approaches 1 decreasing. Our problem then amounts to evaluating

$$\int_0^\pi \log(z - \cos \theta)\, d\theta, \tag{2}$$

which we integrate by parts to get a rational expression. This yields

$$[\theta \log(z - \cos \theta)]_0^\pi - \int_0^\pi \frac{\theta \sin \theta}{z - \cos \theta}\, d\theta.$$

The first term in this expression is a continuous function of z.

Let us study the integral in the above expression and let us show that, for any positive ϵ, there exists a number α, independent of z, such that

$$0 < \int_0^\alpha \frac{\theta \sin \theta}{z - \cos \theta}\, d\theta < \epsilon.$$

Since $z \geqslant 1$, it follows that

$$\frac{1}{z - \cos \theta} \leqslant \frac{1}{1 - \cos \theta}$$

and

$$\int_0^\alpha \frac{\theta \sin \theta}{z - \cos \theta}\, d\theta \leqslant \int_0^\alpha \frac{\theta \sin \theta}{1 - \cos \theta}\, d\theta.$$

Now, the function that is equal to $\theta \sin \theta/(1 - \cos \theta)$ when $0 < \theta \leqslant \alpha$ and equal to 2 when $\theta = 0$ is a continuous function even at $\theta = 0$. Let us denote its least upper bound by M. The integral $\int_0^\alpha [\theta \sin \theta/(1 - \cos \theta)]\, d\theta$ is less than $M\alpha$. The inequality $M\alpha < \epsilon$ determines a number α that is independent of z. Therefore, the integral (2) is uniformly convergent in any closed interval beginning at $z = 1$. From this it follows that this integral is continuous at $z = 1$.

B. The zeros of the polynomial $x^{2n} - 1$ are the $n - 1$ complex numbers

$$\omega_p = \exp[(2i\pi/2n)p] \qquad (p = 1, 2, ..., n - 1),$$

their conjugates

$$\overline{\omega_p} = \exp[(-i\pi/n)p],$$

and the two real numbers 1 and -1. Its factorization is represented by

$$x^{2n} - 1 = (x^2 - 1) \prod_{p=1}^{n-1} (x - \omega_p)(x - \overline{\omega_p})$$

$$= (x^2 - 1) \prod_{p=1}^{n-1} \left(1 - 2x \cos \frac{p}{n} \pi + x^2\right)$$

or

$$\prod_{p=1}^{n-1} \left(1 - 2x \cos \frac{p}{n} \pi + x^2\right) = \frac{x^{2n} - 1}{x^2 - 1}. \tag{3}$$

We note that the function $P_n(x) = (x^{2n} - 1)/(x^2 - 1)$, being a polynomial in x, is certainly continuous. Consequently,

$$P_n(+1) = P_n(-1) = n. \tag{4}$$

This being the case, we can define the integral $I(x)$ as the limit of the Riemann sum obtained by partitioning the interval $[0, \pi]$ at the point $\pi/n, 2\pi/n, ..., (n - 1)\pi/n$:

$$I(x) = \lim_{n\to\infty} \frac{\pi}{n} \sum_{p=1}^{n-1} \log \left(1 - 2x \cos p \frac{\pi}{n} + x^2\right),$$

that is,

$$I(x) = \lim_{n\to\infty} \frac{\pi}{n} \log P_n(x). \tag{5}$$

This limit is easily calculated:

Case 1. $|x| < 1$. Here, $P_n(x)$ tends to $(1 - x^2)^{-1}$ as $n \to \infty$. Therefore, we see from (5) that $I(x) = 0$.

Case 2. $|x| = 1$. We have

$$I(\pm 1) = \lim_{n\to\infty} \frac{\pi}{n} \log n \qquad \text{or} \qquad I(\pm 1) = 0.$$

Case 3. $|x| > 1$. As n increases without bound, $\log P_n(x)$ is equivalent to

$$\log x^{2n} = 2n \log |x|.$$

Therefore,

$$I(x) = 2\pi \log |x|.$$

Summarizing, we have

$$I(x) = \begin{cases} 0 & \text{if} \quad |x| \leqslant 1, \\ 2\pi \log |x| & \text{if} \quad |x| \geqslant 1. \end{cases} \tag{6}$$

C. Let us find these same results by a different kind of reasoning. Let us first prove the following:

$$I(-x) = I(x), \tag{7}$$

$$I\left(\frac{1}{x}\right) = I(x) - 2\pi \log |x|. \tag{8}$$

First of all,

$$I(-x) = \int_0^\pi \log(1 + 2x \cos\theta + x^2)\, d\theta.$$

The change of variable $\theta = \pi - \varphi$ leads to

$$I(-x) = \int_0^\pi \log(1 - 2x \cos\varphi + x^2)\, d\varphi = I(x).$$

On the other hand,

$$\begin{aligned} I\left(\frac{1}{x}\right) &= \int_0^\pi \log\left(1 - \frac{2}{x}\cos\theta + \frac{1}{x^2}\right) d\theta \\ &= \int_0^\pi \log(x^2 - 2x\cos\theta + 1)\, d\theta + \int_0^\pi \log\frac{1}{x^2}\, d\theta, \end{aligned}$$

from which (8) follows.

Let us now show that

$$I(x) = \tfrac{1}{2} I(x^2). \tag{9}$$

If we add the equations

$$I(x) = \int_0^\pi \log(1 - 2x\cos\theta + x^2)\, d\theta,$$

$$I(-x) = \int_0^\pi \log(1 + 2x\cos\theta + x^2)\, d\theta,$$

remembering that $I(-x) = I(x)$, we obtain

$$\begin{aligned}2I(x) &= \int_0^{\pi} \log(1 - 2x^2 \cos 2\theta + x^4)\, d\theta \\ &= \tfrac{1}{2}\int_0^{2\pi} \log(1 - 2x^2 \cos\varphi + x^4)\, d\varphi \\ &= \tfrac{1}{2}\int_0^{\pi} \log(1 - 2x^2 \cos\varphi + x^4)\, d\varphi + \tfrac{1}{2}\int_{\pi}^{2\pi} \log(1 - 2x^2 \cos\varphi + x^4)\, d\varphi.\end{aligned}$$

If we set $\varphi = \pi + \psi$, we obtain

$$2I(x) = \tfrac{1}{2} I(x^2) + \tfrac{1}{2} I(-x^2) = I(x^2)$$

by virtue of (7). This proves (9).

From this, we deduce immediately that $I(x) = I(x^4)/4$ and, more generally, that

$$I(x) = \frac{1}{2^n} I(x^{2^n}) \qquad (n \text{ a positive integer}) \tag{10}$$

Suppose that $|x| \leqslant 1$. Then $|x|^{2^n} \leqslant 1$. We know that $I(x)$ is a continuous function. Therefore, $|I(x)|$ is bounded by some number M on the interval $[-1, 1]$. Hence, it follows from (10) that

$$|I(x)| \leqslant \frac{1}{2^n} M.$$

As we let n increase without bound, we see that

$$I(x) = 0 \qquad \text{if} \qquad |x| \leqslant 1.$$

Therefore, it follows from (8) that under these conditions

$$I\left(\frac{1}{x}\right) = 0 + 2\pi \log \frac{1}{|x|}, \qquad \left|\frac{1}{x}\right| \geqslant 1$$

and hence

$$I(y) = 2\pi \log |y|, \qquad |y| \geqslant 1.$$

Thus we have again found relations (6) by a different method.

D. We have seen that $I(1) = 0$. If we set $\theta = 2t$, we can show easily that $\int_0^{\pi/2} \log \sin t\, dt$ is equal to $(-\pi/2) \log 2$. Let us re-evaluate this integral by integrating the function

$$f(x) = \int_0^{\pi/2} \frac{1}{1 + x^2 \tan^2 t}\, dt$$

under the integral sign. The integrand is discontinuous at $x = 0$, $t = \pi/2$. However, it is majorized by the constant 1, which does not depend on x. Therefore, the integral converges uniformly and we may integrate from $x = 0$ to $x = 1$ under the integral sign:

$$\int_0^1 f(x)\,dx = \int_0^{\pi/2} dt \int_0^1 \frac{dx}{1 + x^2 \tan^2 t}.$$

Now,

$$\int_0^1 \frac{dx}{1 + x^2 \tan^2 t} = \frac{1}{\tan t} \times [\arctan (x \tan t)]_0^1 = \frac{t}{\tan t}.$$

Therefore,

$$\int_0^1 f(x)\,dx = \int_0^{\pi/2} \frac{t\,dt}{\tan t} = [t \log \sin t]_0^{\pi/2} - \int_0^{\pi/2} \log \sin t\,dt = -\int_0^{\pi/2} \log \sin t\,dt.$$

But we can also calculate $f(x)$ first. If we set $\tan t = u$, we have

$$f(x) = \int_0^\infty \frac{du}{(1 + u^2)(1 + x^2u^2)} = \frac{1}{1 - x^2}\left(\int_0^\infty \frac{du}{1 + u^2} - \int_0^\infty \frac{x^2\,du}{1 + x^2u^2}\right)$$
$$= \frac{1}{1 - x^2}\left(\frac{\pi}{2} - x\frac{\pi}{2}\right) = \frac{\pi}{2}\frac{1 - x}{1 - x^2} = \frac{\pi}{2}\frac{1}{1 + x}.$$

Therefore,

$$\int_0^1 f(x)\,dx = \frac{\pi}{2}\int_0^1 \frac{dx}{1 + x} = \frac{\pi}{2}[\log (1 + x)]_0^1 = \frac{\pi}{2} \log 2.$$

Thus

$$\boxed{\int_0^{\pi/2} \log \sin t\,dt = -\frac{\pi}{2} \log 2.}$$

The integral from 0 to π is of course twice this, namely, $-\pi \log 2$.

Remark. The formula

$$\int_0^1 dx \int_0^{\pi/2} \frac{dt}{1 + x^2 \tan^2 t} = \int_0^{\pi/2} dt \int_0^1 \frac{dx}{1 + x^2 \tan^2 t}$$

has been interpreted as an integration under the integration sign. The two sides correspond to two ways of evaluating the double integral

$$\iint \frac{dx\,dt}{1 + x^2 \tan^2 t}$$

over the region

$$0 < x < 1, \qquad 0 < t < \frac{\pi}{2}.$$

11. Fourier series, power series, and their applications

In this exercise we shall study the two series

$$\sum_{n=1}^{\infty} x^n \cos n\theta \qquad \text{and} \qquad \sum_{n=1}^{\infty} x^n \sin n\theta.$$

We shall evaluate their limits, examine their convergence, and treat them both as power series in x and Fourier series in θ. We shall study the series that result from term-by-term integration of the above series. The series $\Sigma n^{-1} x^n \cos n\theta$ has certain interesting features for $x = 1$.

We shall use these series to re-evaluate the two integrals studied in Exercise 10.

Problems. A. Show that, if $|x| < 1$,

$$A = x \cos \theta + \cdots + x^n \cos n\theta + \cdots = \frac{x \cos \theta - x^2}{1 - 2x \cos \theta + x^2},$$

$$B = x \sin \theta + \cdots + x^n \sin n\theta + \cdots = \frac{x \sin \theta}{1 - 2x \cos \theta + x^2},$$

$$C = x \cos \theta + \cdots + x^n \frac{\cos n\theta}{n} + \cdots = -\frac{1}{2} \log (1 - 2x \cos \theta + x^2),$$

$$D = x \sin \theta + \cdots + x^n \frac{\sin n\theta}{n} + \cdots = \arctan \frac{x \sin \theta}{1 - x \cos \theta}.$$

Can we derive A and B from C and D or C and D from A and B? If so, by what operations?

B. With the aid of the series C, evaluate the integral

$$\int_0^{\pi} \log (1 - 2x \cos \theta + x^2)\, d\theta \qquad \text{for} \qquad |x| < 1.$$

Extend the result to the case in which $|x| > 1$.

C. Use Abel's theorem on power series to show that the series C and D are well defined for $x = 1$ and from this deduce that

$$\cos\theta + \cdots + \frac{\cos n\theta}{n} + \cdots = -\log\left(2\sin\frac{\theta}{2}\right) \qquad (0 < \theta < 2\pi),$$

$$\sin\theta + \cdots + \frac{\sin n\theta}{n} + \cdots = \frac{\pi - \theta}{2} \qquad (0 < \theta < 2\pi).$$

D. Show that the series

$$S = \cos\theta + \cdots + \frac{\cos n\theta}{n} + \cdots$$

converges uniformly in the interval $\alpha \leqslant \theta \leqslant \pi$, where α is a given number between 0 and π.
Evaluate the integral

$$\int_0^{\pi/2} \log \sin t \, dt$$

by integrating the above series term by term from α to π and then letting α approach 0.

Solutions. **A.** Let us evaluate $A + iB$ and $C + iD$. The sum $A + iB$ is a geometric series:

$$A + iB = \sum_{n=1}^{\infty} x^n e^{in\theta} = \frac{xe^{i\theta}}{1 - xe^{i\theta}} = \frac{xe^{i\theta}(1 - xe^{-i\theta})}{1 - 2x\cos\theta + x^2},$$

so that

$$A = \frac{x\cos\theta - x^2}{1 - 2x\cos\theta + x^2}, \qquad B = \frac{x\sin\theta}{1 - 2x\cos\theta + x^2}.$$

This result is valid for $|x| < 1$ and for all θ.

Similarly,

$$C + iD = \sum_{n=1}^{\infty} \frac{(xe^{i\theta})^n}{n} = -\log(1 - xe^{i\theta}).$$

This result is valid for $|x| < 1$ and for $|x| = 1$ if $\theta \neq 2k\pi$, where k is an integer. We know that if $\{a_n\}$ is a decreasing sequence of positive numbers that tends to 0, the series

$$\sum a_n \sin n\theta$$

converges and that the series

$$\sum a_n \cos n\theta$$

converges except when θ is a multiple of 2π.

Now,

$$1 - xe^{i\theta} = (1 - 2x \cos \theta + x^2)^{1/2} e^{-i\varphi},$$

where

$$\tan \varphi = \frac{x \sin \theta}{1 - x \cos \theta}$$

so that

$$\log(1 - xe^{i\theta}) = \tfrac{1}{2} \log (1 - 2x \cos \theta + x^2) - i\varphi$$

and

$$C = -\tfrac{1}{2} \log (1 - 2x \cos \theta + x^2), \qquad D = \arctan \frac{x \sin \theta}{1 - x \cos \theta},$$

where this arctangent must vanish for $\theta = 0$.

We may obtain C and D by dividing the series A and B by x and then integrating them with respect to x from 0 to a given value X whose absolute value is less than 1. We can do this because the series A and B, being power series, are uniformly convergent. We can also integrate with respect to θ for a fixed value of x such that $|x| < 1$. Under these conditions we may derive A and B from C and D by differentiating termwise with respect to x or θ, since the series obtained in this manner are uniformly convergent.

B. The series whose general term u_n is $n^{-1}x^n \cos n\theta$ is the Fourier-series expansion of the function $-\frac{1}{2}\log(1 - 2x \cos \theta + x^2)$. The expansion converges uniformly with respect to x for $|x| \leqslant h < 1$.

For every fixed x whose absolute value is less than 1, this expansion converges uniformly in every θ-interval.

Now, the constant term in the expansion is 0. Consequently,

$$I(x) = \int_0^\pi \log (1 - 2x \cos \theta + x^2)\, d\theta = 0 \qquad \text{for} \qquad |x| < 1.$$

For $|x| > 1$, we set $y = 1/x$. Then

$$\begin{aligned} I(x) &= \int_0^\pi \log \left(1 - \frac{2 \cos \theta}{y} + \frac{1}{y^2}\right) d\theta \\ &= \int_0^\pi \log (1 - 2y \cos \theta + y^2)\, d\theta - \int_0^\pi \log y^2\, d\theta, \end{aligned}$$

so that

$$I(x) = \pi \log x^2 \qquad \text{for} \qquad |x| > 1.$$

C. The series C is a power series in x. If $x = 1$, the series

$$\cos\theta + \cdots + \frac{\cos n\theta}{n} + \cdots$$

converges for $\theta \neq 2k\pi$.

From Abel's theorem on power series, the series $n^{-1}x^n \cos n\theta$ converges uniformly for $0 \leqslant x \leqslant 1$ where θ has a fixed value not a multiple of 2π.

Thus the function representing the sum of the series is continuous from the left at $x = 1$. We obtain in the limit

$$\begin{aligned}\cos\theta + \cdots + \frac{\cos n\theta}{n} + \cdots & \\ &= -\tfrac{1}{2}\lim_{x\to 1}\log(1 - 2x\cos\theta + x^2)\\ &= -\tfrac{1}{2}\log(2 - 2\cos\theta)\\ &= -\log\left(2\sin\frac{\theta}{2}\right).\end{aligned}$$

This result is valid for $\theta \neq 2k\pi$. We note that the function representing the sum of the series is not bounded.

Similarly, consider the power series

$$D = x\sin\theta + \cdots + \frac{x^n \sin n\theta}{n} + \cdots.$$

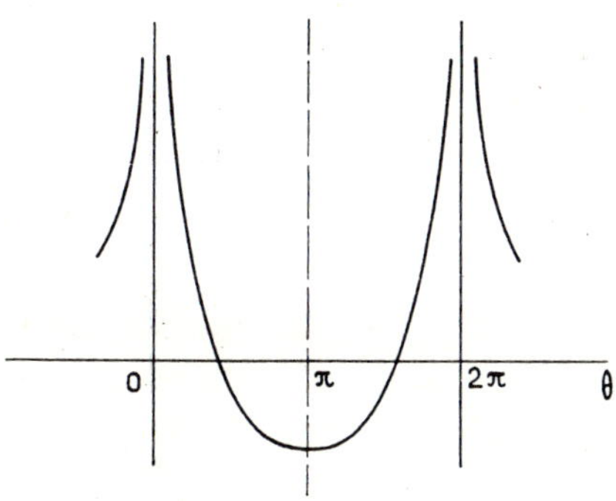

FIGURE 16

This series converges for $x = 1$ and any fixed value of θ. Therefore, its sum is continuous from the left at $x = 1$. If $\theta = 2k\pi$, $D = 0$. Therefore, the sum of the series

$$\sin\theta + \cdots + \frac{\sin n\theta}{n} + \cdots$$

vanishes for $\theta = 2k\pi$, as is obvious.

If $\theta \neq 2k\pi$, the limit of the series is

$$\arctan\frac{\sin\theta}{1-\cos\theta} = \arctan\left(\tan\frac{\pi-\theta}{2}\right).$$

The member on the right should be interpreted as a periodic function with period 2π that is equal to $(\pi - \theta)/2$ if $0 < \theta < 2\pi$. Therefore,

$$\sin\theta + \cdots + \frac{\sin n\theta}{n} + \cdots = \begin{cases} \dfrac{\pi-\theta}{2}, & (0<\theta<2\pi), \\ 0 & (\theta = 0). \end{cases}$$

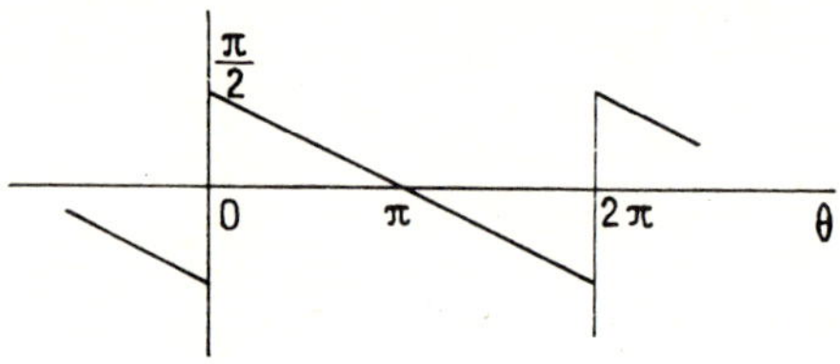

FIGURE 17

D. Let us return to the series

$$S = \cos\theta + \cdots + \frac{\cos n\theta}{n} + \cdots = -\log\left(2\sin\frac{\theta}{2}\right).$$

This series converges uniformly in any interval of the form $[a, \pi]$, where $0 < \alpha < \pi$. To see this, note that

$$2\sin\frac{\theta}{2}\cos n\theta = \sin\left(n+\frac{1}{2}\right)\theta - \sin\left(n-\frac{1}{2}\right)\theta.$$

Thus if we set $S_n = \cos\theta + \cdots + n^{-1}\cos n\theta$, we have

$$2S_n\sin\frac{\theta}{2} = -\sin\frac{\theta}{2} + \frac{1}{2}\sin\frac{3\theta}{2} + \frac{1}{6}\sin\frac{5\theta}{2} + \frac{1}{n(n-1)}\sin\left(n-\frac{1}{2}\right)\theta + \frac{1}{n}\sin\left(n+\frac{1}{2}\right)\theta.$$

The term $(1/n)\sin(n + \frac{1}{2})\theta$ tends to 0. Therefore, the series S is equivalent to the series whose general term is

$$\frac{1}{2n(n-1)}\,\frac{\sin\,(n-\frac{1}{2})\,\theta}{\sin\,(\theta/2)}\,.$$

If $\alpha \leqslant \theta \leqslant \pi$, we have $\sin(\theta/2) \geqslant \sin(\alpha/2)$. Therefore, the series is majorized by the numerical series whose general term is

$$\frac{1}{2\sin\,(\alpha/2)}\,\frac{1}{n(n-1)}$$

and which is convergent.

Therefore, we may integrate the series S from α to π:

$$\sin\alpha + \cdots + \frac{\sin n\alpha}{n^2} + \cdots = + \int_\alpha^\pi \log\left(2\sin\frac{\theta}{2}\right) d\theta.$$

Now, let us let α approach 0. The left-hand member is a uniformly convergent series majorized by the series $\Sigma\, 1/n^2$. Its sum tends to 0 as $\alpha \to 0$. The right-hand member tends to

$$\int_0^\pi \log\left(2\sin\frac{\theta}{2}\right) d\theta,$$

which is a convergent integral (see Exercise 10). Therefore,

$$\int_0^\pi \log\left(2\sin\frac{\theta}{2}\right) d\theta = 0.$$

If we set $\theta/2 = t$, we obtain

$$\int_0^{\pi/2} \log\sin t\; dt = -\frac{\pi}{2}\log 2,$$

which is the result obtained in Exercise 10 by several other methods.

12. Study of a linear integral equation

In this exercise we shall consider the equation

$$\lambda y(t) = \int_0^{2\pi} k(r, t-s)\, y(s)\, ds + g(t), \tag{E}$$

where the kernel $k(r, t)$ is equal to

$$\frac{1 - r^2}{1 - 2r \cos t + r^2}.$$

Here r is a *given* number whose absolute value is less than 1, g is a given function, $y(t)$ is the unknown function, and λ is a parameter.

The kernel $k(r, t - s)$ has the property of remaining unchanged when we permute s and t. The operation

$$Ay(s) = \int_0^{2\pi} k(r, t - s)\, y(s)\, ds$$

is a linear operation on y *provided that y is chosen in a certain vector space V*. Equation (E) may be written

$$\lambda y = \mathbf{A}y + g \qquad \text{or} \qquad (\lambda\mathbf{I} - \mathbf{A})y = g,$$

where $\mathbf{I}$ is the identity operator.

Solving (E) amounts to finding the inverse of the operator $\lambda\mathbf{I} - \mathbf{A}$. This inversion is possible in the space V except possibly for certain values of λ, which, just as for analogous problems in linear algebra, constitute the spectrum of the operator $\mathbf{A}$ and are called *eigenvalues*. When λ is an eigenvalue, the equation has no solution unless g satisfies certain additional requirements that are always satisfied for $g = 0$.

All these results, which we shall verify for the example chosen, are of a rather general nature. In particular, they hold if V is the space of square-summable functions (integrable in the sense of Lebesgue) on $[0, 2\pi]$, in which case the solutions are defined in the same space.

Here, the kernel k is continuous and, hence, square-summable. The elementary theory of uniformly convergent Fourier series enables us to prove directly that there exist solutions that can be expanded in uniformly convergent Fourier series. These solutions are indeed square-summable, but they have more precise local properties than the general theory provides.

It turns out that the eigenvalues are of the form $\lambda = 2\pi r^p$, where p is an integer. Their existence and the fact that they constitute a denumerable set follow from the general theorem on symmetric kernels (which are analogous to real symmetric matrices) and particular cases of Hermitian kernels.

Problems. A. Show that, for $|r| < 1$,

$$\frac{1 - r^2}{1 - 2r \cos\theta + r^2} = 1 + 2r\cos\theta + \cdots + 2r^n \cos n\theta + \cdots$$

and use this expansion to evaluate the integral

$$\int_0^{2\pi} \frac{\cos n\theta}{1 - 2r\cos\theta + r^2}\, d\theta.$$

B. Set

$$k(r, \theta) = \frac{1 - r^2}{1 - 2r\cos\theta + r^2}.$$

and evaluate the integrals

$$a_n = \frac{1}{\pi}\int_0^{2\pi} k(r, t - s)\cos ns\, ds,$$

and

$$b_n = \frac{1}{\pi}\int_0^{2\pi} k(r, t - s)\sin ns\, ds$$

C. Consider the linear integral equation

$$\lambda y(t) = \int_0^{2n} k(r, t - s)\, y(s)\, ds \tag{E}$$

where $y(t)$ is the unknown function, assumed to be defined and bounded on the interval $[0, 2\pi]$, r is a given parameter between -1 and 1, and λ is a real nonzero unknown parameter.

Show that if λ is of the form $\lambda = 2\pi r^p$, where p is an integer, this equation possesses solutions that can be expanded in Fourier series on the interval $[0, 2\pi]$. Exhibit these solutions.

D. Consider the nonhomogeneous equation

$$\lambda y(t) = \int_0^{2\pi} k(r, t - s)\, y(s)\, ds + g(t) \tag{E$'$}$$

where $g(t)$ is a function that can be expanded in a Fourier series for $0 \leqslant t \leqslant 2\pi$.

Show that the solution is determined if λ is not of the form $2\pi r^p$, where p is an integer.

If there exists an integer p such that $\lambda = 2\pi r^p$, (E′) has no solution except when $g(t)$ is orthogonal to $\cos pt$ and $\sin pt$ on the interval $[0, 2\pi]$. In this last case, the solution is the sum of a definite Fourier series plus the term $a\cos pt + b\sin pt$, where a and b are arbitrary constants.

Solutions. **A.** In the preceding exercise we have shown that, for $|r| < 1$,

$$r \cos \theta + \cdots + r^n \cos n\theta + \cdots = \frac{r \cos \theta - r^2}{1 - 2r \cos \theta + r^2}.$$

Therefore, for $|r| < 1$,

$$1 + 2r \cos \theta + \cdots + \cdots + 2r^n \cos n\theta + \cdots = \frac{1 - r^2}{1 - 2r \cos \theta + r^2}.$$

This series converges uniformly with respect to θ in any interval whatever.

From this, we conclude that $2r^n/(1 - r^2)$ is the nth Fourier coefficient of the function

$$\frac{1}{1 - 2r \cos \theta + r^2}.$$

In other words,

$$\int_0^{2\pi} \frac{\cos n\theta}{1 - 2r \cos \theta + r^2}\, d\theta = 2\pi \frac{r^n}{1 - r^2} \qquad \text{if} \qquad |r| < 1.$$

B. Consider the function

$$k(r, t - s) = \frac{1 - r^2}{1 - 2r \cos (t - s) + r^2}.$$

From Problem A, this function is the sum of the series

$$\begin{aligned} 1 + 2r \cos (t - s) + \cdots + 2r^n \cos n(t - s) + \cdots \\ = 1 + 2r \cos t \cos s + 2r \sin t \sin s + \cdots \\ + 2r^n \cos nt \cos ns + 2r^n \sin nt \sin ns + \cdots. \end{aligned}$$

This is a Fourier series with respect to the variable s, with t playing the role of a parameter. According to the expression for the Fourier coefficients, we have

$$a_n = \frac{1}{\pi} \int_0^{2\pi} k(r, t - s) \cos ns\, ds = 2r^n \cos nt,$$

$$b_n = \frac{1}{\pi} \int_0^{2\pi} k(r, t - s) \sin ns\, ds = 2r^n \sin nt.$$

C. Let us see whether the linear homogeneous integral equation

$$\lambda y(t) = \int_0^{2\pi} k(r, t - s) y(s)\, ds \tag{E}$$

has bounded solutions on the interval $[0, 2\pi]$. From the expansion of $k(r, t - s)$, we have

$$\lambda y(t) = \int_0^{2\pi} y(s)[1 + 2r \cos(t - s) + \cdots + 2r^n \cos n(t - s) + \cdots]\, ds.$$

Since y is bounded, the series in the integrand converges uniformly since it is majorized by a convergent geometric series.

Therefore, we may integrate term by term, getting

$$\begin{aligned}\lambda y(t) &= \int_0^{2\pi} y(s)\, ds + \cdots + 2r^n \cos nt \int_0^{2\pi} y(s) \cos ns\, ds \\ &\quad + 2r^n \sin nt \int_0^{2\pi} y(s) \sin ns\, ds + \cdots . \end{aligned} \tag{1}$$

This formula indicates that $y(t)$ can be expanded in a uniformly convergent Fourier series (again because the series $\sum r^n$ converges). The coefficients are given by Fourier's formulas. Thus

$$\lambda \frac{1}{\pi} \int_0^{2\pi} y(t) \cos nt\, dt = 2r^n \int_0^{2\pi} y(s) \cos ns\, ds,$$

$$\lambda \frac{1}{\pi} \int_0^{2\pi} y(t) \sin nt\, dt = 2r^n \int_0^{2\pi} y(s) \sin ns\, ds.$$

It follows from these formulas that, in general,

$$\int_0^{2\pi} y(t) \cos nt\, dt = 0, \qquad \int_0^{2\pi} y(t) \sin nt\, dt = 0.$$

The function $y(t)$, being orthogonal to all the functions in a complete series, vanishes for almost every t. However, the integral in the right-hand member of (E) vanishes for every t. Therefore, $y(t) = 0$ is the only solution of the integral equation.

An exception occurs if there exists a value p of n such that

$$\lambda = 2\pi r^p.$$

Then $y(t)$ is orthogonal to all the functions $\cos nt$, $\sin nt$ except $\cos pt$ and $\sin pt$. The representation (1) of $y(t)$ by a Fourier series has two nonzero terms, corresponding to $n = p$. Therefore, $y(t)$ is of the form

$$y(t) = a \cos pt + b \sin pt, \tag{2}$$

where a and b are arbitrary constants.

D. Suppose that $g(t)$ in the nonhomogeneous equation

$$\lambda y(t) = \int_0^{2\pi} k(r, t - s)\, y(s)\, ds + g(t) \tag{E'}$$

can be expanded in a Fourier series:

$$g(t) = \frac{\alpha_0}{2} + \alpha_1 \cos t + \beta_1 \sin t + \cdots + \alpha_n \cos nt + \beta_n \sin nt + \cdots .$$

Then

$$\lambda y(t) = \int_0^{2\pi} y(s)\, ds + \cdots + 2r^n \cos nt \int_0^{2\pi} y(s) \cos ns\, ds$$
$$+ 2r^n \sin nt \int_0^{2\pi} y(s) \sin ns\, ds + \cdots + \frac{\alpha_0}{2} + \cdots + \alpha_n \cos nt + \beta_n \sin nt + \cdots.$$

Now $y(t)$ can be expanded in a Fourier series, and Fourier's formulas yield

$$\begin{aligned} \frac{\lambda}{\pi} \int_0^{2\pi} y(t) \cos nt\, dt &= 2r^n \int_0^{2\pi} y(s) \cos ns\, ds + \alpha_n , \\ \frac{\lambda}{\pi} \int_0^{2\pi} y(t) \sin nt\, dt &= 2r^n \int_0^{2\pi} y(s) \sin ns\, ds + \beta_n . \end{aligned} \tag{3}$$

If λ is not of the form $2\pi r^p$, equations (3) yield the integral $\int_0^{2\pi} y(t) \cos nt\, dt$ and $\int_0^{2\pi} y(t) \sin nt\, dt$, which, except for a factor $1/\pi$, are the Fourier coefficients of $y(t)$. Therefore, we have a definite solution, given by

$$\begin{aligned} y(t) &= \frac{1}{\lambda} \sum \frac{2\pi r^n}{\lambda - 2\pi r^n} (\alpha_n \cos nt + \beta_n \sin nt) + \frac{1}{\lambda} g(t) \\ &= \sum \frac{1}{\lambda - 2\pi r^n} (\alpha_n \cos nt + \beta_n \sin nt). \end{aligned}$$

If $\lambda = 2\pi r^p$, the system (3) is in general impossible, because two of the equations are incompatible.

An exception occurs if α_p and β_p are both zero, that is, if

$$\int_0^{2\pi} g(t) \cos pt\, dt = 0, \qquad \int_0^{2\pi} g(t) \sin pt\, dt = 0.$$

If these conditions are satisfied, the Fourier coefficients of $y(t)$ for $n = p$ are indeterminate. Then

$$y(t) = \sum_{n \neq p} \frac{1}{\lambda - 2\pi r^n} (\alpha_n \cos nt + \beta_n \sin nt) + a \cos pt + b \sin pt,$$

where a and b are arbitrary constants.

13. Approximation of a continuous function by a polynomial

A famous theorem of Weierstrass asserts that every *continuous* function on a closed interval can be uniformly approximated by a polynomial within any desired margin ϵ of error. The difficulty with this theorem stems from the fact that the continuous function in question is not necessarily differentiable or even of bounded variation.

Various proofs of this theorem have been given. One of them consists in replacing the curve representing the function with an inscribed polygonal line. By several easy reductions, the problem is reduced to the following one: Show that $|x|$ can be uniformly approximated by a polygon on an interval containing the origin.

The purpose of the present exercise is to give several proofs of this result using both power series and trigonometric series.

We note that all these procedures, although excellent from a theoretical standpoint, are not satisfactory from a practical one. They represent $|x|$ by means of polynomials of high degree and, to obtain a convenient approximation, very lengthy calculations are necessary.

Problems. We wish to show that, for any $\epsilon > 0$ and any interval containing the origin, there exists a polynomial $P(x)$ such that

$$||x| - P(x)| < \epsilon$$

for every x in that interval.

A. Write

$$|x| = [1 - (1 - x^2)]^{1/2}$$

and expand the radical in powers of $1 - x^2$. Show that the series obtained converges for $-\sqrt{2} < x < \sqrt{2}$, including the value $x = 0$. Then replace the series with a partial sum.

B. Expand the function $|\sin \pi t|$, which is periodic with period 1, in a Fourier series. Then, set $\sin \pi t = x$ and show that the sum of the first n terms of the Fourier series is a polynomial in x.

C. Expand $|x|$ in a Fourier series on the interval $[-1, 1]$. Show that the series converges uniformly, replace the series with the sum of its first n terms, expand each of these terms in powers of x, and approximate them with partial sums.

Solutions. **A.** If we expand

$$|x| = [1 - (1 - x^2)]$$

in powers of $1 - x^2$, we get

$$| x | = 1 - \frac{1}{2}(1 - x^2) - \frac{1}{8}(1 - x^2)^2 - \cdots \\ - \frac{1 \cdot 3 \cdots (2p - 3)}{2 \cdot 4 \cdots 2p}(1 - x^2)^p + \cdots . \qquad (1)$$

The radius of convergence of this series in powers of $z = 1 - x^2$ is 1. In fact, $(1 - z)^{1/2}$ is holomorphic for $| z | < 1$ and it has a critical point at $z = 1$. Therefore, it converges for

$$| 1 - x^2 | < 1$$

whether

$$-1 < 1 - x^2 < 1 \qquad \text{or} \qquad 0 < x^2 < 2.$$

The interval of convergence therefore is the union of the two intervals

$$-\sqrt{2} < x < 0 \qquad \text{and} \qquad 0 < x < \sqrt{2}\,.$$

The elementary theory of power series offers no information for $x = 0$ or $x = \pm\sqrt{2}$. Let us examine the general term. The coefficient in the general term is

$$\frac{1 \cdot 3 \cdots (2p - 3)}{2 \cdot 4 \cdots 2p} = \frac{(2p)!}{2^{2p}(p!)^2(2p - 1)}\,.$$

The most natural method of studying the order of magnitude of this term is to apply Stirling's formula to the factorials. The quantity $p!$ is of the same order of infinity as $(2\pi)^{1/2}p^{p+(1/2)}e^{-p}$. Therefore, up to equivalence,

$$\frac{(2p)!}{2^{2p}(p!)^2(2p - 1)} = \frac{\sqrt{2\pi}(2p)^{2p+(1/2)}\, e^{-2p}}{2^{2p}2\pi p^{2p+1}\, e^{-2p}2p} = \frac{1}{2\sqrt{\pi}}\frac{1}{p^{3/2}}\,.$$

This amounts to saying that there exists a constant M such that the coefficient of the general term of the series (1) is less than $M/p^{3/2}$. Therefore, the general term itself is less than $M(1 - x^2)^p/p^{3/2}$. The series converges absolutely for $0 \leqslant | 1 - x^2 | \leqslant 1$. Hence it is uniformly convergent for $-\sqrt{2} \leqslant x \leqslant \sqrt{2}$, including the value $x = 0$.

Now let $S_n(x)$ denote the sum of the first n terms in the series (1). For given ϵ, there exists a number $n_0(\epsilon)$, independent of x, such that, for $-\sqrt{2} \leqslant x \leqslant \sqrt{2}$ and $n > n_0(\epsilon)$,

$$|\, | x | - S_n(x) \,| < \epsilon.$$

Since $S_n(x)$ is a polynomial in x, we see that it is possible to approximate the continuous function $|x|$ uniformly by a polynomial even though $|x|$ is not differentiable at the origin.

Remark 1. There is nothing special about the interval $|x| \leqslant \sqrt{2}$, for which the above reasoning is valid. More generally, we may set

$$|x| = \lambda\left[1 - \left(1 - \frac{x^2}{\lambda^2}\right)\right]^{1/2},$$

where λ is any positive number. The interval then becomes

$$|x| \leqslant \lambda\sqrt{2}.$$

Remark 2. The essential feature of the method we have just used consists in the fact that we employ a power series in the closed interval of convergence, a fact that necessitates certain precautions. We may avoid this difficulty by proceeding as follows:

As a positive number a tends to 1, the expression $[1 - a^2(1 - x^2)]^{1/2}$ tends to $|x|$ uniformly on every interval.

We therefore first replace the angular curve $y = |x|$ with the semihyperbola

$$y = [y - a^2(1 - x^2)].$$

Again we expand the radical in powers of $1 - x^2$.

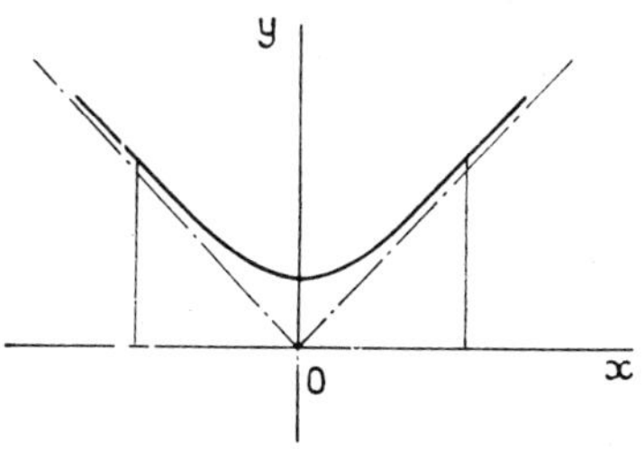

FIGURE 18

For $a < 1$, the series converges uniformly in the closed interval $-1 \leqslant x \leqslant 1$, which is contained in the interior of the interval of convergence ($|x| = 1$ corresponds to $1 - x^2 = 0$). If we let a approach 1 from below, the sum of the first n terms of the series approaches the sum of the series. Now the sum of the first n terms is a polynomial. We need only choose n sufficiently large to get the desired approximation.

B. First, we expand the function $|\sin \pi t|$, which is periodic with period 1 (see Exercise 9), in a Fourier series, obtaining

$$|\sin \pi t| = \frac{2}{\pi} - \frac{4}{\pi}\left(\frac{\cos 2\pi t}{4-1} + \cdots + \frac{\cos 2\pi pt}{4p^2-1} + \cdots\right). \tag{2}$$

This series converges uniformly on every interval and does indeed represent $|\sin \pi t|$ for every t. In other words, for any given ϵ, there exists a number $N(\epsilon)$ such that, for $n > N(\epsilon)$ and for all t,

$$\left| |\sin \pi t| - \frac{2}{\pi} + \frac{4}{\pi}\sum_{p=1}^{n} \frac{\cos 2\pi pt}{4p^2-1}\right| < \epsilon.$$

Let us make the change of variable $\sin \pi t = x$. We know that, if m is an integer, $\cos 2\pi mt$ is a polynomial in x. Specifically, if we set $\pi t = \alpha$, we obtain

$$\begin{aligned}\cos 2m\alpha &= \cos^{2m}\alpha - C_{2m}^{2}\cos^{2m-2}\alpha\sin^2\alpha \\ &\quad + C_{2m}^{4}\cos^{2m-4}\alpha\sin^4\alpha + \cdots + (-1)^m C_{2m}^{2m}\sin^{2m}\alpha \\ &= (1-x^2)^m - C_{2m}^{2}(1-x^2)^{m-1}x^2 + \cdots + (-1)^m C_{2m}^{2m}x^{2m}.\end{aligned}$$

Therefore, the partial sum

$$\frac{2}{\pi} - \frac{4}{\pi}\sum_{p=1}^{n}\frac{\cos 2\pi pt}{4p^2-1}$$

is a polynomial in x that differs from $|x|$ by less than ϵ.

C. The periodic function with period 2 that is equal to $|x|$ on the interval $[-1, 1]$ is continuous, and it has a continuous derivative except at points at which its argument is an integer. Therefore, for every value of x, it is the sum of an even Fourier series that converges uniformly on every interval. Specifically,

$$|x| = \frac{a_0}{2} + a_1 \cos \pi x + \cdots + a_n \cos \pi n x + \cdots,$$

where

$$a_n = 2\int_0^1 |x| \cos \pi n x\, dx = \begin{cases} 0 & \text{for } n \text{ even,} \\ -\dfrac{4}{\pi^2 n^2} & \text{for } n \text{ odd.}\end{cases}$$

We can choose n in such a way that, for any x in the interval $[-1, 1]$,

$$\left| |x| - \left(\frac{a_0}{2} + a_1\cos \pi x + \cdots + a_n \cos \pi n x\right)\right| < \frac{\epsilon}{2}.$$

When we have thus chosen n, we expand $\cos \pi x, \ldots, \cos \pi n x$ in powers of x:

$$\cos \pi p x = 1 - \tfrac{1}{2}(\pi p x)^2 + \cdots + \frac{(-1)^k}{k!}(\pi p x)^k + \cdots .$$

Then we choose k so that

$$\left| \cos \pi p x - \left[1 - \frac{1}{2}(\pi p x)^2 + \cdots + \frac{(-1)^k}{k!}(\pi p x)^k\right] \right| < \frac{\epsilon}{2n} .$$

We thus obtain a polynomial $P(x)$ such that

$$|\,|x| - P(x)\,| < \epsilon.$$

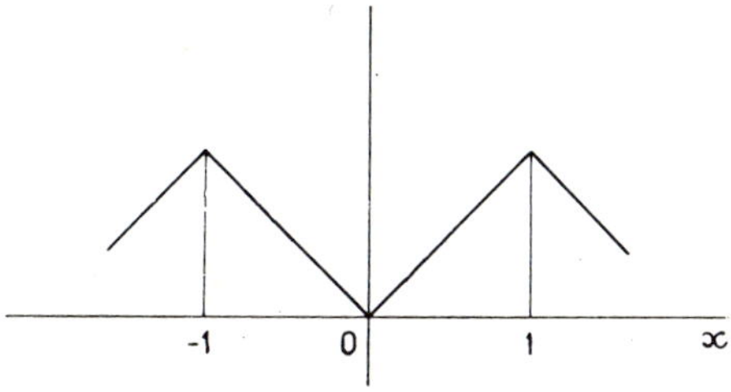

FIGURE 19

14. The Fourier transform of 1/cosh t

The purposes of this exercise are (1) to give an example of a function that is its own Fourier transform [infinitely many such functions exist, the classical one being the function $\exp(-t^2/2)$], (2) to give an example of the law of probability and of the calculation of a characteristic function, (3) to expand a discontinuous function in a Fourier series, and (4) to integrate term by term a nonuniformly convergent series. We point out that the integral $\int_0^\infty (\cos xt/\cosh t)\, dt$ can also be evaluated by an integration in the complex plane (see Exercise 4 of Chapter 4).

Problems. A. Expand in a Fourier series the odd periodic function of period 2π that is equal to $\cosh tx$ for $0 < x < \pi$ (where t is a fixed parameter). What happens to this expansion when $x = \pi/2$?

B. Expand $1/\cosh t$ in powers of e^{-t}. Multiply this expansion by $\cos xt$ and show that the resulting series can be integrated term by term from 0 to ∞. By comparison with the result of Problem 1, evaluate the integral

$$\int_0^\infty \frac{\cos xt}{\cosh t}\,dt.$$

Solutions. **A.** We wish to expand in a Fourier series the function $f(x)$ that is odd, periodic of period 2π, and equal to $\cosh tx$ for $0 < x < \pi$.

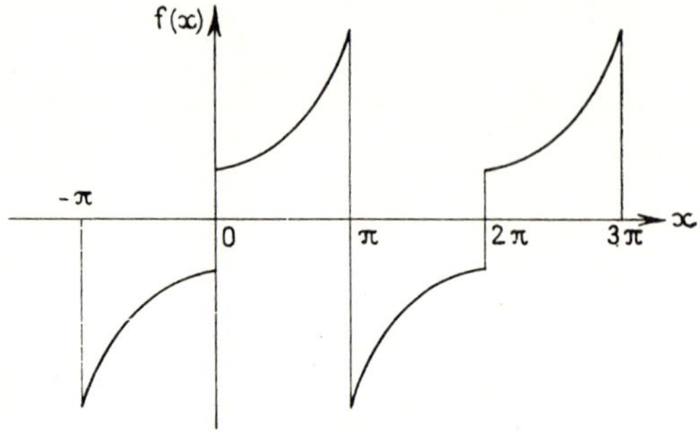

FIGURE 20

The coefficients a_n all vanish. The coefficients b_n are given by

$$b_n = \frac{1}{\pi}\int_{-\pi}^{\pi} f(x)\sin nx\,dx = \frac{2}{\pi}\int_0^\pi \sin nx \cosh tx\,dx.$$

Evaluation of this integral is easy (by using the exponential forms of $\sin nx$ and $\cosh tx$). We obtain

$$b_n = [1 - (-1)^n \cosh t\pi]\,\frac{2n}{\pi(t^2 + n^2)}\,.$$

According to Jordan's theorem, the series $\sum_{n=1}^\infty b_n \sin nx$ converges uniformly for all x in every closed interval contained in the open interval $(0, \pi)$.

At every point at which $f(x)$ is continuous, the sum of the series is $f(x)$. Thus, in particular, for $0 < x < \pi$,

$$\cosh tx = \sum_{n=1}^\infty [1 - (-1)^n \cosh t\pi]\,\frac{2n}{\pi(t^2 + n^2)}\sin nx.$$

Let us set $x = \pi/2$. Noting that

$$\sin nx = \begin{cases} 0 & \text{if} \quad n = 2p, \\ (-1)^p & \text{if} \quad n = 2p + 1, \end{cases}$$

$$1 + \cosh t\pi = 2\cosh^2 \frac{t\pi}{2}\,,$$

we obtain

$$\boxed{\frac{1}{\cosh(\pi/2)\,t} = \frac{4}{\pi}\sum_{p=0}^{\infty}(-1)^p \frac{2p+1}{(2p+1)^2+t^2}\,.}$$

B. Let us expand $1/\cosh t$ in powers of e^{-t}. For $t > 0$,

$$\frac{1}{\cosh t} = \frac{2e^{-t}}{1+e^{-2t}} = 2e^{-t} - 2e^{-3t} + \cdots + (-1)^n 2e^{-(2n+1)t} + \cdots.$$

Therefore,

$$\frac{\cos xt}{\cosh t} = 2e^{-t}\cos xt - 2e^{-3t}\cos xt + \cdots + (-1)^n 2e^{-(2n+1)t}\cos xt + \cdots.$$

If we wish to integrate this series term by term from $t = 0$ to $t = \infty$, we encounter two difficulties when we try to use the uniform convergence of the series: (a) The interval of integration is infinite [this difficulty can be surmounted (see Exercise 7)]. (b) The series does not converge uniformly in the interval of integration, one end point of which is the point $t = 0$.

We shall show directly that the integral of the remainder tends to 0, that is, that

$$A = \int_0^\infty \left| \frac{\cos xt}{\cosh t} - \sum_{p=0}^{n} (-1)^p\, 2e^{-(2p+1)t} \cos tx \right| dt$$

tends to 0 as $n \to \infty$.

Since we are dealing with a geometric series, the remainder is easily calculated. We find

$$A = \int_0^\infty \left| \cos xt\, \frac{e^{-(2n+3)t}}{1+e^{-2t}} \right| dt \leqslant \int_0^\infty e^{-(2n+3)t}\, dt.$$

The last integral is equal to $(2n+1)^{-1}$ and it does indeed tend to 0 as $n \to \infty$.

Finally,

$$\int_0^\infty \frac{\cos xt}{\cosh t}\, dt = 2\sum_{p=0}^{\infty}(-1)^p \int_0^\infty \cos xt\, e^{-(2p+1)t}\, dt$$

$$= \sum_{p=0}^{\infty}(-1)^p \frac{2(2p+1)}{(2p+1)^2+x^2}\,.$$

We saw in Problem A that

$$\sum_{p=0}^{\infty}(-1)^p \frac{2p+1}{(2p+1)^2+x^2} = \frac{\pi}{4}\frac{1}{\cosh(\pi x/2)}.$$

Therefore,

$$\int_0^{\infty}\frac{\cos xt}{\cosh t}\,dt = \frac{\pi}{2}\frac{1}{\cosh(\pi x/2)}.$$

This formula gives the Fourier transform of $1/\cosh t$. In probability-theory language, this means that $(\pi \cosh t)^{-1}$ is a probability density whose characteristic function is $1/\cosh(\pi x/2)$.

If we replace t with $t\,(\pi/2)^{1/2}$ and x with $x\,(2/\pi)^{1/2}$, the above formula becomes

$$\int_0^{\infty}\frac{\cos xt}{\cosh t\,(\pi/2)^{1/2}}\,dt = \left(\frac{\pi}{2}\right)^{1/2}\frac{1}{\cosh x\,(\pi/2)^{1/2}}.$$

It takes a symmetric form: The function $1/\cosh t\,(\pi/2)^{1/2}$ is its own Fourier cosine transform.

15. Gibbs' phenomenon

If $f(x)$ is a function of bounded variation, its Fourier series converges for every value of x. However, in a neighborhood of a point at which $f(x)$ is not continuous, the convergence may not be uniform. [We recall that it is uniform in every interval in which $f(x)$ is continuous.]

"Gibbs' phenomenon" brings out the fact that the partial sums $S_n(x)$ of the series do not converge uniformly in a neighborhood of a point of discontinuity x_0 of $f(x)$. They tend to $f(x)$ at every point x different from x_0 but sufficiently close to x_0. However, the distance between $f(x)$ and $S_n(x)$ does not tend to zero.

We can verify this result with the example of the function that is periodic with period 2π, odd, and equal to $\pi/4$ when $0 < x < \pi$. This function has a saltus at the origin.

Problems. Consider the function $f(x)$ that is odd, periodic with period 2π, and equal to $\pi/4$ when $0 < x < \pi$.

A. Expand $f(x)$ in a Fourier series.

B. Denote by $S_n(x)$ the sum of the first n terms of the series. Show that

$$S_n(x) = \int_0^x \frac{\sin 2nt}{2 \sin t}\, dt.$$

Evaluate the ordinate y_n of the first maximum of positive abscissa of the function $S_n(x)$. Show that y_n tends to a limit y as $n \to \infty$ and show by numerical calculation that $y > \pi/4$.

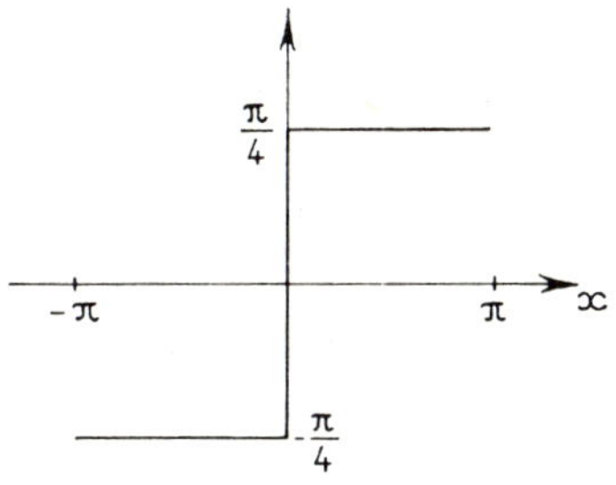

FIGURE 21

Solutions. **A.** Let

$$f(x) = b_1 \sin x + \cdots + b_n \sin nx + \cdots$$

denote the Fourier expansion of the function $f(x)$ defined above. Here

$$b_n = \frac{1}{\pi} \int_0^{2\pi} f(x) \sin nx\, dx = \frac{2}{\pi} \int_0^{\pi} \frac{\pi}{4} \sin nx\, dx$$
$$= -\frac{1}{2n} (\cos nx)_0^{\pi}.$$

Therefore,

$$b_n = \begin{cases} 0 & \text{if} \quad n \text{ is even}, \\ \dfrac{1}{n} & \text{if} \quad n \text{ is odd}. \end{cases}$$

This series is convergent in the closed interval $0 \leqslant x \leqslant \pi$. According to Jordan's theorem, its sum is equal to $f(x)$ in the open interval $0 < x < \pi$:

$$\frac{\pi}{4} = \sin x + \frac{\sin 3x}{3} + \cdots + \frac{\sin (2n-1)x}{2n-1} + \cdots.$$

For $x = 0$ and $x = \pi$, the sum is 0.

B. The convergence is not uniform in a closed interval containing $x = 0$, since $f(x)$ is *discontinuous* at that point. In a neighborhood of $x = 0$, no matter how large n is, we can find points at which the partial sum $S_n(x)$ differs from $\pi/4$ or from $-\pi/4$ by a fixed amount. This fact constitutes Gibbs' phenomenon.

Let us find the maxima of $S_n(x)$ by setting the derivative $S_n'(x)$ equal to zero. According to a classical formula,

$$S_n'(x) = \cos x + \cos 3x + \cdots + \cos (2n - 1)x = \frac{1}{2} \frac{\sin 2nx}{\sin x}.$$

From this, we see immediately that $S_n(x)$ has maxima at the points

$$x_k = \frac{(2k + 1)\pi}{2n}$$

and minima at the points

$$x_k' = \frac{2k\pi}{2n} = \frac{k\pi}{n}.$$

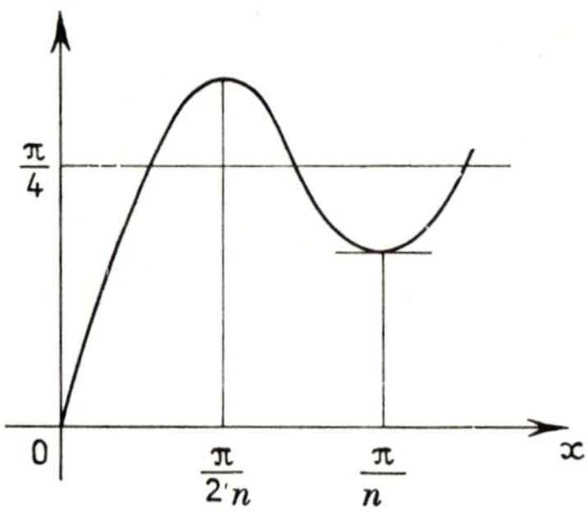

FIGURE 22

The curve $y = S_n(x)$ thus oscillates about the value

$$y = \frac{\pi}{4}.$$

Let us integrate $S_n'(x)$ to get a different expression for $S_n(x)$:

$$S_n(x) = \int_0^x \frac{\sin 2nt}{2 \sin t}\, dt.$$

The ordinate of the first maximum is equal to

$$y_n = \int_0^{\pi/2n} \frac{\sin 2nt}{2 \sin t}\, dt.$$

Let us find the limit of y_n as $n \to \infty$. We have

$$2y_n = \int_0^{\pi/2n} \frac{\sin 2nt}{t}\,dt + \int_0^{\pi/2n} \sin 2nt \left(\frac{1}{\sin t} - \frac{1}{t}\right) dt.$$

In the first integral, we set $2nt = s$. The integral then becomes

$$\int_0^{\pi} \frac{\sin s\,ds}{s},$$

which is independent of n.

The second integral is less in absolute value than

$$\int_0^{\pi/2n} \left| \frac{1}{\sin t} - \frac{1}{t} \right| dt.$$

Now, $|(1/\sin t) - (1/t)|$ is a continuous function in a neighborhood of the origin. If M is its least upper bound, the above integral is less than $\pi M/2n$, which tends to 0. Therefore, the limit of y_n is

$$Y = \frac{1}{2}\int_0^{\pi} \frac{\sin s}{s}\,ds.$$

To find the numerical value of Y, we expand $(\sin s)/s$ in a power series. Since this series converges uniformly on every finite interval, we may integrate term by term. We obtain

$$\frac{\sin s}{s} = 1 - \frac{s^2}{3!} + \frac{s^4}{5!} + \cdots, \qquad 2Y = \pi - \frac{\pi^3}{3 \cdot 3!} + \frac{\pi^5}{5 \cdot 5!} - \cdots.$$

Thus Y is the sum of an alternating series whose general term decreases in absolute value and approaches zero. If we cut the series off at the term

$$\frac{\pi^{2n-1}}{(2n-1)(2n-1)!},$$

the remainder of the series will be less in absolute value than

$$\frac{\pi^{2n+1}}{(2n+1)(2n+1)!}.$$

For $n = 3$,

$$\frac{\pi^7}{7 \cdot 7!} < 0.1.$$

Thus

$$\pi - \frac{\pi^3}{18} + \frac{\pi^5}{600} - 0.1 < 2Y < \pi - \frac{\pi^3}{18} + \frac{\pi^5}{600}.$$

This inequality shows that Y is of the order of 0.9. If we were to include more terms of the series, we would obtain

$$Y = 0.92$$

with digits discarded after the second decimal place. Therefore,

$$Y > \frac{\pi}{4} = 0.78.$$

*16. Uniform distributions

In Exercise 7 of Chapter 1, we defined the concept of a sequence $\{x_n\}$ that is equally distributed on the interval [0, 1]. We showed that for a sequence $\{x_n\}$ to be equally distributed on [0, 1], it is necessary and sufficient that

$$\int_0^1 f(x)\,dx = \lim_{N\to\infty} \frac{f(x_1) + \cdots + f(x_N)}{N}$$

for every function $f(x)$ that is integrable on that interval. In the present case, we prove and use a form reduced to the condition stated below. We show that it is not necessary that all integrable functions $f(x)$ satisfy it but only that the function $f(x) = e^{2i\pi kx}$, where k is any nonzero integer, satisfy it.

We shall apply this result to the sequence $\{x_n\}$ of numbers $x_n = \theta n$, mod 1, that is, to the sequence of decimal portions of the multiples of a number θ. If θ is irrational, this sequence is equally distributed over [0, 1]. From this it results, in particular, that the sequence is dense in [0, 1]. This last result, less precise and more elementary, can of course be proved directly.

Problems. Let us denote by L an interval $a < x < b$ in the interior of the interval [0, 1]. Let ϵ be a sufficiently small positive number. Consider two functions $g_1(x)$ and $g_2(x)$ both continuous and periodic with period 1 such that

$$\begin{cases} g_2(x) = 0 \quad \text{if} \quad 0 < x < a - \epsilon \quad \text{and} \quad b + \epsilon < x < 1. \\ g_2(x) = 1 \quad \text{if} \quad a < x < b. \\ g_2(x) \text{ varies linearly when } a - \epsilon < x < a \quad \text{and} \quad b < x < b + \epsilon. \end{cases}$$

$$\begin{cases} g_1(x) = 0 \quad \text{if} \quad 0 < x < a \quad \text{and} \quad b < x < 1. \\ g_1(x) = 1 \quad \text{if} \quad a + \epsilon < x < b - \epsilon. \\ g_1(x) \text{ varies linearly when } a < x < a + \epsilon \quad \text{and} \quad b - \epsilon < x < b. \end{cases}$$

Finally, let $f(x)$ denote a function that is periodic with period 1 and that is equal to 1 when $a < x < b$ but equal to 0 when $0 < x < a$ and $b < x < 1$.

A. Construct the curves for g_1, g_2, and f. Show that g_1 and g_2 can be expanded in absolutely and uniformly convergent Fourier series. Is the same true of f?

B. If $x_1, \ldots, x_q, \ldots$ is an evenly distributed sequence, the theorem referred to above shows immediately that

$$\lim_{N\to\infty} \frac{1}{N} [\exp(zi\pi kx_1) + \cdots + \exp(2i\pi kx_N)] = 0 \tag{1}$$

for every nonzero integer k.

Show that, conversely, if condition (1) is satisfied, the sequence $\{x_q\}$ is evenly distributed. To do this, show first that, for $i = 1$ and $i = 2$,

$$\lim_{N\to\infty} \frac{1}{N} [g_i(x_1) + \cdots + g_i(x_N)] = \int_0^1 g_i(x)\, dx.$$

By using the inequalities $g_1 \leqslant f \leqslant g_2$ and the equation

$$\int_0^1 g_2\, dx - \int_0^1 g_1\, dx = 2\epsilon,$$

show that

$$\lim_{N\to\infty} \frac{1}{N} [f(x_1) + \cdots + f(x_N)] = \int_0^1 f(x)\, dx,$$

Interpret this condition with the aid of the definition of an evenly distributed sequence (see Exercise 7 of Chapter 1).

C. Let θ denote an irrational number. Consider the sequence $\{x_q\}$, where $x_q = \theta q$, mod 1 (in other words, x_q is the difference between the

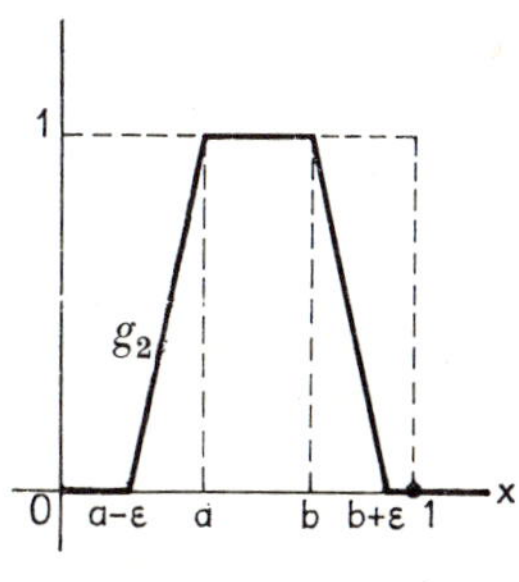

FIGURE 23

product θq and the integral part of it, that is, x_q is the decimal part of θq). Show that the sequence $\{x_q\}$ is evenly distributed.

Solutions. **A.** The function g_2 with period 1 shown in Figure 23 is continuous. Its derivative exists and is continuous except at four points in each period, at which it has discontinuities of the first kind. Thus we know that $g_2(x)$ has a Fourier expansion of the form

$$g_2(x) = \sum_k c_k e^{2i\pi kx}$$

and that there exists a number M such that

$$| c_k | < \frac{M}{k^2}.$$

Therefore, the Fourier series for this function converges absolutely and uniformly. The same reasoning holds for g_1. Since f has points of discontinuity, its expansion converges only in $1/k$ and is not absolutely convergent. It converges uniformly only in the intervals in which f is continuous.

B. The functions g_1, f, g_2 are such that

$$g_1 \leqslant f \leqslant g_2, \qquad \int_0^1 g_2\, dx - \int_0^1 g_1\, dx = 2\epsilon.$$

We set

$$S_N(k) = \frac{1}{N}[\exp(2i\pi kx_1) + \cdots + \exp(2i\pi kx_N)].$$

By hypothesis, $S_N(k)$ tends to 0 as $N \to \infty$, except, of course, in the case of $S_N(0)$, which remains equal to 1. If $g(x)$ represents one of the functions $g_1(x)$ or $g_2(x)$, we have

$$\frac{1}{N}[g(x_1) + \cdots + g(x_N)] = \sum_k c_k S_N(k).$$

Since $| S_N(k)| < 1$, the series on the right, which is majorized by the series M/k^2, is uniformly convergent. Therefore,

$$\lim_{N\to\infty} \sum_k c_k S_N(k) = \sum_k c_k [\lim_{N\to\infty} S_N(k)] = c_0 .$$

We note that this passage to the limit, which is possible with g_1 and g_2, cannot be done directly with f, whose Fourier series is not uniformly convergent over the entire interval. Since c_0, the first Fourier coefficient of g, is equal to $\int_0^1 g(x)\,dx$, we have

$$\lim_{N\to\infty} \frac{1}{N}[g(x_1) + \cdots + g(x_N)] = \int_0^1 g(x)\,dx.$$

We set

$$A_1 = \frac{1}{N}[g_1(x_1) + \cdots + g_1(x_N)], \qquad A_2 = \frac{1}{N}[g_2(x_1) + \cdots + g_2(x_N)],$$

$$A = \frac{1}{N}[f(x_1) + \cdots + f(x_N)].$$

Since $g_1 \leqslant f \leqslant g_2$, we have $A_1 \leqslant A \leqslant A_2$, so that

$$0 \leqslant A - A_1 \leqslant A_2 - A_1 .$$

Similarly, since

$$\int_0^1 g_2\,dx - \int_0^1 g_1\,dx = 2\epsilon,$$

we have

$$0 < \int_0^1 f\,dx - \int_0^1 g_1\,dx < \int_0^1 g_2\,dx - \int_0^1 g_1\,dx = 2\epsilon.$$

Finally, we may choose N sufficiently large that

$$-\epsilon < A_1 - \int_0^1 g_1\,dx < \epsilon, \qquad -\epsilon < A_2 - \int_0^1 g_2\,dx < \epsilon.$$

We then have

$$A - \int_0^1 f\,dx = (A - A_1) + \left(A_1 - \int_0^1 g_1\,dx\right) + \left(\int_0^1 g_1\,dx - \int_0^1 f\,dx\right)$$

and

$$\left| A - \int_0^1 f\,dx \right| < A_2 - A_1 + \epsilon + 2\epsilon.$$

But

$$A_2 - A_1 < \int_0^1 g_2\,dx + \epsilon - \left(\int_0^1 g_1\,dx - \epsilon\right) < 2\epsilon + 2\epsilon = 4\epsilon.$$

Finally,

$$\left| A - \int_0^1 f\,dx \right| < 7\epsilon.$$

Thus we may choose N so great that

$$\left| \frac{1}{N}[f(x_1) + \cdots + f(x_N)] - \int_0^1 f(x)\,dx \right| < 7\epsilon.$$

But $f(x_q)$ is equal to 1 if x_q belongs to L and is otherwise equal to 0. Therefore, the sum $f(x_1) + \cdots + f(x_N)$ is equal to the number N' of points x_q among the points $x_1, ..., x_N$ that belong to the interval L. Since $\int_0^1 f(x)\, dx$ is the length l of L, we see that N'/N tends to a limit equal to l. Since this result holds for every L, we have the definition of a uniform distribution.

C. *Example.* $x_q = \theta q$, mod 1. Let us evaluate the sum

$$S_N = \frac{1}{N}(e^{2i\pi k\theta} + \cdots + e^{2i\pi kN\theta}).$$

This is a geometric progression. We obtain

$$S_N = \frac{1}{2iN}\left\{\frac{\exp[2i\pi k(N + \frac{1}{2})\theta] - e^{i\pi k\theta}}{\sin \pi k\theta}\right\}.$$

If θ is irrational, the denominator $\sin \pi k\theta$ is never 0 except for $k = 0$. Therefore,

$$| S_N | < \frac{1}{N \sin \pi k\theta}.$$

S_N tends to 0 as $N \to \infty$ and the sequence of the x_q (modulo 1) is evenly distributed.

If θ is rational, we already know that this sequence assumes only a finite number of values. It is not dense on the segment [0, 1]. Also, if θ is a fraction (reduced to its simplest form) s/p, then S_N assumes the value 1 whenever k is a multiple of p. Thus there are nonzero values of k for which S_N does not tend to 0 and the criterion of Problem B is not applicable.

17. Volumes of the interior of cubes in *n*-dimensional spaces

Exercise 7 of Chapter 2 is stated in probability-theory language, but it is immediately reduced to the evaluation of multiple integrals. In the present exercise, on the other hand, we use probability-theory techniques to obtain a geometric result. Of course, we are still dealing with measure-theoretic methods. However, in this case, probability-theory language is advantageous and shortens the calculations.

Let us recall certain properties that will be of use to us. Let X denote a real random variable and let $f(x)$ denote its probability density. Then $f(x)$ is the probability that $x \leqslant X < x + dx$.

We say that X is **uniformly distributed** over an interval $x_0 < x < x_1$ if the probability density of X is equal to $(\alpha_1 - x_0)^{-1}$ when x belongs to the interval but is 0 when it does not.

The characteristic function of X is the Fourier transform of the density $f(x)$. If several random variables are independent, the characteristic function of their sum is the product of their individual characteristic functions.

If the position of a point is uniformly distributed throughout a cube in n-dimensional space, its n coordinates are independent variables uniformly distributed along the sides of the cube. The probability of a subset E of the cube is the volume of E divided by the volume of the cube.

$$\frac{V^2}{4\pi^2 a} p'(a) = 4a \int_0^{R-a} \rho^2 \, d\rho + \int_{R-a}^{R} [R^2 - (a-\rho)^2]\rho \, d\rho.$$

After performing the integrations, we find

$$\frac{V^2}{4\pi^2} p'(a) = \frac{a^5}{12} - a^3R^2 + \frac{4}{3} a^2R^3 = \frac{a^2}{12}(2R-a)^2(4R+a).$$

Knowing that $p(0) = 0$, we conclude by integrating the above equation that

$$\frac{V^2}{4\pi^2} p(a) = \frac{6a^2}{72} - \frac{a^4R^2}{4} + \frac{4}{9} a^3R^3,$$

$$p(a) = \frac{1}{32}\left(\frac{a}{R}\right)^6 - \frac{9}{16}\left(\frac{a}{R}\right)^4 + \left(\frac{a}{R}\right)^3.$$

The probability $p(a)$ vanishes for $a = 0$ and is equal to 1 for $a = 2R$. The probability density $p'(a)$ vanishes at $a = 0$ and $a = 2R$.

Problems. A. Let X be a random variable uniformly distributed between -1 and 1. Calculate the characteristic function of X.

B. Calculate the characteristic function $\varphi(t)$ of the sum Y of n independent random variables that are equal to X. From this, derive the distribution function of Y.

C. *An application.* Show that the volume V of the interior of the cube C in n-dimensional space defined by the inequalities

$$-1 < x_1 < 1, \ldots, -1 < x_n < 1$$

and lying between the two planes

$$x_1 + \cdots + x_n = -a \qquad \text{and} \qquad x_1 + \cdots + x_n = a$$

is equal to

$$V = \frac{2^{n+1}}{\pi} \int_0^\infty \left(\frac{\sin t}{t}\right)^n \frac{\sin at}{t}\, dt.$$

Verify this result for $n = 1$ and $n = 2$.

Solutions. **A.** Let X be a random variable uniformly distributed on $[-1, 1]$. Its probability density is $\frac{1}{2}$ and its characteristic function is

$$\frac{1}{2}\int_{-1}^{1} e^{itx}\, dx = \frac{\sin t}{t}.$$

B. The sum Y of n independent random variables uniformly distributed on $[-1, 1]$ has the characteristic function

$$\varphi(t) = \left(\frac{\sin t}{t}\right)^n.$$

According to Fourier's reciprocity formulas, the probability density of Y is

$$f(y) = \frac{1}{2\pi}\int_{-\infty}^{\infty} \left(\frac{\sin t}{t}\right)^n e^{-itx}\, dt = \frac{1}{2\pi}\int_{-\infty}^{\infty} \left(\frac{\sin t}{t}\right)^n \cos ty\, dt.$$

The probability that $-a < Y < a$ is therefore equal to

$$p = \int_{-a}^{a} f(y)\, dy = \frac{1}{\pi}\int_{-\infty}^{\infty} \left(\frac{\sin t}{t}\right)^n \frac{\sin at}{t}\, dt.$$

We also know that $p = 1$ for all $a > n$. Since Y varies from $-n$ to n, this probability can be interpreted as follows.

Let C be the n-dimensional cube defined by the inequalities

$$-1 < x_i < 1 \qquad \text{for} \qquad i = 1, 2, \ldots, n.$$

Let V be the volume inside this cube and between the two planes

$$\sum x_i = \pm a.$$

The probability p is the ratio of this volume V to the volume 2^n of the cube; that is,

$$V = 2^n p = \frac{2^{n+1}}{\pi}\int_0^\infty \left(\frac{\sin t}{t}\right)^n \frac{\sin at}{t}\, dt.$$

Example 1. $n = 1$. Here

$$V = \frac{4}{\pi} \int_0^\infty \frac{\sin t}{t} \frac{\sin at}{t} \, dt.$$

If $a < 1$, then V is the length of the segment $[-a, a]$, namely, $2a$. If $a > 1$, then $V = 2$. The verification is easy:

$$2 \sin t \sin at = \cos (a - 1)t - \cos (a + 1)t.$$

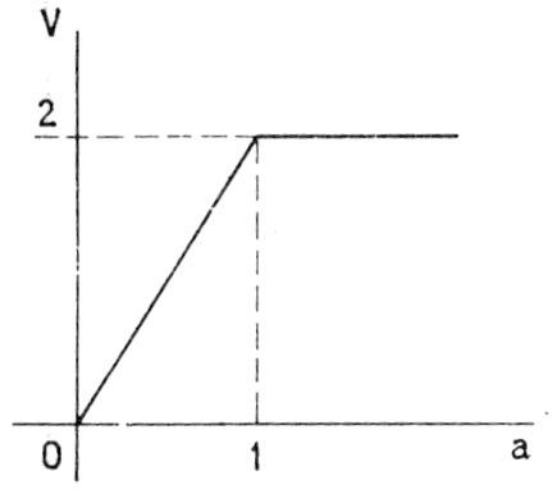

FIGURE 24

From this we get

$$V = \frac{2}{\pi} \int_0^\infty \frac{1 - \cos (a + 1)t}{t^2} \, dt - \frac{2}{\pi} \int_0^\infty \frac{1 - \cos (a - 1)t}{t} \, dt.$$

Let us calculate $(2/\pi) \int_0^\infty [(1 - \cos \lambda t)/t^2] \, dt$ by differentiating under the integral sign. The derivative is equal to 1 if $\lambda > 0$ and is equal to -1 if $\lambda < 0$. From this we conclude that

$$\frac{2}{\pi} \int_0^\infty \frac{1 - \cos \lambda t}{t^2} \, dt = \begin{cases} \lambda & \text{if} \quad \lambda > 0, \\ -\lambda & \text{if} \quad \lambda < 0. \end{cases}$$

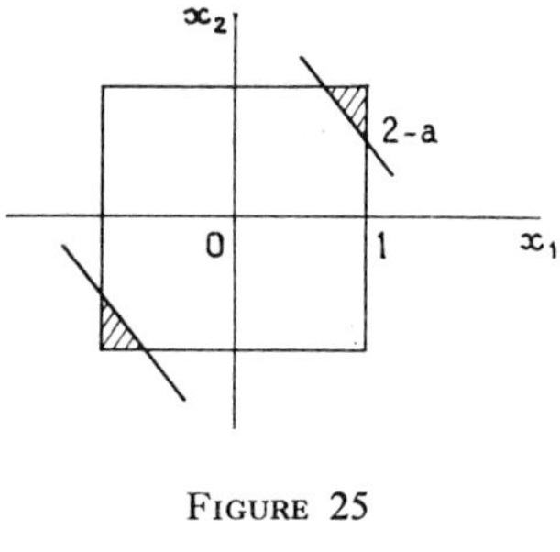

FIGURE 25

Consequently,

$$V = \begin{cases} 2a & \text{if} \quad 0 < a < 1, \\ 2 & \text{if} \quad a > 1, \end{cases}$$

which is what we expected to find.

Example 2. $n = 2$. The integral

$$V = \frac{8}{\pi} \int_0^\infty \left(\frac{\sin t}{t}\right)^2 \frac{\sin at}{t} \, dt$$

represents the shaded area in Figure 25 if $a < 2$, and it represents the area of the square $V = 4$ for all $a > 2$.

If $a < 2$, its value is therefore (see Figure 26)

$$V = 4 - (2 - a)^2 = 4a - a^2.$$

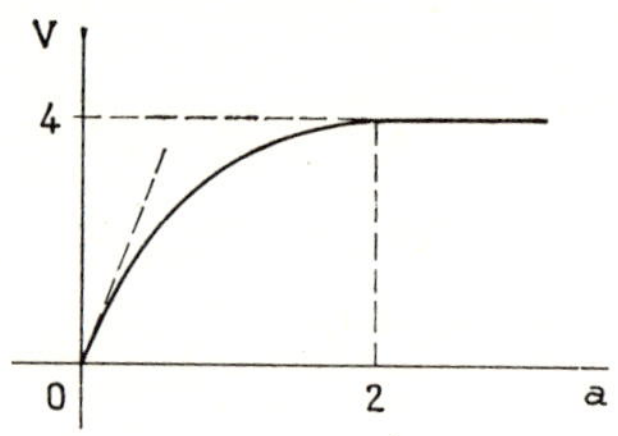

FIGURE 26

CHAPTER 3

Line Integrals and Multiple Integrals

The exercises of this chapter are devoted to integrations involving the following concepts:

Line integrals.

Double, triple, and higher-order integrals (proper and improper).

Surface integrals in three-dimensional space.

Transformation of integrals. The Jacobian determinant. Change of variables.

The formulas of Stokes and Ostrogradskiy. Integration of outer differential forms (in three-dimensional space).

The gamma and beta functions.

Not many exercises are devoted to the pure calculation of double and triple integrals. However, numerous occasions for making such calculations will appear in the course of more advanced exercises. Certain of the exercises in this chapter are indirect exercises on uniform convergence, integration under the integral sign, and even Fourier series and integrals.

In connection with line integrals, we have gone slightly afield with the integration of exact differentials around a closed curve (we shall encounter problems of this sort again in connection with analytic functions) and the elementary theory of integrating factors. We need to know that, for a differential form $\omega = P\,dx + Q\,dy + R\,dz$ to have an integrating factor, it is necessary and sufficient that the vector (P, Q, R) be orthogonal to its curl and that, for the differential form

$\omega = \Sigma P_i \, dx_i$ to be an exact differential, it is necessary and sufficient that its outer differential $d\omega$ vanish. We wish to be able to state Stokes' and Ostrogradskiy's theorems both in classical language and with the aid of the formalism of Grassman algebra. We do not, however, include Frobenius' theorem.

In connection with the area of a surface, we use the concept of a linear element in three-dimensional space and a linear element on a surface. However, we do not include problems on the differential geometry of surfaces and, of course, we omit any study of Riemann surfaces.

The point of view adapted for improper multiple integrals is somewhat different from that for simple integrals. Since a multiple integral can converge only if it converges absolutely, we turn our attention to the integration of positive functions and the convergence of the integral follows from the validity of a particular method of calculation. In general, it is not expedient to test for convergence on the basis of comparison of local properties (equivalence of infinitesimals or variables that become infinite). As an application, we shall give some examples of integration under the integral sign which do not follow from the properties of uniform convergence.

With regard to the gamma and beta functions, we have confined ourselves to the evaluation of certain improper integrals by means of their use and the complement formula. We recall that

$$B(x, y) = \int_0^1 t^{x-1} (1 - t)^{y-1} \, dt, \qquad \Gamma(s) = \int_0^\infty t^{s-1} e^{-1} \, dt,$$

and that

$$B(x, y) = \frac{\Gamma(x)\Gamma(y)}{\Gamma(x + y)}, \qquad \Gamma(s)\Gamma(1 - s) = \int_0^\infty \frac{t^{s-1}}{1 + t} \, dt = \frac{\pi}{\sin s\pi}.$$

B is defined if x and y are both positive; Γ is defined if s is positive.

1. Evaluation of a line integral

Problems. Consider the line integral

$$I = \int_\Gamma P \, dx + Q \, dy,$$

where

$$Q = \frac{e^x}{x^2 + y^2} (x \cos y + y \sin y)$$

and

$$P = \frac{e^x}{x^2 + y^2} (x \sin y - y \cos y)$$

and where Γ is a simple closed curve encircling the origin.

A. Verify that $\partial P/\partial y = \partial Q/\partial x$. Show that the functions

$$A = P + y \qquad \text{and} \qquad B = Q - x$$

are continuous even at the origin. Show from this that $I = 2\pi$.

B. Obtain the same result by converting I to polar coordinates and integrating over a circle with center at the origin, letting the radius tend to 0.

C. Set $z = x + iy$ and $\bar{z} = x - iy$. Show that I is the imaginary part of an integral of a meromorphic function and evaluate this integral by use of Cauchy's formula.

Solutions. **A.** Consider the line integral

$$I = \int_\Gamma \frac{e^x}{x^2 + y^2} (x \cos y + y \sin y)\, dy + \frac{e^x}{x^2 + y^2} (x \sin y - y \cos y)\, dx.$$

If we set

$$Q = \frac{e^x(x \cos y + y \sin y)}{x^2 + y^2}, \qquad P = \frac{e^x(x \sin y - y \cos y)}{x^2 + y^2},$$

we can verify immediately that

$$\frac{\partial Q}{\partial x} = \frac{\partial P}{\partial y}.$$

Therefore, I is the integral of the differential of some function $\varphi(x, y)$.

On the other hand, P and Q are continuous functions of x and y except at the point $x = y = 0$. Therefore, if Γ is a closed curve, $I = 0$, except possibly if Γ encircles the origin.

In what follows we assume that Γ is a simple closed curve with positive orientation that does encircle the origin.

Compare I with the integral

$$J = \int_\Gamma \frac{x\, dy - y\, dx}{x^2 + y^2},$$

the value of which is 2π. Let us set

$$I - J = \int_\Gamma A\,dx + B\,dy.$$

Then

$$A = \frac{e^x}{x^2 + y^2}(x \sin y - y \cos y) + \frac{y}{x^2 + y^2},$$
$$B = \frac{e^x}{x^2 + y^2}(x \cos y + y \sin y) - \frac{x}{x^2 + y^2}.$$

Let us now show that A and B can be defined in such a manner that they are continuous even at the origin. This will imply that $I - J = 0$ and that $I = J = 2\pi$.

Let us look at A. In a neighborhood of the origin, we have

$$e^x = 1 + x + \frac{x^2}{2} + x^2\epsilon_1,$$
$$\cos y = 1 - \frac{y^2}{2} + y^2\epsilon_2,$$
$$\sin y = y + y^2\epsilon_3,$$

where ϵ_1, ϵ_2, and ϵ_3 are functions that tend to 0 as x and y tend to 0.

With the aid of these expressions, we can put A in the form

$$A = \frac{y}{2} + \frac{xy}{2} + y\,\frac{\alpha x^2 + \beta xy + \gamma y^2}{x^2 + y^2},$$

where α, β, and γ are three functions that tend to 0 as x and y tend to 0.

As x and y tend to 0, the first two terms tend to 0. The third is the product of two factors. The first of these factors, y, tends to 0. The second, $(\alpha x^2 + \beta xy + \gamma y^2)/(x^2 + y^2)$, is bounded, as is easily seen when it is written in polar coordinates:

$$\alpha \cos^2\theta + \beta \cos\theta \sin\theta + \gamma \sin^2\theta.$$

Therefore, the last term of A also tends to 0. Thus A tends to 0 as x and y tend to 0, so that A is continuous at the origin. We can show in a similar way that B is continuous at the origin. Therefore,

$$I = J = 2\pi.$$

B. We have

$$I = \int_\Gamma e^x \cos y\,\frac{x\,dy - y\,dx}{x^2 + y^2} + \int_\Gamma e^x \sin y\,\frac{x\,dx + y\,dy}{x^2 + y^2}.$$

In polar coordinates,

$$I = \int_{\Gamma} e^{r\cos\theta} \cos(r \sin\theta)\, d\theta + \int_{\Gamma} e^{r\cos\theta} \sin(r \sin\theta)\, \frac{dr}{r}.$$

The value of I is unknown in advance, but it does not change when we deform Γ arbitrarily in a continuous manner (replacing Γ with a homologous curve in the doubly connected region obtained by deleting the origin from the plane). Let us take as Γ the circle of radius r with center at the origin. Since r is constant, the second integral vanishes and

$$I = \int_0^{2\pi} e^{r\cos\theta} \cos(r \sin\theta)\, d\theta.$$

Let us let r (upon which I does not depend) approach 0. The integrand in the above expression is continuous. Therefore, we may take the limit under the integral sign. Now, for $r = 0$,

$$I = \int_0^{2\pi} d\theta = 2\pi,$$

which is the value found in Problem A.

C. If we set $z = x + iy$ and $\bar{z} = x - iy$, we have

$$I = \int_{\Gamma} \frac{e^z}{2iz}\, dz - \int_{\Gamma} \frac{e^{\bar{z}}}{2i\bar{z}}\, d\bar{z}.$$

Thus I is the imaginary part of the integral

$$\int_{\Gamma} \frac{e^z}{z}\, dz.$$

According to Cauchy's formula, this integral is equal to $2i\pi$, so that $I = 2\pi$.

2. An example of an integrable differential form

Problems. Let $P(x, y, z)$, $Q(x, y, z)$, and $R(x, y, z)$ be three homogeneous functions of degree m.

A. If $Px + Qy + Rz$ is not identically 0, the differential form

$$\omega = P\,dx + Q\,dy + R\,dz$$

has the integrating factor

$$\frac{1}{Px + Qy + Rz}.$$

Show this both by direct verification and by changing to spherical coordinates.

B. If $Px + Qy + Rz = 0$, the surfaces representing the solutions of the total differential equation $\omega = 0$ are cones and ω has an integrating factor of the form $r^\alpha f(\theta, \varphi)$.

Solutions. **A.** Suppose that the differential form

$$\omega = P\,dx + Q\,dy + R\,dz$$

is integrable. Then the vector (P, Q, R) is orthogonal to its curl.

We can verify by direct calculation that

$$\frac{P\,dx + Q\,dy + R\,dz}{Px + Qy + Rz}$$

is an exact differential since P, Q, and R satisfy Euler's equation for homogeneous functions of degree m. However, it is more interesting to make a change of variables, since the condition of integrability is invariant.

Let us change to spherical coordinates. We have

$$\begin{aligned} dx &= \sin\theta\cos\varphi\,dr + r(\cos\theta\cos\varphi\,d\theta - \sin\theta\sin\varphi\,d\varphi), \\ dy &= \sin\theta\sin\varphi\,dr + r(\cos\theta\sin\varphi\,d\theta + \sin\theta\cos\varphi\,d\varphi), \\ dz &= \cos\theta\,dr - r\sin\theta\,d\theta. \end{aligned}$$

We then have

$$\omega = (P\sin\theta\cos\varphi + Q\sin\theta\sin\varphi + R\cos\theta)\,dr + rH,$$

where H is a differential form in $d\theta$ and $d\varphi$. Now

$$Px + Qy + Rz = r(P\sin\theta\cos\varphi + Q\sin\theta\sin\varphi + R\cos\theta).$$

On the other hand, P, Q, and R are homogeneous of degree m. Therefore, they may be expressed as the product of r^m and functions of θ and φ. It follows that

$$\frac{rH}{Px + Qy + Rz}$$

is independent of r. Hence there exist functions A and B of θ and φ such that

$$\frac{\omega}{Px + Qy + Rz} = \frac{dr}{r} + A\,d\theta + B\,d\varphi.$$

Let us express the condition that ω is integrable in terms of the variables r, θ, and φ, that ω is integrable. If this is the case, so is

$$\frac{\omega}{Px + Qy + Rz}$$

integrable. Thus

$$\begin{vmatrix} \frac{1}{r} & A & B \\ \frac{\partial}{\partial r} & \frac{\partial}{\partial \theta} & \frac{\partial}{\partial \varphi} \\ \frac{1}{r} & A & B \end{vmatrix} = \frac{1}{r}\left(\frac{\partial B}{\partial \theta} - \frac{\partial A}{\partial \varphi}\right) = 0.$$

This condition means that $A\,d\theta + B\,d\varphi$ is an exact differential. Thus $\omega/(Px + Qy + Rz)$ is an exact differential, and $1/(Px + Qy + Rz)$ is an integrating factor of ω.

Example. $\omega = y^2\,dx + x^2\,dy$ is not an exact differential. But

$$\frac{y^2\,dx + x^2\,dy}{y^2x + x^2y}$$

is the differential of $\log xy/(x + y)$.

B. The above calculations are not valid when $Px + Qy + Rz$ is identically 0. In this case, we have

$$\omega = r^{m+1}(M\,d\theta + N\,d\varphi),$$

where M and N are functions of θ and φ. The condition for integrability is verified. The expression ω/r^{m+1} is a differential form in two variables. Therefore, it admits integrating factors in the form $f(\theta, \varphi)$, and ω has an integrating factor $f(\theta, \varphi)/r^{m+1}$.

The solutions of the equation $\omega = 0$ are of the form

$$g(\theta, \varphi) = 0.$$

The variable r does not appear in their equation. Therefore, they are cones with apex at the origin.

*3. An exercise on integrating factors

Problems. A. Consider two real differential forms

$$\omega_1 = M_1\,dx + N_1\,dy, \qquad \omega_2 = M_2\,dx + N_2\,dy,$$

where M_1, N_1, M_2, and N_2 are real functions of x and y that are twice continuously differentiable in some region. For ω_1 and ω_2 to have a common integrating factor μ, it is necessary and sufficient that a certain differential form $P\,dx + Q\,dy$ be an exact differential.

B. Let f denote a given function of a single variable. Find the condition under which the two differential forms

$$\omega_1 = dx + f(x)\,dy, \qquad \omega_2 = f(y)\,dx + dy$$

have a common integrating factor μ. Show that $f(x)$ must be of the form

$$f(x) = Ae^{\omega x} + Be^{-\omega x} + C,$$

where A, B, and C are constants satisfying the condition

$$C^2 = 1 + 4AB.$$

(Here, ω may be real or purely imaginary. In the second case, A and B must be complex conjugates.)

C. Let us set $f(x) = \sinh x$. Show that

$$\mu = \frac{\cosh x \cosh y + \sinh x + \sinh y}{1 - \sinh x \sinh y}$$

is a common integrating factor of ω_1 and ω_2.

Solutions. **A.** Let us seek the condition under which the two differential forms

$$\omega_1 = M_1\,dx + N_1\,dy, \qquad \omega_2 = M_2\,dx + N_2\,dy$$

will have a common integrating factor μ. This integrating factor is the common solution of the two partial differential equations

$$\frac{\partial}{\partial y}(\mu M_1) = \frac{\partial}{\partial x}(\mu N_1), \qquad \frac{\partial}{\partial y}(\mu M_2) = \frac{\partial}{\partial x}(\mu N_2)$$

or

$$\begin{aligned}\frac{\partial M_1}{\partial y} - \frac{\partial N_1}{\partial x} &= \left(N_1 \frac{\log \mu}{\partial x} - M_1 \frac{\partial \log \mu}{\partial y}\right), \\ \frac{\partial M_2}{\partial y} - \frac{\partial N_2}{\partial x} &= \left(N_2 \frac{\partial \log \mu}{\partial x} - M_2 \frac{\partial \log \mu}{\partial y}\right).\end{aligned} \tag{1}$$

Now, two first-order partial differential equations do not in general have a common solution. For this there is a compatibility condition. If the coefficients of ω_1 and ω_2 are not proportional, equations (1) express $\partial \log \mu/\partial x$ and $\partial \log \mu/\partial y$ with the aid of the given conditions

$$\frac{\partial \log \mu}{\partial x} = P, \qquad \frac{\partial \log \mu}{\partial y} = Q.$$

Here P and Q are known functions which depend on M_1, M_2, N_1, and N_2 and their first derivatives. The compatibility condition is that

$$\frac{\partial P}{\partial y} = \frac{\partial Q}{\partial x},$$

which is the same as the requirement that

$$P\,dx + Q\,dy$$

be an exact differential.

If $M_2 = kM_1$ and $N_2 = kN_1$, then $\omega_2 = k\omega_1$. If μ is an integrating factor of ω_1, so must μk be. This is possible only if k is a first integral of the differential equation

$$M_1\,dx + N_1\,dy = 0.$$

We know that every integrating factor of ω_1 is the product of one of these integrating factors and any first integral of this differential equation.

B. We choose, for example,

$$\begin{aligned} M_1 &= 1, & N_1 &= f(x), \\ M_2 &= f(y), & N_2 &= 1. \end{aligned}$$

Then

$$-f'(x) = f(x)\frac{\partial \log \mu}{\partial x} - \frac{\partial \log \mu}{\partial y}, \qquad f'(y) = \frac{\partial \log \mu}{\partial x} - f(y)\frac{\partial \log \mu}{\partial y},$$

so that

$$[1-f(x)f(y)]\,\frac{\partial \log \mu}{\partial x}=f'(y)+f'(x)f(y), \qquad P=\frac{f'(y)+f'(x)f(y)}{1-f(x)f(y)},$$
$$[1-f(x)f(y)]\,\frac{\partial \log \mu}{\partial y}=f'(x)+f(x)f'(y), \qquad Q=\frac{f'(x)+f(x)f'(y)}{1-f(x)f(y)}.$$

The condition $\partial P/\partial y - \partial Q/\partial x = 0$ may be written

$$[1-f(x)f(y)][f''(y)-f''(x)]+f(x)f'^2(y)-f(y)f'^2(x)=0$$

or

$$f''(y)-f''(x)+f(x)[f'^2(y)-f(y)f''(y)]-f(y)[f'^2(x)-f(x)f''(x)]=0. \quad (2)$$

In four-dimensional space, consider the two vectors whose components are

$$1, \qquad -f''(x), \qquad f(x), \qquad -[f'^2(x)-f(x)f''(x)]$$

and

$$f''(y), \qquad 1, \qquad f'^2(y)-f(y)f''(y), \qquad f(y).$$

Condition (2) states that these vectors are "orthogonal." Now, their end points describe two curves parametrized respectively by the two *independent* parameters x and y. Therefore, these two curves are in two orthogonal subspaces. Two situations may exist:

(a) One of these vectors has a fixed direction. In this case, the second vector is in the (three-dimensional) hyperplane orthogonal to that direction.

(b) The two vectors are in two supplementary orthogonal two-dimensional subspaces (planes).

Let us examine these two situations. They reduce to two because of the symmetry of x and y in the problem.

(a) There exist four constants α, β, γ, and δ such that

$$\frac{1}{\alpha}=-\frac{f''(x)}{\beta}=\frac{f(x)}{\gamma}=-\frac{f'^2(x)-f(x)f''(x)}{\delta},$$
$$\alpha f''(y)+\beta+\gamma[f'^2(y)-f(y)f''(y)]+\delta f(y)=0.$$

It follows from the first line of equations that f is a *constant*. Therefore, $f'=f''=0, \beta=0$, and $\delta=0$, and the differential forms are exact differentials. This case is banal.

(b) In the second case, the components of the vector

$$1, \qquad -f''(x), \qquad f(x), \qquad -[f'^2(x)-f(x)f''(x)]$$

are connected by *two* independent linear relations. It follows that $1, f''$, and f are connected by a relation, namely,

$$af'' + bf = c. \tag{3}$$

If $a = 0$, f is a constant. In the opposite case, we may set $a = 1$, and f'' is a solution of a second-order linear differential equation with constant coefficients.

If $b = 0$, we have

$$f = \frac{c}{2}x^2 + c_1 x + c_2 .$$

We can verify directly that this is a solution only if $c = c^1 = 0$, that is, if f is constant.

If $b \neq 0$, there exists a constant ω, real or purely imaginary, such that

$$f(x) = Ae^{\omega x} + Be^{-\omega x} + C, \tag{4}$$

where A, B, C, and ω are constants representing combinations of the constants b, c and the constants of integration.

Remark. If we assume that the function f has at least one zero y_0, we can get this result more rapidly. By assigning to y the value y_0 in (2), we obtain

$$f_0'' - f''(x) + f_0'^2 f(x) = 0,$$

where

$$f_0'' = f''(y_0) \qquad \text{and} \qquad f_0'^2 = f'^2(y_0)$$

are constants. This is (3).

It remains to show that the solution (4) is valid. We substitute it in (2), remembering that

$$f'' = \omega^2(f - c), \qquad f'^2 = \omega^2(f - c)^2 - 4AB\omega^2.$$

A brief calculation shows that $C^2 = 1 + 4AB$.

The form chosen for f is valid when ω, A, and B are real constants. However, ω can also be purely imaginary.

To get a real function f, we therefore set

$$\omega = i\lambda, \qquad A + B = \alpha, \qquad i(A - B) = \beta,$$

where A and B are imaginary conjugates. We then have

$$f(x) = \alpha \cos \lambda x + \beta \sin \lambda x + C, \tag{5}$$

with

$$C^2 = 1 + \alpha^2 + \beta^2.$$

C. If $A = \frac{1}{2}$, $B = -\frac{1}{2}$, and $C = 0$, we have $C^2 = 1 + 4AB$ and

$$f(x) = \sinh x.$$

The differential forms

$$\omega_1 = dx + \sinh x\, dy \qquad \text{and} \qquad \omega_2 = \sinh y\, dx + dy$$

thus have a common integrating factor μ, which is a common solution of the two equations

$$\frac{\partial \mu}{\partial y} = \frac{\partial}{\partial x}(\mu \sinh x), \qquad \frac{\partial}{\partial y}(\mu \sinh y) = \frac{\partial \mu}{\partial x} \tag{6}$$

or

$$\sinh x \frac{\partial \log \mu}{\partial x} - \frac{\partial \log \mu}{\partial y} = -\cosh x, \quad \frac{\partial \log \mu}{\partial x} - \sinh y \frac{\partial \log \mu}{\partial y} = \cosh y.$$

Therefore,

$$\frac{\partial \log \mu}{\partial x} = \frac{\cosh y + \sinh y \cosh x}{1 - \sinh x \sinh y}, \qquad \frac{\partial \log \mu}{\partial y} = \frac{\cosh x + \sinh x \cosh y}{1 + \sinh x \sinh y}.$$

And we know that these equations are compatible. There exists a function $\log \mu$ such that

$$d \log \mu = \frac{(\cosh y + \sinh y \cosh x)\, dx + (\cosh x + \sinh x \cosh y)\, dy}{1 - \sinh x \sinh y}.$$

We set

$$A = \cosh x \cosh y + \sinh x + \sinh y, \qquad B = 1 - \sinh x \sinh y.$$

Let us first verify the following two identities:

$$\begin{aligned}(\cosh x + \sinh x \cosh y)B &= (\cosh y - \cosh x \sinh y)A,\\ (\cosh y + \sinh y \cosh x)B &= (\cosh x - \cosh y \cosh x)A.\end{aligned}$$

Now,

$$\begin{aligned} A_x' &= \sinh x \cosh y + \cosh x, \qquad & A_y' &= \cosh x \sinh y + \cosh y,\\ B_x' &= -\cosh x \sinh y, & B_y' &= -\sinh x \cosh y.\end{aligned}$$

Therefore,

$$A_x'B = (A_y' + 2B_x')A, \qquad A_y'B = (A_x' + 2B_y')A.$$

From this, we conclude that

$$B\,dA = 2A\,dB + A(A_y'\,dx + A_x'\,dy).$$

Now,

$$d\log\mu = \frac{A_y'\,dx + A_x'\,dy}{B}.$$

Therefore,

$$d\log\mu = \frac{B\,dA - 2A\,dB}{AB} = \frac{dA}{A} - 2\frac{dB}{B},$$

so that

$$\log\mu = \log\frac{A}{B^2}, \qquad \mu = \frac{A}{B^2} \text{ (up to a constant factor),}$$

$$\mu = \frac{\cosh x \cosh y + \sinh x + \sinh y}{1 - \sinh x \sinh y}.$$

*4. Compatibility conditions in elasticity theory

Let u_i denote a vector field in three- (or n-) dimensional space. We assign to this field the derived tensor $\partial u_i/\partial x_j$, which consists of a symmetric part

$$s_{ij} = \frac{\partial u_i}{\partial x_j} + \frac{\partial u_j}{\partial x_i}$$

and an antisymmetric part

$$p_{ij} = \frac{\partial u_i}{\partial x_j} - \frac{\partial u_j}{\partial x_i}$$

(the curl of u_i if $n = 3$).

Suppose that the s_{ij} are given. Can we derive the u_i? The purpose of this problem is to show that this is possible only if the s_{ij} verify certain conditions known as *compatibility conditions*.

In the mechanics of continuous media, we consider a system in an initial state and then deform that system. The point of the coordinates x_i undergoes a displacement represented by the vector u_i. The tensor s_{ij} is the deformation tensor and the tensor p_{ij} is the rotation tensor. The question posed is then: Find the displacement when the deformation tensor is given. Is every field of symmetric tensors a deformation tensor? If not, what conditions must the field of symmetric tensors satisfy to be a deformation tensor?

Problems. A. Suppose that $a(x, y)$, $b(x, y)$, and $c(x, y)$ are two twice continuously differentiable functions of variables x and y. Find two functions $u(x, y)$ and $v(x, y)$ such that

$$\frac{\partial u}{\partial x} = \frac{a}{2}, \qquad \frac{\partial u}{\partial y} + \frac{\partial v}{\partial x} = b, \qquad \frac{\partial v}{\partial y} = \frac{c}{2}. \tag{1}$$

Show that this problem has a solution if and only if

$$\frac{\partial^2 a}{\partial y^2} - 2\frac{\partial^2 b}{\partial x\,\partial y} + \frac{\partial^2 c}{\partial x^2} = 0. \tag{2}$$

B. Suppose that a symmetric tensor s_{ij} is a given function of $x_1, x_2, \ldots, x_n$ in n-dimensional space. Find a vector field u_i, which is a function of $x_1, \ldots, x_n$, such that

$$\frac{\partial u_i}{\partial x_j} + \frac{\partial u_j}{\partial x_i} = s_{ij}. \tag{3}$$

Show that this problem has a solution if and only if the s_{ij} verify the conditions

$$\frac{\partial^2 s_{ki}}{\partial x_j\,\partial x_l} - \frac{\partial^2 s_{kj}}{\partial x_i\,\partial x_l} = \frac{\partial^2 s_{li}}{\partial x_j\,\partial x_k} - \frac{\partial^2 s_{lj}}{\partial x_i\,\partial x_k}. \tag{4}$$

(We may introduce, as auxiliary unknowns, the antisymmetric tensor $p_{ij} = \partial u_i/\partial x_j - \partial u_j/\partial x_i$.)

C. We choose $n = 3$ and use the following notations:

$$\begin{array}{llllll} 2e_1 = s_{11}, & 2e_2 = s_{22}, & 2e_3 = s_{33}, & 2g_1 = s_{23}, & 2g_2 = s_{31}, & 2g_3 = s_{12} \\ u_1 = u, & u_2 = v, & u_3 = w, & x_1 = x, & x_2 = y, & x_3 = z. \end{array}$$

Write the compatibility relations (4) in these notations.

D. With n still equal to 3, suppose that

$$s_{ij} = 2t\delta_{ij}, \tag{5}$$

where t is an unknown function of x, y, and z. Show that

$$t = ax + by + cz + h,$$

where a, b, c, and h are constants, and calculate u, v, and w.

Solutions. **A.** Find two functions u and v such that

$$\frac{\partial u}{\partial x} = \frac{a}{2}, \qquad \frac{\partial u}{\partial y} + \frac{\partial v}{\partial x} = b, \qquad \frac{\partial v}{\partial y} = \frac{c}{2}, \tag{1}$$

where a, b, and c are given. Define the function

$$b' = \frac{\partial u}{\partial y} - \frac{\partial v}{\partial x}.$$

Then

$$\frac{\partial u}{\partial y} = \frac{b + b'}{2}, \qquad \frac{\partial v}{\partial x} = \frac{b - b'}{2},$$

and u and v are defined by their differentials

$$du = \tfrac{1}{2}[a\,dx + (b + b')\,dy], \qquad dv = \tfrac{1}{2}[(b - b')\,dx - c\,dy].$$

This is possible only if

$$\frac{\partial a}{\partial y} = \frac{\partial}{\partial x}(b + b'), \qquad \frac{\partial}{\partial y}(b - b') = \frac{\partial c}{\partial x}.$$

If a, b, b', and c were all given, these two relations would constitute the conditions under which the problem has a solution. But b' is not given. Therefore, we need to choose a, b, and c in such a way that the equations

$$\frac{\partial b'}{\partial x} = \frac{\partial a}{\partial y} - \frac{\partial b}{\partial x}, \qquad \frac{\partial b'}{\partial y} = \frac{\partial b}{\partial y} - \frac{\partial c}{\partial x}$$

are compatible. The function b' is defined by its partial derivatives. The condition under which the above equation is possible is

$$\frac{\partial}{\partial y}\left(\frac{\partial b}{\partial x} - \frac{\partial a}{\partial y}\right) = \frac{\partial}{\partial x}\left(\frac{\partial c}{\partial x} - \frac{\partial b}{\partial y}\right) \qquad \text{or} \qquad \frac{\partial^2 c}{\partial x^2} - 2\frac{\partial^2 b}{\partial x\,\partial y} + \frac{\partial^2 a}{\partial y^2} = 0.$$

B. Let us generalize to n dimensions the preceding problem. Suppose that we are given a symmetric tensor s_{ij}, which is a function of $x_1, \ldots, x_n$. We seek a vector field u_i such that

$$\frac{\partial u_i}{\partial x_j} + \frac{\partial u_j}{\partial x_i} = s_{ij}. \tag{3}$$

We take as auxiliary unknowns the functions

$$p_{ij} = \frac{\partial u_i}{\partial x_j} - \frac{\partial u_j}{\partial x_i},$$

which constitute a field of antisymmetric tensors ($p_{ii} = 0$, $p_{ij} = -p_{ji}$). Then

$$2\frac{\partial u_i}{\partial x_j} = s_{ij} + p_{ij}.$$

The u_i are defined by the differential formulas

$$2du_i = \sum_j (s_{ij} + p_{ij})\, dx_j\,,$$

subject to the condition that

$$\frac{\partial}{\partial x_k}(s_{ij} + p_{ij}) = \frac{\partial}{\partial x_j}(s_{ik} + p_{ik}),$$

or

$$\frac{\partial p_{ij}}{\partial x_k} - \frac{\partial p_{ik}}{\partial x_j} = \frac{\partial s_{ik}}{\partial x_j} - \frac{\partial s_{ij}}{\partial x_k}. \tag{6}$$

How should we choose the s_{ij} for these conditions to define the p_{ij}? If $k = i$, we have $p_{ii} = 0$ and

$$\frac{\partial p_{ij}}{\partial x_i} = \frac{\partial s_{ii}}{\partial x_j} - \frac{\partial s_{ij}}{\partial x_i}, \tag{7}$$

which determines the partial derivatives $\partial p_{ij}/\partial x_i$.

Suppose now that i, j, and k are three distinct integers. Let us write the relations we have obtained by circular permutations of the indices in (6):

$$\frac{\partial p_{ij}}{\partial x_k} - \frac{\partial p_{ik}}{\partial x_j} = \frac{\partial s_{ik}}{\partial x_j} - \frac{\partial s_{ij}}{\partial x_k},$$

$$\frac{\partial p_{jk}}{\partial x_i} - \frac{\partial p_{ji}}{\partial x_k} = \frac{\partial s_{ji}}{\partial x_k} - \frac{\partial s_{jk}}{\partial x_i},$$

$$\frac{\partial p_{ki}}{\partial x_j} - \frac{\partial p_{kj}}{\partial x_i} = \frac{\partial s_{kj}}{\partial x_i} - \frac{\partial s_{ki}}{\partial x_j}.$$

Since $p_{ji} = -p_{ij}$, these relations define the three unknowns

$$\frac{\partial p_{ij}}{\partial x_k}, \qquad \frac{\partial p_{jk}}{\partial x_i}, \qquad \frac{\partial p_{ki}}{\partial x_j}.$$

By adding, we get

$$2\left(\frac{\partial p_{ij}}{\partial x_k} + \frac{\partial p_{jk}}{\partial x_i} + \frac{\partial p_{ki}}{\partial x_j}\right) = 0.$$

Therefore,

$$\frac{\partial p_{ij}}{\partial x_k} = \frac{\partial s_{ik}}{\partial x_j} - \frac{\partial s_{kj}}{\partial x_i}. \tag{8}$$

Thus equations (8) define all the partial derivatives of the p_{ij}, since the formula already obtained,

$$\frac{\partial p_{ij}}{\partial x_j} = \frac{\partial s_{ij}}{\partial x_j} - \frac{\partial s_{jj}}{\partial x_i} \tag{7}$$

is contained in (8) and is not independent of it.

Therefore, the desired condition under which the problem has a solution is that, for every i,

$$\sum_k \left(\frac{\partial s_{ki}}{\partial x_j} - \frac{\partial s_{kj}}{\partial x_i}\right) dx_k$$

be the differential of some function p_{ij}. This condition may be written

$$\frac{\partial^2 s_{ki}}{\partial x_j \partial x_l} - \frac{\partial^2 s_{kj}}{\partial x_i \partial x_l} = \frac{\partial^2 s_{li}}{\partial x_j \partial x_k} - \frac{\partial^2 s_{lj}}{\partial x_i \partial x_k}. \tag{4}$$

C. The case in which $n = 3$. We set

$$\begin{aligned} &s_{11} = 2e_1, \quad s_{22} = 2e_2, \quad s_{33} = 2e_3; \qquad u_1 = u, \quad u_2 = v, \quad u_3 = w, \\ &s_{23} = 2g_1, \quad s_{31} = 2g_2, \quad s_{12} = 2g_3; \qquad x_1 = x, \quad x_2 = y, \quad x_3 = z. \end{aligned}$$

Let us again get the solution of our problem for the three-dimensional case, using these notations. We have

$$\frac{\partial u}{\partial x} = e_1, \qquad \frac{\partial v}{\partial y} = e_2, \qquad \frac{\partial w}{\partial z} = e_3,$$

$$\frac{\partial v}{\partial z} + \frac{\partial w}{\partial y} = 2g_1, \qquad \frac{\partial w}{\partial x} + \frac{\partial u}{\partial z} = 2g_2, \qquad \frac{\partial u}{\partial y} + \frac{\partial v}{\partial x} = 2g_3.$$

We set

$$\frac{\partial v}{\partial z} - \frac{\partial w}{\partial y} = 2p_1, \qquad \frac{\partial w}{\partial x} - \frac{\partial u}{\partial z} = 2p_2, \qquad \frac{\partial u}{\partial y} - \frac{\partial v}{\partial x} = 2p_3.$$

The matrix of the partial derivatives of u, v, and w is then

$$\begin{pmatrix} e_1 & g_3 + P_3 & g_2 - p_2 \\ g_3 - p_3 & e_2 & g_1 + p_1 \\ g_2 + p_2 & g_1 - p_1 & e_3 \end{pmatrix}.$$

Therefore, the conditions under which the problem has a solution are

$$\frac{\partial e_1}{\partial y} = \frac{\partial g_3}{\partial x} + \frac{\partial p_3}{\partial x}, \qquad \frac{\partial e_1}{\partial z} = \frac{\partial g_2}{\partial x} - \frac{\partial p_2}{\partial x},$$

$$\frac{\partial g_3}{\partial y} - \frac{\partial p_3}{\partial y} = \frac{\partial e_2}{\partial x}, \qquad \frac{\partial g_3}{\partial z} - \frac{\partial p_3}{\partial z} = \frac{\partial g_1}{\partial x} + \frac{\partial p_1}{\partial x},$$

$$\frac{\partial g_2}{\partial y} + \frac{\partial p_2}{\partial y} = \frac{\partial g_1}{\partial x} - \frac{\partial p_1}{\partial x}, \qquad \frac{\partial g_2}{\partial z} + \frac{\partial p_2}{\partial z} = \frac{\partial e_3}{\partial x},$$

$$\frac{\partial g_3}{\partial z} + \frac{\partial p_3}{\partial z} = \frac{\partial g_2}{\partial y} - \frac{\partial p_2}{\partial y},$$

$$\frac{\partial e_2}{\partial z} = \frac{\partial g_1}{\partial y} + \frac{\partial p_1}{\partial y},$$

$$\frac{\partial g_1}{\partial z} - \frac{\partial p_1}{\partial z} = \frac{\partial e_3}{\partial y}.$$

These equations constitute a linear system enabling us to evaluate the nine partial derivatives of p_1, p_2, and p_3:

$$\frac{\partial p_1}{\partial x} = \frac{\partial g_3}{\partial z} - \frac{\partial g_2}{\partial y}, \qquad \frac{\partial p_1}{\partial y} = \frac{\partial e_2}{\partial z} - \frac{\partial g_1}{\partial y}, \qquad \frac{\partial p_1}{\partial z} = \frac{\partial g_1}{\partial z} - \frac{\partial e_3}{\partial y},$$

$$\frac{\partial p_2}{\partial x} = \frac{\partial g_2}{\partial x} - \frac{\partial e_1}{\partial z}, \qquad \frac{\partial p_2}{\partial y} = \frac{\partial g_1}{\partial x} - \frac{\partial g_3}{\partial z}, \qquad \frac{\partial p_2}{\partial z} = \frac{\partial e_3}{\partial x} - \frac{\partial g_2}{\partial z},$$

$$\frac{\partial p_3}{\partial x} = \frac{\partial e_1}{\partial y} - \frac{\partial g_3}{\partial x}, \qquad \frac{\partial p_3}{\partial y} = \frac{\partial g_3}{\partial y} - \frac{\partial e_2}{\partial x}, \qquad \frac{\partial p_3}{\partial z} = \frac{\partial g_2}{\partial y} - \frac{\partial g_1}{\partial x}.$$

Thus we have the compatibility conditions

$$\frac{\partial^2 e_1}{\partial y^2} + \frac{\partial^2 e_2}{\partial x^2} = 2\frac{\partial^2 g_3}{\partial x\, \partial y}; \quad \frac{\partial^2 e_2}{\partial z^2} + \frac{\partial^2 e_3}{\partial y^2} = 2\frac{\partial^2 g_1}{\partial y\, \partial z}; \quad \frac{\partial^2 e_3}{\partial x^2} + \frac{\partial^2 e_1}{\partial z^2} = 2\frac{\partial^2 g_2}{\partial z\, \partial x};$$

$$\frac{\partial^2 e_1}{\partial y\, \partial z} = \frac{\partial h}{\partial x} - 2\frac{\partial^2 g_1}{\partial x^2}; \qquad \frac{\partial^2 e_2}{\partial z\, \partial x} = \frac{\partial h}{\partial y} - 2\frac{\partial^2 g_2}{\partial y^2}; \qquad \frac{\partial^2 e_3}{\partial x\, \partial y} = \frac{\partial h}{\partial z} - 2\frac{\partial^2 g_3}{\partial z^2};$$

with

$$h = \frac{\partial g_1}{\partial x} + \frac{\partial g_2}{\partial y} + \frac{\partial g_3}{\partial z}.$$

There are six independent relations connecting the s_{ij}, that is, connecting e_1, e_2, e_3, g_1, g_2, and g_3.

D. *An application.* We are still working in three-dimensional space. We choose $s_{ij} = 2t\delta_{ij}$, where t is an unknown function of x, y, and z. Then

$$e_1 = e_2 = e_3 = t, \qquad g_1 = g_2 = g_3 = 0.$$

Let us write the compatibility conditions:

$$\frac{\partial^2 t}{\partial y^2} + \frac{\partial^2 t}{\partial x^2} = 0 \qquad \frac{\partial^2 t}{\partial y^2} + \frac{\partial^2 t}{\partial z^2} = 0, \qquad \frac{\partial^2 t}{\partial x^2} + \frac{\partial^2 t}{\partial z^2} = 0,$$
$$\frac{\partial^2 t}{\partial y\, \partial z} = 0, \qquad \frac{\partial^2 t}{\partial z\, \partial x} = 0, \qquad \frac{\partial^2 t}{\partial x\, \partial y} = 0.$$

Thus the six second partial derivatives of t are zero, so that t is of the form

$$t = ax + by + cz + h,$$

where a, b, c, and h are constants.

Let us now calculate p_1, p_2, and p_3. The matrix of the nine partial derivatives is

$$\begin{pmatrix} 0 & c & -b \\ -c & 0 & a \\ b & -a & 0 \end{pmatrix}.$$

Therefore,

$$dp_1 = c\, dy - b\, dz, \qquad dp_2 = -c\, dx + a\, dz, \qquad dp_3 = b\, dx - a\, dy,$$

so that

$$p_1 = cy - bz + \alpha, \qquad p_2 = -cx + az + \beta, \qquad p_3 = bx - ay + \gamma,$$

where α, β, and γ are three new constants.

From this we derive the matrix of the partial derivatives of u, v, and w:

$$\begin{pmatrix} ax + by + cz + h & bx - ay + \gamma & cx - az - \beta \\ -bx + ay - \gamma & ax + by + cz + h & cy - bz + \alpha \\ -cx + az + \beta & -cy + bz - \alpha & ax + by + cz + h \end{pmatrix}.$$

Then, for example,

$$\begin{aligned} du &= (ax + by + cz + h)\, dx + (bx - ay + \gamma)\, dy + (cx - az - \beta)\, dz \\ &= a(x\, dx - y\, dy - z\, dz) + b(y\, dx + x\, dy) + c(z\, dx + x\, dz) \\ &\qquad + h\, dx + \gamma\, dy - \beta\, dz. \end{aligned}$$

We can verify that du is the differential of a function

$$u = \frac{a}{2}(x^2 - y^2 - z^2) + bxy + cxz + hx + \gamma y - \beta z + u_0,$$

where u_0 is a new constant. We may also write

$$u = -\frac{a}{2}(x^2 + y^2 + z^2) + x(ax + by + cz + h) + \gamma y - \beta z + u_0 .$$

Let us denote by

Φ	the vector whose components are	α	β	γ
X	the vector whose components are	x	y	z
U	the vector whose components are	u	v	w
U_0	the vector whose components are	u_0	v_0	w_0
A	the vector whose components are	a	b	c

and let us set

$$r^2 = x^2 + y^2 + z^2, \qquad t = ax + by + cz + h,$$

where $t - h$ is the scalar product of X and the vector A. We have

$$U = U_0 - \frac{A}{2} r^2 + tX + X \times \Phi.$$

5. An example of Green's function

Consider the differential equation

$$y'' = f(x).$$

The *Cauchy problem* consists in finding a function y that satisfies this equation such that $f'(a) = f(a) = 0$. The *boundary-value problem* consists in finding a solution that vanishes at $x = a$ and $x = b$, where $b > a$. The solution of the Cauchy problem may be written

$$y = \int_a^x (x - s)f(s)\, ds.$$

If we let x vary on a chosen interval $a \leqslant x \leqslant b$, we have

$$y = \int_a^b C(x - s)y(s)\, ds,$$

where

$$C(s) = \begin{cases} 0 & \text{if} \quad s < 0, \\ s & \text{if} \quad s > 0. \end{cases}$$

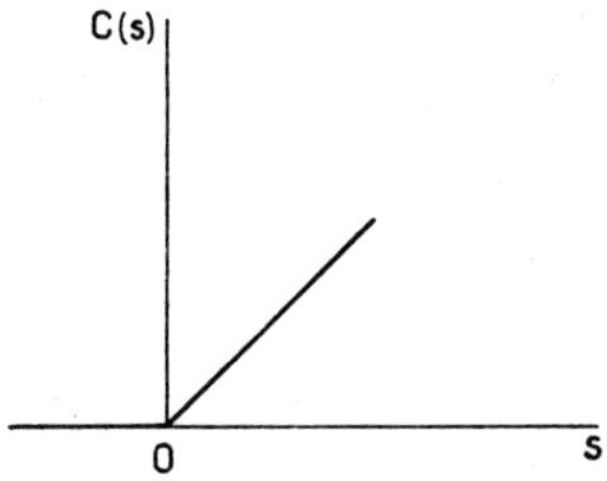

FIGURE 1

Of course, the value of b is conventional and plays no precise role in the expression for y.

The purpose of this exercise is to put the solution of the boundary-value problem in a comparable form:

$$y = \int_a^b G(x, s)f(s)\, ds$$

where $G(x, s)$, known as Green's function, plays a role analogous to $C(x - s)$, but, of course, it has a quite different structure.

Problem. Solve the differential equation

$$y'' = f(x),$$

where $f(x)$ is a given integrable function satisfying the conditions

$$y(a) = 0, \qquad y(b) = 0,$$

where a and b are given numbers such that $a < b$.

Put the solution in the form

$$y = \int_a^b G(x, s)f(s)\, ds,$$

where $G(x, s)$ is a function defined in the square $a < x < b$, $a < s < b$.

Study the surface $z = G(x, s)$.

Solution. Consider the differential equation

$$y'' = f(x).$$

If we integrate twice, we obtain

$$y' = \int_a^x f(s)\,ds + A, \qquad y = \int_a^x dt \int_a^t f(s)\,ds + Ax + B,$$

where A and B are constants.

This iterated integral is equal to a double integral over the triangle

$$a < t < x, \qquad a < s < t$$

in the st-plane. By reversing the order of the integrations, we obtain

$$\begin{aligned} y &= \int_a^x f(s)\,ds \int_s^x dt + Ax + B \\ &= \int_a^x (x - s)f(s)\,ds + Ax + B. \end{aligned}$$

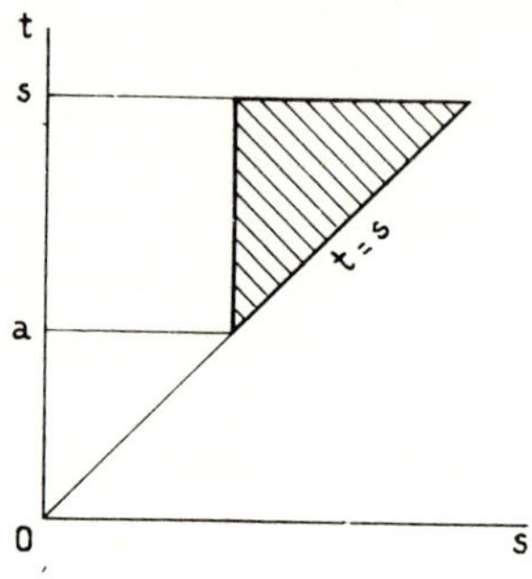

FIGURE 2

For $x = a$, we have $y = 0$, so that

$$B = -Aa.$$

For $x = b$, we again have $y = 0$, so that

$$\int_a^b (b - s)f(s)\,ds + A(b - a) = 0.$$

Therefore, we obtain

$$y = \int_a^x (x - s)f(s)\,ds - \frac{x - a}{b - a} \int_a^b (b - s)f(s)\,ds.$$

We set $\int_a^b = \int_a^x + \int_x^b$ and regroup the integrals:

$$y = \int_a^x \left[(x - s) - \frac{x - a}{b - a}(b - s)\right] f(s)\,ds - \frac{x - a}{b - a} \int_x^b (b - s)f(s)\,ds$$

or

$$y = \int_a^x \frac{(b-x)(a-s)}{b-a} f(s)\, ds + \int_x^b \frac{(a-x)(b-s)}{b-a} f(s)\, d\,.$$

We set

$$y = \int_a^b G(x, s)\, f(s)\, ds,$$

where the function $G(x, s)$ is defined on the square

$$a < x < b, \qquad a < s < b$$

by

$$G(x, s) = \begin{cases} \dfrac{(a-s)(b-x)}{b-a} & \text{if} \quad a \leqslant s \leqslant x, \\ \dfrac{(a-x)(b-s)}{b-a} & \text{if} \quad x \leqslant s \leqslant b. \end{cases}$$

This function G is symmetric in x and s and is continuous at $x = s$, but its partial derivative $\partial G/\partial x$ (for example) varies from $-(a-s)/(b-a)$ to $-(b-s)/(b-a)$, that is, by one unit, as x passes the value s.

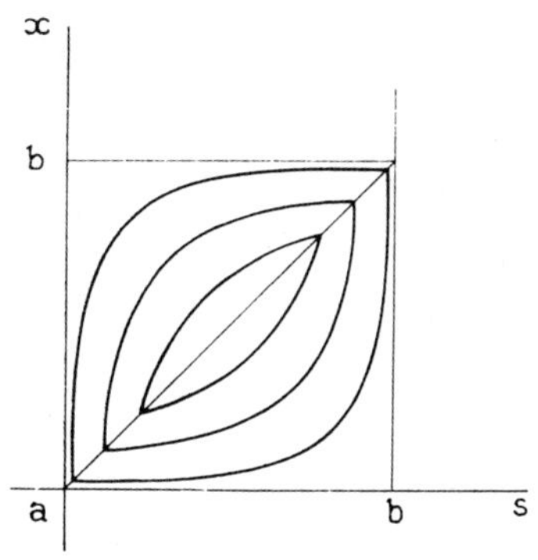

FIGURE 3

The surface $z = G(x, s)$ is composed of two portions of hyperbolic paraboloids that are symmetric about the plane $x = s$. It is entirely below the plane $z = 0$ and it rests on the square $x = a$, $x = b$, $s = a$, $s = b$ in that plane.

The planes $x =$ const and $s =$ const cut each of the two hyperbolic paraboloids along their rectilinear generators. The intersection with

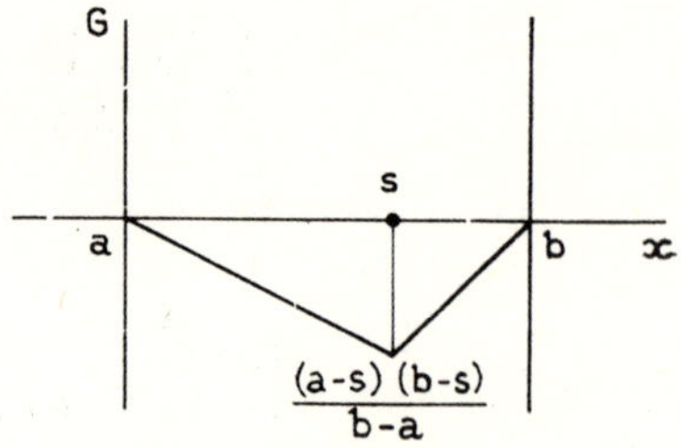

FIGURE 4

the plane $x = s$ is a parabola with vertex at $z = -(b - a)/4$ $x = s = (a + b)/2$. This parabola is an edge of the surface.

Finally, the planes $x =$ const intersect the two hyperbolic paraboloids along portions of hyperbolas that are asymptotic (in projection) to the sides of the square above.

6. Evaluation of a multiple integral

Problems. A. Suppose that α, β, and γ are three given numbers. Reduce the evaluation of the integral

$$I = \iiint_{\Omega} f(\alpha x + \beta y + \gamma z)\, dx\, dy\, dz,$$

where Ω is the sphere of radius 1 with center at the origin, to the evaluation of a simple integral. Complete the evaluation with $f(u) = \cos u$.

B. Do the same for the integral

$$J = \iint_{C} f(\alpha x + \beta y)\, dx\, dy$$

over the circle C of radius 1 with center at the origin. The expression for J then introduces the Bessel function $J_1(h)$.

Solutions. **A.** Let us make the change of variables defined as follows: Let the plane $Z = 0$ be the plane $\alpha x + \beta y + \gamma z = 0$ and let the X- and Y-axes be two perpendicular axes in that plane. By a suitable orientation of the Z-axis, we can have

$$Z = \frac{\alpha x + \beta y + \gamma z}{(\alpha^2 + \beta^2 + \gamma^2)^{1/2}} = \frac{\alpha x + \beta y + \gamma z}{h}.$$

We know that in a rectangular change of axes,

$$\left| \frac{D(x, y, z)}{D(X, Y, Z)} \right| = 1,$$

so that

$$\begin{aligned} I &= \iiint_{\Omega} f(\alpha x + \beta y + \gamma z)\, dx\, dy\, dz \qquad (\Omega: x^2 + y^2 + z^2 < 1), \\ &= \iiint_{\Omega'} f(Zh)\, dX\, dY\, dZ \qquad (\Omega': X^2 + Y^2 + Z^2 < 1), \\ &= \int_{-1}^{+1} f(Zh)\, dZ \iint_{S_z} dX\, dY, \end{aligned}$$

where S_Z is the circle constituting the intersection of Ω' with the plane $Z = 0$.

We have

$$\iint_{S_z} dX\, dY = \pi(1 - Z^2),$$

so that

$$I = \pi \int_{-1}^{+1} (1 - Z^2) f(Zh)\, dZ.$$

Example. If $f(u) = \cos u$,

$$I = \pi \int_{-1}^{+1} (1 - Z^2) \cos Zh\, dZ.$$

This integral is easily evaluated by successive integrations by parts. We find

$$I = \frac{4\pi}{h^2} \left(\frac{\sin h}{h} - \cos h \right).$$

To construct the curve for $I(h)$, we shall confine ourselves to positive values of $h[= (\alpha^2 + \beta^2 + \gamma^2)^{1/2}]$. We note the following:

(1) $I(0) = 4\pi/3$ [this result is obvious when we write $I(0)$ in the form

$$I(0) = f(0) \iiint_{\Omega} dx\, dy\, dz].$$

$I(h)$ is continuous from 0 to $+\infty$.

$I(h) = 0$ when $\tan h = h$, that is, at an infinite number of points which, when numbered (beginning with 1) in order of increasing abscissas, have abscissas closer and closer to $(n + 1)\pi/2$.

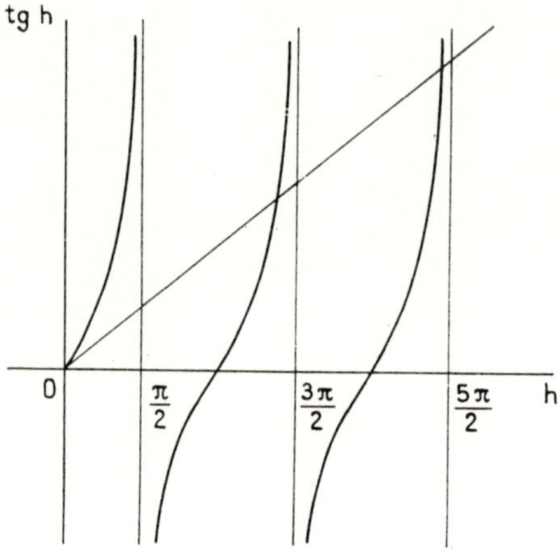

FIGURE 5

The maxima and minima of $I(h)$ are given by study of the derivative

$$\frac{dI}{dh} = \frac{4\pi}{h^3}\left[h \sin h - 3\left(\frac{\sin h}{h} - \cos h\right)\right].$$

The bracketed expression vanishes and changes sign at $\tan h = 3h/(3 - h^2)$, that is, at points whose abscissas become closer and closer to $n\pi$ as $h \to \infty$.

The derivative of the function $\sin h - h \cos h$ is $h \sin h$. This function vanishes at $h = n\pi$, so that its maxima and minima are equal to $(-1)^{n+1} n\pi$, and the corresponding values of I are

$$(-1)^{n+1} \frac{4\pi}{n^2\pi^2} = (-1)^{n+1} \frac{4}{\pi} \frac{1}{n^2}.$$

They constitute a sequence of remarkable points lying alternately on the two curves $y = \pm 4/\pi h^2$, and they become closer to the maxima and minima of I at increasing distances from the origin.

In some we see that the curve $I(h)$ represents an approximately periodic phenomenon but one that is rapidly attenuated. We point out that $I(h)$ presents certain analogies with the Bessel function

$$J_{3/2}(h) = \left(\frac{2}{\pi h}\right)^{1/2} \left(\frac{\sin h}{h} - \cos h\right).$$

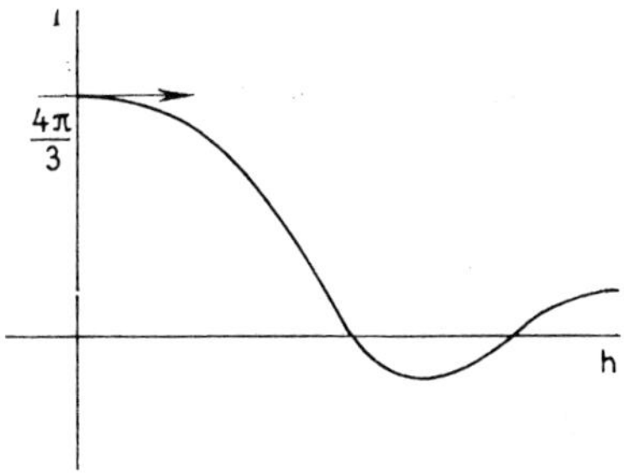

FIGURE 6

More precisely,

$$I(h) = \left(\frac{2\pi}{h}\right)^{3/2} J_{3/2}(h).$$

Thus the curve of $I(h)$ can be obtained from that of $J_{3/2}(h)$. The factor $(2\pi/h)^{3/2}$ indicates the attenuation of the function.

B. By an analogous procedure, we can see, by setting

$$Y = \frac{\alpha x + \beta y}{(\alpha^2 + \beta^2)^{1/2}},$$

that, $h = (\alpha^2 + \beta^2)^{1/2}$,

$$\begin{aligned} J &= \iint_C f(\alpha x + \beta y)\, dx\, dy = \iint_{C'} f(hY)\, dX\, dY \qquad \begin{pmatrix} C\colon x^2 + y^2 \leqslant 1 \\ C'\colon X^2 + Y^2 \leqslant 1 \end{pmatrix} \\ &= \int_{-1}^{+1} f(hY) \times 2(1 - Y^2)^{1/2}\, dY \\ &= 2\int_{-1}^{+1} (1 - Y^2)^{1/2} f(hY)\, dY. \end{aligned}$$

If $f(u) = \cos u$, we have

$$J = 2\int_{-1}^{+1} (1 - Y^2)^{1/2} \cos(hY)\, dY = 4\int_0^1 (1 - Y^2)^{1/2} \cos(hY)\, dY = \frac{2\pi}{h} J_1(h),$$

where $J_1(h)$ denotes the Bessel function

$$J_1(h) = \frac{2h}{\pi} \int_0^1 \cos \omega h (1 - \omega^2)^{1/2}\, d\omega.$$

The curve of $J_1(x)$ resembles an attenuated sinusoid. It is easily obtained from the curve of $J = (2\pi/h)J_1(h)$, where the factor $1/h$ indicates the attenuation of the function.

Remark. In n-dimensional space, we can calculate

$$I = \int_\Omega f(\alpha_1 x_1 + \cdots + \alpha_n x_n)\, dx_1 \cdots dx_n\,,$$

where Ω is the interior of the sphere

$$x_1^2 + \cdots + x_n^2 < 1.$$

in the same way. If

$$h = (\alpha_1^2 + \cdots + \alpha_n^2)^{1/2}$$

we have

$$I = \int_{\Omega'} f(hX_1)\, dX_1\, dX_2 \cdots dX_n = \int_{-1}^{1} f(hX_1)\, dX_1 \int dX_2 \cdots dX_n\,,$$

where the second integral denotes the area of the intersection of the unit sphere with the plane $X_1 = 0$. This integral is equal to

$$(1 - X_1^2)^{(n-1)/2}\, \omega,$$

where ω is the measure (volume) of the $(n - 1)$-dimensional unit sphere. Thus, we have

$$I = \omega \int_{-1}^{1} (1 - X^2)^{(n-1)/2} f(hX)\, dX,$$

where $\omega = 2$ if $n = 2$, $\omega = \pi$ if $n = 3$, and, in general,

$$\omega = \frac{\pi^{(n-1)/2}}{\Gamma[(n+1)/2]}\,.$$

When $f(hX) = \cos hX$, we need to evaluate the integral

$$\int_{-1}^{1} (1 - X^2)^{(n-1)/2} \cos hX\, dX,$$

which is equal to

$$\sqrt{\pi}\,\Gamma\left(\frac{n+1}{2}\right)\left(\frac{2}{h}\right)^{n/2} J_{n/2}(h),$$

where J_λ is the Bessel function of order λ. Thus we have

$$I = \left(\frac{2\pi}{h}\right)^{n/2} J_{n/2}(h).$$

If n is an odd number $2p + 1$, then $J_{(2p+1)}(h)$ is an elementary function, hence I is an elementary expression. If n is even, the expression for I involves Bessel functions of even orders.

7. The probability of the distance between two points in the interior of a sphere

This exercise is worded in the language of probability theory. However, probability-theory techniques play a completely negligible role in it, and it is simply a matter of evaluating multiple integrals. We need only make the following correspondences:

The probability that a point A belongs to a set E that is interior to a sphere S and that is defined by certain properties	Volume (Lebesgue measure) of the set E divided by the volume V of the sphere S
Elementary probability	$\dfrac{dx\,dy\,dz}{V}$
Elementary probability of a pair of independent points A and B (product of the elementary probabilities of A and B)	$\dfrac{dx\,dy\,dz\,dx'\,dy'\,dz'}{V^2}$

Let us recall finally that if $p(a)$ is the probability that a particular random variable U will be less than a, the corresponding density (if it exists) is the derivative $p'(a)$. The differential $p'(a)\,da$ is the probability that

$$a \leqslant U < a + da.$$

If U is a positive variable, we have

$$p(a) = \int_{-\infty}^{a} p'(s)\,ds.$$

Here, $p(a)$ and $p'(a)$ appear as "product measures." The desired probability $p(a)$ is the measure of a certain set in six-dimensional space.

Problem. Find the probability that the distance l between two points A and B picked at random in the interior of a sphere S of radius R will be less than a given number a.

Solutions. We choose the elementary probabilities *a priori*. The simplest hypothesis is that A and B have uniform probability distributions in the sphere and that these probabilities are independent. If $V = 4\pi R^3/3$ is the volume of the sphere, the elementary probability of the pair (A, B) is

$$\frac{dx\, dy\, dz\, dx'\, dy'\, dz'}{V^2}.$$

We need to integrate this in the six-dimensional region defined by

$$x^2 + y^2 + z^2 < R^2, \qquad x'^2 + y'^2 + z'^2 < R^2$$
$$(x - x')^2 + (y - y')^2 + (z - z')^2 < a^2.$$

When $a \geqslant 2R$, we naturally find that the probability p is 1. Let us now suppose that $a < 2R$. With A assumed fixed, let us evaluate the integral

$$\Omega = \iiint dx'\, dy'\, dz'$$

in that part of space common to the sphere S and the sphere of radius a with center at A (inside S).

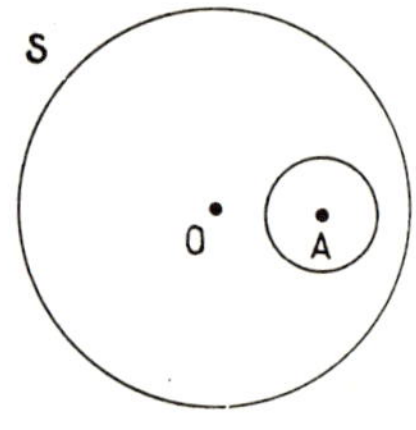

FIGURE 7

We then have

$$p(A) = \frac{1}{V^2} \iiint \Omega\, dx\, dy\, dz,$$

this integral being taken over the sphere S.

This method leads to rather lengthy calculations. It is faster to calculate first the probability density $p'(a)$, that is, the probability $p'(a)\, da$ that the distance l will lie between a and $a + da$.

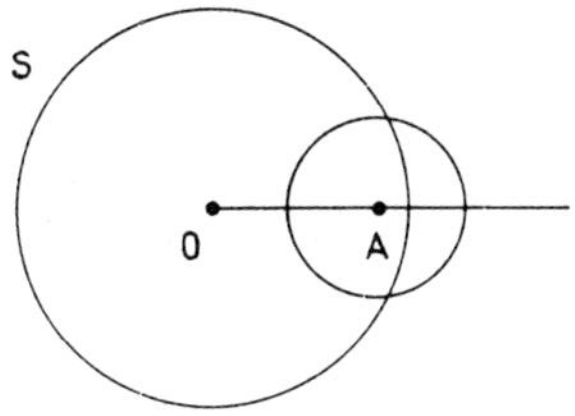

FIGURE 8

With A fixed, we calculate the volume between two concentric spheres with center at A and with radii respectively equal to a and $a + da$ in the interior of the sphere S.

Let us set $OA = \rho$. If $\rho + a < R$, the sphere of radius a with center at A lies entirely in the interior of S, and the desired volume is $4\pi a^2\, da$. If $\rho + a > R$, this volume is the product of da and the area of the spherical dome of radius a in the interior of S. Let h denote the altitude of this dome. The area in question is $2\pi ah$, where

$$h = OP - OA + a = \frac{\rho^2 + R^2 - a^2}{2\rho} + a - \rho = \frac{R^2 - (a-\rho)^2}{2\rho}.$$

In sum,

$$V^2 p'(a) = \iiint\limits_{\rho<R-a} 4\pi a^2\, dx\, dy\, dz + \iiint\limits_{R-a<\rho<R} 2\pi a \frac{R^2 - (a-\rho)^2}{2\rho}\, dx\, dy\, dz.$$

Let us shift to spherical coordinates:

$$\frac{V^2}{4\pi^2 a} p'(a) = 4a \int_0^{R-a} \rho^2\, d\rho + \int_{R-a}^{R} [R^2 - (a-\rho)^2]\rho\, d\rho.$$

After performing the integrations, we find

$$\begin{aligned} \frac{V^2}{4\pi^2} p'(a) &= \frac{a^5}{12} - a^3R^2 + \frac{4}{3} a^2R^3 \\ &= \frac{a^2}{12}(2R-a)^2(4R+a). \end{aligned}$$

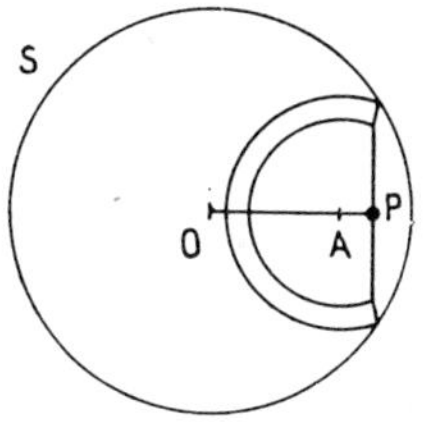

FIGURE 9

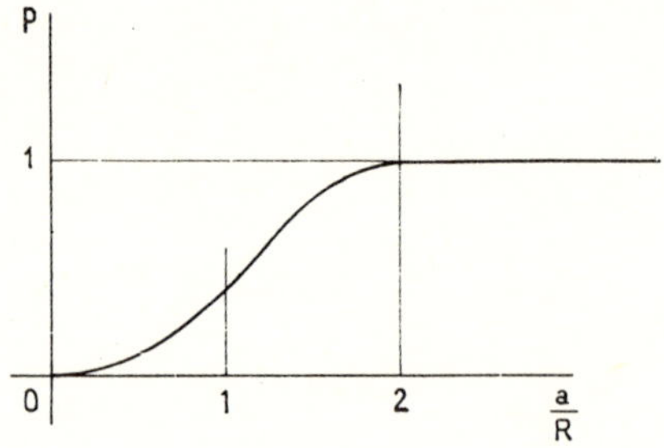

FIGURE 10

Knowing that $p(0) = 0$, we conclude by integrating the above equation that

$$\frac{V^2}{4\pi^2} p(a) = \frac{6a^2}{72} - \frac{a^4R^2}{4} + \frac{4}{9} a^3R^3,$$

$$p(a) = \frac{1}{32}\left(\frac{a}{R}\right)^6 - \frac{9}{16}\left(\frac{a}{R}\right)^4 + \left(\frac{a}{R}\right)^3.$$

The probability $p(a)$ vanishes for $a = 0$ and is equal to 1 for $a = 2R$. The probability density $p'(a)$ vanishes at $a = 0$ and $a = 2R$.

8. Multiple integrals, power series, and Fourier series

In this exercise we shall expand one function in a Fourier series and a second function in a power series. For particular values of the variables, these two series coincide, which yields the value of a certain definite integral.

Problems. A. Expand the odd function $\varphi(\theta)$ that is periodic with period 2π and that is equal to $\theta(\pi - \theta)$ in the interval $[0, \pi]$. What numerical series do we obtain for $\theta = \pi/2$?

B. Prove the formula

$$\int_0^x \frac{\arctan s}{s} \log \frac{x}{s}\, ds = x - \frac{x^3}{3^3} + \cdots + \frac{(-1)^n x^{2n+1}}{(2n+1)^3} + \cdots.$$

C. From the results obtained, derive the formula

$$\int_0^1 (\log s)^2 \frac{ds}{1+s^2} = \frac{\pi^3}{16}.$$

Solutions. **A.** Consider the odd function $\varphi(\theta)$ that is periodic with period 2π and that is equal to $\theta(\pi - \theta)$ in the interval $[0, \pi]$.

The coefficients β_n in its Fourier-series expansion

$$\varphi(\theta) = \beta_1 \sin \theta + \beta_2 \sin 2\theta + \cdots + \beta_n \sin n\theta + \cdots$$

are given by

$$\beta_n = \frac{2}{\pi} \int_0^\pi \theta(\pi - \theta) \sin n\theta \, d\theta.$$

Integrating by parts, we obtain

$$\begin{aligned} \beta_n &= -\frac{2}{\pi} \left[\frac{\theta(\pi - \theta) \cos n\theta}{n} \right]_0^\pi + \frac{2}{n\pi} \int_0^\pi (\pi - 2\theta) \cos n\theta \, d\theta \\ &= \frac{2}{n\pi} \int_0^\pi (\pi - 2\theta) \cos n\theta \, d\theta. \end{aligned}$$

Integrating by parts again, we obtain

$$\begin{aligned} \beta_n &= \frac{2}{n\pi} \left[\left(\frac{\pi - 2\theta}{n} \sin n\theta \right)_0^\pi + \frac{2}{n} \int_0^\pi \sin n\theta \, d\theta \right] \\ &= \frac{4}{\pi n^3} (\cos n\pi)_\pi^0 . \end{aligned}$$

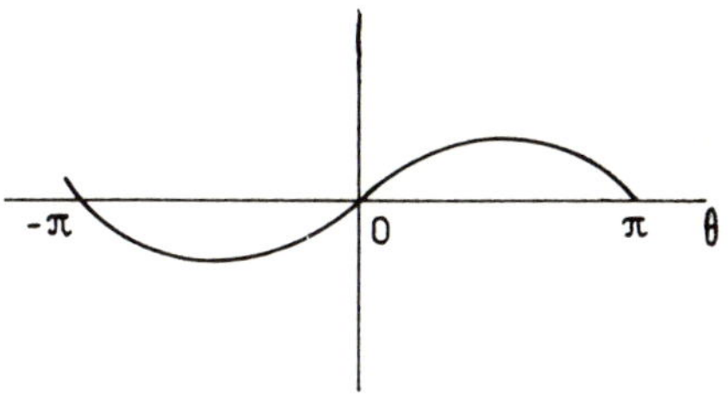

FIGURE 11

Now, $\cos n\pi = (-1)^n$, so that

$$\beta_n = \begin{cases} 0 & \text{if} \quad n = 2h, \\ \dfrac{8}{\pi(2h + 1)^3} & \text{if} \quad n = 2h + 1. \end{cases}$$

Finally,

$$\theta(\pi - \theta) = \sum_{h=0}^{\infty} \frac{8}{\pi} \frac{\sin (2h + 1)\theta}{(2h + 1)^3} \qquad (0 \leqslant \theta \leqslant \pi). \tag{1}$$

That $\beta_{2h} = 0$ is only to be expected, since the line $\theta = \pi/2$ is an axis of symmetry of the curve representing $\varphi(\theta)$.

For $\theta = \pi/2$, we obtain

$$\frac{\pi^2}{4} = \frac{8}{\pi} \sum_{h=0}^{\infty} \frac{(-1)^h}{(2h+1)^3}$$

or

$$\frac{\pi^3}{32} = \sum_{h=0}^{\infty} \frac{(-1)^h}{(2h+1)^3} = 1 - \frac{1}{3^3} + \frac{1}{5^3} + \cdots + \frac{(-1)^h}{(2h+1)^3} + \cdots. \tag{2}$$

This result is numerically plausible. We can verify by use of the properties of alternating series that

$$1 - \frac{1}{27} < \frac{\pi^3}{32} < 1 - \frac{1}{27} + \frac{1}{125}.$$

B. Let us set

$$f(x) = \int_0^x \frac{\arctan s}{s} \log \frac{x}{s}\, ds.$$

Then

$$f'(x) = \frac{1}{x} \int_0^x \frac{\arctan s}{s}\, ds.$$

Now, we know that

$$\arctan s = s - \frac{s^3}{3} + \frac{s^5}{5} + \cdots + \frac{(-1)^n s^{2n+1}}{2n+1} + \cdots \qquad (|\, s \,| \leqslant 1),$$

$$\int_0^x \frac{\arctan s}{s}\, ds = x - \frac{x^3}{3^2} + \frac{x^5}{5^2} + \cdots + (-1)^n \frac{x^{2n+1}}{(2n+1)^2} + \cdots \qquad (|\, x \,| < 1),$$

since term-by-term integration is permissible at every point x of the interval of convergence $[-1, 1]$.

Similarly, a second term-by-term integration yields

$$\int_0^x \frac{dt}{t} \int_0^t \frac{\arctan s}{s}\, ds = x - \frac{x^3}{3^3} + \frac{x^5}{5^3} + \cdots + \frac{(-1)^n x^{2n+1}}{(2n+1)^3} + \cdots \qquad (|\, x \,| < 1).$$

Now

$$\int_0^x \frac{dt}{t} \int_0^t \frac{\arctan s}{s}\, ds = \iint_{\mathscr{D}} \frac{\arctan s}{ts}\, dt\, ds,$$

where $\mathscr{D}$ is the region in the st-plane shown in Figure 12.

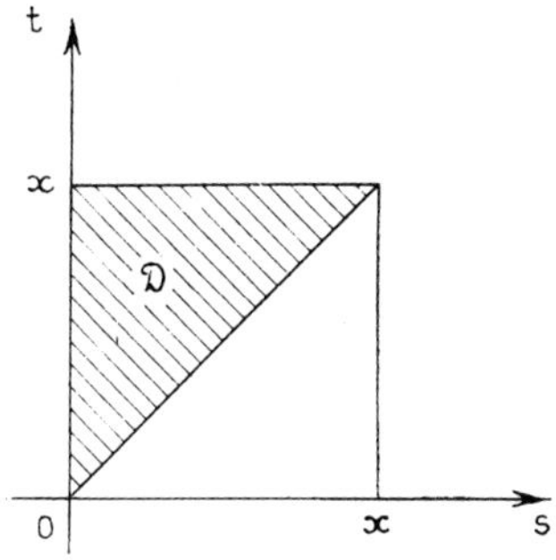

FIGURE 12

By reversing the order of the integrations, we obtain

$$\int_0^x \frac{\arctan s}{s}\, ds \int_s^x \frac{dt}{t} = \int_0^x \frac{\arctan s}{s} \log \frac{x}{s}\, ds.$$

Therefore,

$$\int_0^x \frac{\arctan s}{s} \log \frac{x}{s}\, ds = x + \cdots + \frac{(-1)^n x^{2n+1}}{(2n+1)^3} + \cdots \tag{3}$$

C. The series

$$\sum \frac{(-1)^n x^{2n+1}}{(2n+1)^3}$$

converges absolutely and uniformly for $-1 \leqslant x \leqslant 1$, since it is majorized by the series $\sum 1/(2n+1)^3$. Therefore, its sum is continuous from the left at $x = 1$, so that (3) holds also for $x = 1$.

If we set $x = 1$ in (3), we obtain

$$-\int_0^1 \frac{\arctan s}{s} \log s\, ds = 1 - \frac{1}{3^3} + \frac{1}{5^3} + \cdots + \frac{(-1)^n}{(2n+1)^3} + \cdots = \frac{\pi^3}{32}.$$

Integration by parts enables us to transform the integral on the left:

$$-\int_0^1 \frac{\arctan s}{s} \log s\, ds = -\left[\frac{\arctan s\,(\log s)^2}{2}\right]_0^1 + \frac{1}{2}\int_0^1 \frac{(\log s)^2}{1+s^2}\, ds$$
$$= \frac{1}{2}\int_0^1 \frac{(\log s)^2}{1+s^2}\, ds,$$

in which the bracketed expression vanishes. Therefore, the desired result[†] is

$$\int_0^1 (\log s)^2 \frac{ds}{1+s^2} = \frac{\pi^3}{16}.$$

9. Hölder's inequality

In connection with simple integrals, we have given several proofs of Hölder's inequality

$$\int_a^b \varphi\psi\, dx \leqslant \left(\int_a^b \varphi^p\, dx\right)^{1/p} \left(\int_a^b \psi^q\, dx\right)^{1/q}, \qquad \frac{1}{p}+\frac{1}{q}=1,$$

where φ and ψ are nonnegative integrable functions (see Exercises 3 and 4 of Chapter 1). Here we shall give a different proof, one that uses elementary calculations of multiple integrals and the inequality

$$na_1a_2 \cdots a_n \leqslant a_1{}^n + a_2{}^n + \cdots + a_n{}^n.$$

This inequality is equivalent to the classical inequality between the geometric and arithmetic means of n positive numbers.

Problems. A. Let a, b, and c denote three nonnegative numbers. Prove that

$$a^3+b^3+c^3 \geqslant 3abc.$$

For what values of a, b, and c does equality hold?

B. Let us set

$$a=f(x)g(y)g(z), \qquad b=f(y)g(z)g(x), \qquad c=f(z)g(x)g(y),$$

where f and g are two nonnegative functions that are integrable from 0 to 1. Use these notations to write the inequality of the first problem. Integrate the inequality obtained over the region $0<x<1, 0<y<1, 0<z<1$. Then, by a change of notation, prove that

$$\int_0^1 \varphi\psi\, dx \leqslant \left(\int_0^1 \varphi^3\, dx\right)^{1/3} \left(\int_0^1 \psi^{3/2}\, dx\right)^{2/3},$$

where φ and ψ are two nonnegative functions that are integrable from 0 to 1. What relation must hold between φ and ψ for equality to hold?

† We might set $\log s = -t$ and derive this result from Exercise 14 of Chapter 2.

C. Write and then verify by direct calculation the inequality of Question B when

$$\varphi(x) = x^{\alpha} \qquad \text{and} \qquad \psi(x) = x^{\beta},$$

where α and β are two given numbers.

D. Let $a_1, a_2, \ldots, a_n$ denote n positive numbers. Show that

$$a_1^n + a_2^n + \cdots + a_n^n \geqslant n a_1 a_2 \cdots a_n .$$

Let us set

$$a_1 = f_1(x_1) f_2(x_2) \cdots f_n(x_n),$$

where $f_1, f_2, \ldots, f_n$ are n functions that are integrable on $[0, 1]$. We define $a_2, a_3, \ldots, a_n$ by circular permutations on the variables $x_1, x_2, \ldots, x_n$ in the expression for a_1. Prove the inequality

$$\left[\int_0^1 f_1(x) f_2(x) \cdots f_n(x)\, dx\right]^n \leqslant \int_0^1 f_1^n(x)\, dx \int_0^1 f_2^n(x)\, dx \cdots \int_0^1 f_n^n(x)\, dx.$$

By a change of notation, use this inequality to show that, if p and q are rational numbers exceeding 1 such that $1/p + 1/q = 1$, then

$$\int_0^1 \varphi\psi\, dx \leqslant \left(\int_0^1 \varphi^p\, dx\right)^{1/p} \left(\int_0^1 \psi^q\, dx\right)^{1/q},$$

where φ and ψ are nonnegative integrable functions.

Solutions. **A.** The classical identity

$$a^3 + b^3 + c^3 - 3abc = \frac{a+b+c}{2}[(b-c)^2 + (c-a)^2 + (a-b)^2] \tag{1}$$

shows that if a, b, and c are real nonnegative numbers, the polynomial $a^3 + b^3 + c^3 - 3abc$ is nonnegative. This polynomial vanishes if and only if $a = b = c$.

This result can also be shown by examining the variations in the function $u(a) = a^3 - 3abc + b^3 + c^3$.

B. The expression

$$\Phi(x, y, z) = [f(x)g(y)g(z)]^3 + [f(y)g(z)g(x)]^3 + [f(z)g(x)g(y)]^3 \\ -3f(x)f(y)f(z)[g(x)g(y)g(z)]^2$$

is identical to the polynomial (1) if we set

$$a = f(x)g(y)g(z), \qquad b = f(y)g(z)g(x), \qquad c = f(z)g(x)g(y).$$

Therefore, it is nonnegative in the cube K defined by $0 < x < 1$, $0 < y < 1$, $0 < z < 1$ because the functions f and g are nonnegative on the closed interval $[0, 1]$ and, hence, $a, b, c \geqslant 0$. The same holds for the integral

$$I = \iiint_K \Phi(x, y, z)\, dx\, dy\, dz.$$

For I to vanish, it is necessary that $a = b = c$ and hence that

$$\frac{f(x)}{g(x)} = \frac{f(y)}{g(y)} = \frac{f(z)}{g(z)}.$$

The common value of these ratios is therefore a constant and the ratio f/g must be constant.

If we introduce the simple integrals

$$J_1 = \int_0^1 f^3(x)\, dx, \qquad J_2 = \int_0^1 g^3(x)\, dx, \qquad J_3 = \int_0^1 f(x)g^2(x)\, dx,$$

we have

$$I = 3J_1J_2^2 - 3J_3^3 = 3(J_1J_2^2 - J_3^3).$$

Since $I \geqslant 0$, we have

$$J_3^3 \leqslant J_1J_2^2$$

or

$$J_3 \leqslant J_1^{1/3}J_2^{2/3}. \tag{2}$$

Let us make a change of notation by setting

$$f(x) = \varphi(x), \qquad g^2(x) = \psi(x).$$

Then inequality (2) may be written

$$\int_0^1 \varphi(x)\psi(x)\, dx \leqslant \left[\int_0^1 \varphi^3(x)\, dx\right]^{1/3} \left[\int_0^1 \psi^{3/2}(x)\, dx\right]^{2/3}. \tag{3}$$

Equality corresponds to

$$f = kg \qquad \text{or} \qquad \varphi(x) = k\psi^{1/2}(x).$$

C. For the function $\varphi^3(x) = x^{2\alpha}$ to be integrable on [0, 1], it is necessary that $3\alpha > -1$ or $\alpha > -\frac{1}{3}$.

Similarly, for $\psi^{3/2}(x) = x^{3\beta/2}$ to be integrable on [0, 1], it is necessary that

$$\frac{3\beta}{2} > -1 \qquad \text{or} \qquad \beta > -\frac{2}{3}.$$

It is easily verified that $\varphi(x)\psi(x) = x^{\alpha+\beta}$ is integrable, since

$$\alpha + \beta > -1.$$

Inequality (3) is then written

$$\int_0^1 x^{\alpha+\beta}\,dx \leqslant \left(\int_0^1 x^{3\alpha}\,dx\right)^{1/3} \left(\int_0^1 x^{3\beta/2}\,dx\right)^{2/3}. \tag{4}$$

When we carry out the calculations, we see that

$$\frac{1}{\alpha+\beta+1} \leqslant \frac{1}{(3\alpha+1)^{1/3}} \frac{1}{[(3\beta/2)+1]^{2/3}}$$

or

$$(\alpha+\beta+1)^3 \geqslant (3\alpha+1)\left(\frac{3\beta}{2}+1\right)^2. \tag{5}$$

Now, if we introduce the polynomial

$$\psi(\alpha, \beta) = (\alpha+\beta+1)^3 - (3\alpha+1)\left(\frac{3\beta}{2}+1\right)^2,$$

we see immediately that $\alpha = \beta/2$ is a double root of $\psi(\alpha, \beta)$ treated as a function of α, and we may write

$$\psi(\alpha, \beta) = \left(\alpha - \frac{\beta}{2}\right)^2 (4\beta + \alpha + 3),$$

Here ψ is positive if $4\beta + \alpha > -3$. Now, the inequalities $\alpha > -1/3$, $\beta > -2/3$ imply

$$4\beta + \alpha \geqslant -\tfrac{8}{3} - \tfrac{1}{3} = -3.$$

Thus we verify inequality (5).

Remark. Inequality (4) is Hölder's inequality for

$$p = \tfrac{1}{3}, \qquad q = \tfrac{3}{2}.$$

Schwarz' inequality can be proved in an analogous fashion by starting with the inequality

$$a^2 + b^2 - 2ab = (a - b)^2 \geqslant 0$$

with

$$a = f(x)g(y) \qquad \text{and} \qquad b = f(y)g(x).$$

D. Let us prove that, more generally, if $a_1, \ldots, a_n$ are nonnegative numbers, then

$$a_1{}^n + \cdots + a_n{}^n \geqslant na_1a_2 \cdots a_n . \tag{6}$$

We set

$$\psi(a) = a^n - naa_2 \cdots a_n + a_2{}^n + \cdots + a_n{}^n.$$

Then

$$\psi'(a) = n(a^{n-1} - a_2a_3 \cdots a_n).$$

Note that $\psi'(a)$ vanishes for $a = (a_2a_3 \cdots a_n)^{1/(n-1)}$ and, at that value of a, $\psi(a)$ assumes the value

$$a_2{}^n + \cdots + a_n{}^n - (n-1)(a_2a_3 \cdots a_n)^{n/(n-1)}.$$

If ψ is nonnegative at this minimum, the inequality will be proved for all values of a. Let us set

$$a_2^{n/(n-1)} = b_2 , \ldots, a_n^{n/(n-1)} = b_n .$$

We then obtain the inequality

$$b_2^{n-1} + \cdots + b_n^{n-1} - (n-1)b_2 \cdots b_n \geqslant 0.$$

This is the same inequality but for $n - 1$ instead of n. If the desired inequality holds for $n - 1$, it must hold also for n. But we know that it does hold for $n = 2$ (and $n = 3$). Therefore, it holds for all n.

This inequality expresses the fact that the arithmetic mean of n positive numbers $a_1, \ldots, a_n$ is at least equal to their geometric mean $\sqrt[n]{(a_1a_2 \cdots a_n)}$.

Let us now consider n functions $f_1, \ldots, f_n$ and n variables $x_1, \ldots, x_n$. We set

$$a_1 = f_1(x_1)f_2(x_2) \cdots f_n(x_n)$$

and then permute the x_i circularly:

$$\begin{aligned} a_2 &= f_1(x_2)f_2(x_3) \cdots f_n(x_1), \\ a_3 &= f_1(x_3)f_2(x_4) \cdots f_n(x_2), \\ &\cdots\cdots\cdots\cdots\cdots\cdots \\ a_n &= f_1(x_n)f_2(x_1) \cdots f_n(x_{n-1}). \end{aligned}$$

Let us write inequality (6) with these notations and then integrate over the region $0 < x_1 < 1, \ldots, 0 < x_n < 1$. The first member will then contain n integrals of the form

$$\int_0^1 f_1^n(x_1)\,dx_1\,, \ldots, \int_0^1 f_n^n(x_n)\,dx_n\,,$$

which differ only by the index of the variable. Therefore, each has the same value as the first one, so that the effect is the same as multiplying that one by n. In the right-hand member, we have

$$n[f_1(x_1)f_2(x_1)\cdots f_n(x_1)][f_1(x_2)f_2(x_2)\cdots f_n(x_2)]\cdots[f_1(x_n)f_2(x_n)\cdots f_n(x_n)].$$

By integrating, we obtain

$$n\left[\int_0^1 f_1(x_1)f_2(x_1)\cdots f_n(x_1)\,dx_1\right]^n.$$

If we divide by n, we obtain the inequality

$$\left[\int_0^1 f_1(x)f_2(x)\cdots f_n(x)\,dx\right]^n \leqslant \int_0^1 f_1^n(x)\,dx\cdots\int_0^1 f_n^n(x)\,dx.$$

Let us now set $n = \alpha + \beta$ and let us take the first α functions f_i identical to a function f and the last β functions identical to a function g. We then have

$$\left(\int_0^1 f^\alpha g^\beta\,dx\right)^{\alpha+\beta} \leqslant \left(\int_0^1 f^{\alpha+\beta}\,dx\right)^\alpha \left(\int_0^1 g^{\alpha+\beta}\,dx\right)^\beta.$$

Finally, if we change our notations and write

$$p = \frac{\alpha+\beta}{\alpha}, \qquad q = \frac{\alpha+\beta}{\beta}, \qquad f^\alpha = \varphi, \qquad g^\beta = \psi,$$

we get

$$\int_0^1 \varphi\psi\,dx \leqslant \left(\int_0^1 \varphi^p\,dx\right)^{1/p}\left(\int_0^1 \psi^q\,dx\right)^{1/q}, \qquad \frac{1}{p}+\frac{1}{q} = 1.$$

This is Hölder's inequality with p and q rational numbers.

If φ is integrable and nonnegative, the integral $(\int_0^1 \varphi^p\,dx)^{1/p}$ is a continuous function of p. Therefore, the inequality that we have obtained remains valid when p and q are irrational.

10. Evaluation of a plane area by several changes of variables

Problems. A. Consider a family F of conics of the form

$$\frac{x^2}{s-1}+\frac{y^2}{s-3}=1 \tag{1}$$

where s is a real parameter. Specify the type of conic defined by this equation according to the different values of s.

Show that, in general, one ellipse and one hyperbola of the family pass through any given point in the plane.

B. One ellipse E_0 and one hyperbola H_0 of the family F pass through the point $x_0 = 5/4, y_0 = 3/4$. Find the corresponding values of s.

We introduce curvilinear coordinate u and v such that

$$x = a \cosh u \cos v \qquad \text{and} \qquad y = b \sinh u \sin v. \tag{2}$$

Show that a and b may be chosen in such a way that the ellipses of the family F are curves of the form $u =$ const and the hyperbolas are curves of the form $v =$ const. Find these constants for E_0 and H_0. By making the change of variable (2) find the area $\mathscr{A}$ of the region $\mathscr{D}$ shown in Figure 13 that is bounded by the coordinate axes and by arcs of the curves E_0 and H_0.

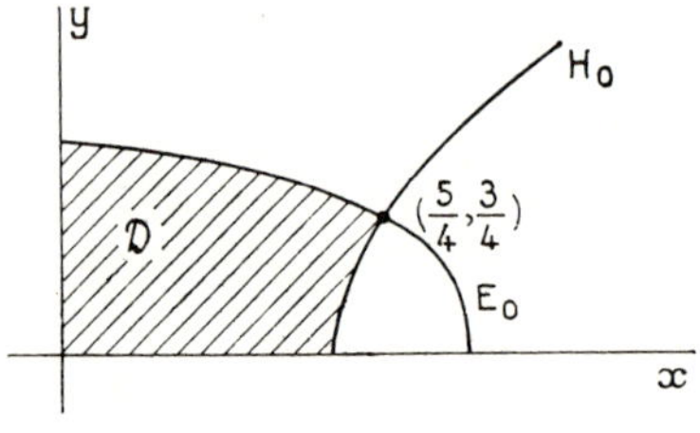

FIGURE 13

C. Denote by λ and μ, respectively, the values of the parameter s corresponding to ellipse E and the hyperbola H that pass through a point (x, y). Here λ and μ may be treated as curvilinear coordinates in the plane, like the formulas for the transformation from x and y to λ and μ.

With the aid of the variables λ and μ, recalculate the area of the domain $\mathscr{D}$ of Problem B.

Solutions. **A.** Consider the conic C whose equation is

$$\frac{x^2}{s-1} + \frac{y^2}{s-3} = 1. \tag{1}$$

If $s > 3$, C is an ellipse; if $1 < s < 3$, C is a hyperbola. If $s < 1$, C is an "imaginary ellipse."

All these conics have the same foci, the abscissas of which are

$$\pm [(s-1) - (s-3)]^{1/2} = \pm \sqrt{2}$$

on the x-axis. They constitute a confocal family.

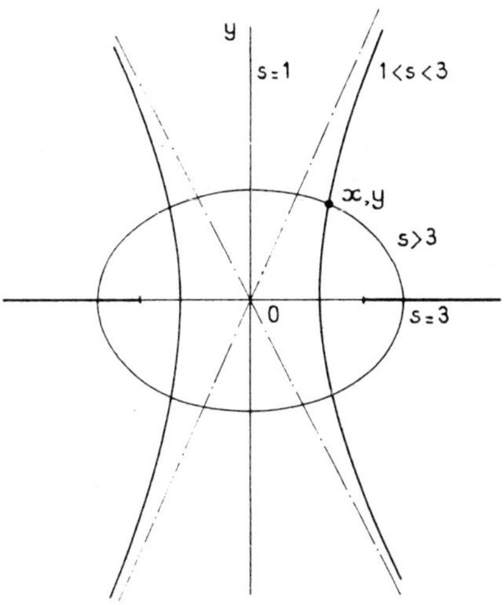

FIGURE 14

As s approaches 3 from above, the length of the major axis of the ellipse tends to $2\sqrt{2}$ and that of the minor axis tends to 0. The ellipse itself approaches the segment $[-\sqrt{2}, +\sqrt{2}]$. As s tends to 3 from below, the length of the major axis of the hyperbola tends to $2\sqrt{2}$ and the asymptotes approach coincidence with the x-axis. The hyperbola itself approaches that portion of the x-axis outside the segment $(-\sqrt{2}, +\sqrt{2})$.

As $s \to 1$ from above, the hyperbola approaches coincidence with the y-axis.

Through any point (x, y) in the plane, there pass one ellipse and one hyperbola of the family F. The parameters s of these two conics are roots of the equation

$$\varphi(s) = \frac{x^2}{s-1} + \frac{y^2}{s-3} - 1 = 0.$$

The function $\varphi(s)$ is a decreasing function and its curve, which is easily constructed, exhibits two roots, one between 1 and 3 and the other greater than 3.

It can be verified immediately that the result remains valid if either x or y or both are 0.

Instead of the equation

$$\varphi(s) = 0,$$

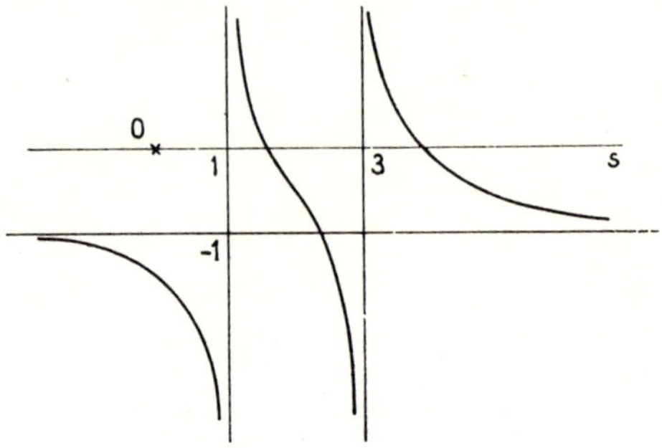

FIGURE 15

we could equally well study the same equation with the fractions eliminated. We obtain a second-degree equation

$$s^2 - (x^2 + y^2 + 4)s + 3x^2 + y^2 + 3 = 0, \tag{1'}$$

the discriminant of which is equal to

$$(x^2 - 2)^2 + y^2(2x^2 + y^2 + 4) \geqslant 0.$$

B. Through the point $x_0 = \frac{5}{4}$, $y_0 = 3/4$ pass the two conics corresponding to the parameters

$$\begin{aligned} s = \lambda_0 &= \frac{1}{2}\{x_0^{\,2} + y_0^{\,2} + 4 + [(x_0^{\,2} + y_0^{\,2} + 4)^2 - 4(3x_0^{\,2} + y_0^{\,2} + 3)]^{1/2}\} \\ &= \frac{33}{8}\,(= 4.125), \end{aligned}$$

$$\begin{aligned} s = \mu_0 &= \frac{1}{2}\{x_0^{\,2} + y_0^{\,2} + 4 - [(x_0^{\,2} + y^2 + 4)^2 - 4(3x_0^{\,2} + y_0^{\,2} + 3)]^{1/2}\} \\ &= 2. \end{aligned}$$

The first ($\lambda_0 = 33/8 > 3$) is an ellipse E_0; the second ($1 < \mu_0 = 2 < 3$) is a hyperbola H_0.

The length of the semiaxes of E_0 are given by

$$\alpha_0^2 = \lambda_0 - 1 = \frac{25}{8}, \qquad \alpha_0 = \frac{5}{4}\sqrt{2},$$

$$\beta_0^2 = \lambda_0 - 3 = \frac{9}{8}, \qquad \beta_0 = \frac{3}{4}\sqrt{2}.$$

The length of the semiaxes of H_0 are given by

$$\alpha_0'^2 = \mu_0 - 1 = 1 \qquad (\alpha_0' = 1),$$

$$\beta_0'^2 = 3 - \mu_0 = 1 \qquad (\beta_0' = 1).$$

The equations of these two conics are given, respectively, by

$$(E_0) \quad \frac{x^2}{25/8} + \frac{y^2}{9/8} = 1, \qquad (H_0) \quad x^2 - y^2 = 1.$$

The ellipse (E_0) can be represented parametrically by the equations

$$(E_0) \quad \begin{cases} x = a \cosh u_0 \cos v, \\ y = b \sinh u_0 \sin v, \end{cases}$$

under the condition that

$$a \cosh u_0 = \alpha_0 = \tfrac{5}{4}\sqrt{2}, \qquad b \sinh u_0 = \beta_0 = \tfrac{3}{4}\sqrt{2}. \tag{3}$$

Similarly, the hyperbola (H_0) can be represented by the equations

$$(H_0) \quad \begin{cases} x = a \cos v_0 \cosh u, \\ y = b \sin v_0 \sinh u, \end{cases}$$

under the condition that

$$a \cos v_0 = \alpha_0' = 1, \qquad b \sin v_0 = \beta_0' = 1. \tag{4}$$

From the system of equations (3) and (4), we conclude that $a^2 = b^2 = 2$. If we take $a = b = \sqrt{2}$, it follows that

$$\cos v_0 = \sin v_0 = \frac{\sqrt{2}}{2},$$

so that

$$v_0 = \frac{\pi}{4}$$

and

$$\cosh u_0 = \tfrac{5}{4}, \qquad \sinh u_0 = \tfrac{3}{4},$$

so that

$$u_0 = \cosh^{-1}(\tfrac{5}{4}) = \log\{\tfrac{5}{4} + [(\tfrac{5}{4})^2 - 1]^{1/2}\} = \log 2.$$

If we make the change of variables

$$\begin{aligned} x &= a \cosh u \cos v = \sqrt{2} \cosh u \cos v, \\ y &= b \sinh u \sin v = \sqrt{2} \sinh u \sin v, \end{aligned} \tag{2'}$$

the curve $u = u_0 = \log 2$ represents the ellipse (E_0) and the curve

$$v = v_0 = \frac{\pi}{4}$$

represents the hyperbola (H_0).

For an ellipse E corresponding to the parameter u to be (in the first quadrant) interior to $\mathscr{D}$, it is necessary that the length of its semiminor axis $\beta = \sqrt{2} \sinh u$ be less than $\beta_0 = (\tfrac{3}{4})\sqrt{2}$:

$$0 < \sqrt{2} \sinh u < \tfrac{3}{4}\sqrt{2} \qquad \text{or} \qquad 0 < u < u_0 = \log 2.$$

Similarly, for a hyperbola H corresponding to the parameter v to cross the region $\mathscr{D}$, it is necessary that the length of its semimajor axis

$$\alpha' = \sqrt{2} \cos v$$

be less than $\alpha_0' = 1$:

$$0 < \sqrt{2} \cos v < 1$$

or

$$v_0 = \frac{\pi}{4} < v < \frac{\pi}{2}.$$

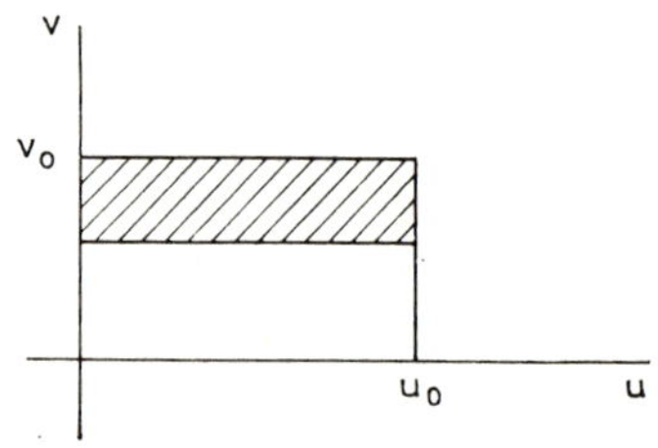

FIGURE 16

In the uv-plane, the region $\mathscr{D}'$ corresponding to $\mathscr{D}$ is therefore defined by

$$0 < u < \log 2, \qquad \frac{\pi}{4} < v < \frac{\pi}{2}.$$

If we differentiate formulas (2), we get

$$dx = \sqrt{2}\sinh u \cos v \, du - \sqrt{2}\cosh u \sin v \, dv,$$
$$dy = \sqrt{2}\cosh u \sin v \, du + \sqrt{2}\sinh u \cos v \, dv,$$

from which we get

$$\left| \frac{D(x, y)}{D(u, v)} \right| = 2(\sinh^2 u \cos^2 v + \cosh^2 u \sin^2 v)$$

and

$$\mathscr{A} = 2 \iint_{\mathscr{D}'} \sinh^2 u \cos^2 v \, du \, dv + 2 \iint_{\mathscr{D}'} \sinh^2 u \sin^2 v \, du \, dv,$$

$$\begin{aligned} \iint \sinh^2 u \cos^2 v \, du \, dv &= \int_0^{\log 2} \sinh^2 u \, du \int_{\pi/4}^{\pi/2} \cos^2 v \, dv \\ &= \left(\frac{\sinh 2u}{4} - \frac{u}{2} \right)_0^{\log 2} \left(\frac{\sin 2v}{4} + \frac{v}{2} \right)_{\pi/4}^{\pi/2} \\ &= \left(\frac{15}{32} - \frac{\log 2}{2} \right) \left(\frac{\pi}{8} - \frac{1}{4} \right), \\ \iint_{\mathscr{D}'} \cosh^2 u \sin^2 v \, du \, dv &= \int_0^{\log 2} \cosh^2 u \, du \int_{\pi/4}^{\pi/2} \sin^2 v \, dv \\ &= \left(\frac{\sinh 2u}{4} + \frac{u}{2} \right)_0^{\log 2} \left(\frac{v}{2} - \frac{\sin 2v}{4} \right)_{\pi/4}^{\pi/2} \\ &= \left(\frac{15}{32} + \frac{1}{2} \log 2 \right) \left(\frac{\pi}{8} + \frac{1}{4} \right), \end{aligned}$$

and, finally,

$$\mathscr{A} = 2 \left(\frac{15}{32} \frac{\pi}{4} + \frac{\log 2}{4} \right) = \frac{15\pi}{64} + \frac{\log 2}{2} = 1.08\ldots.$$

C. Through every point (x, y) of the plane pass an ellipse (E) and a hyperbola (H) of the family (F). We may take the corresponding parameters $s = \lambda$ and $s = \mu$ as curvilinear coordinates.

The sum and products of the roots of (1′) are

$$\lambda + \mu = x^2 + y^2 + 4, \qquad \lambda\mu = 3x^2 + y^2 + 3,$$

from which we conclude that

$$x^2 = \tfrac{1}{2}[(\lambda\mu - (\lambda + \mu) + 1)] = \tfrac{1}{2}(\lambda - 1)(\mu - 1),$$

$$y^2 = \tfrac{1}{2}[(-\lambda\mu + 3(\lambda + \mu) - 9)] = \tfrac{1}{2}(\lambda - 3)(3 - \mu),$$

and, remembering that $\lambda > 3$, $1 < \mu < 3$, in the first quadrant,

$$x = \frac{1}{\sqrt{2}}(\lambda - 1)^{1/2}(\mu - 1)^{1/2}, \qquad y = \frac{1}{\sqrt{2}}(\lambda - 3)^{1/2}(3 - \mu)^{1/2}.$$

We have seen that the semiminor axis β of (E) varies in $\mathscr{D}$ from 0 to $\beta_0 = (\frac{3}{4})\sqrt{2}$. The parameter $\lambda = \beta^2 + 3$ thus varies from 3 to $\lambda_0 = 33/8$.

Similarly, the semimajor axis α' of (H) varies from 0 to 1 and hence $\mu' = \alpha'^2 + 1$ varies from 1 to $\mu_0 = 2$.

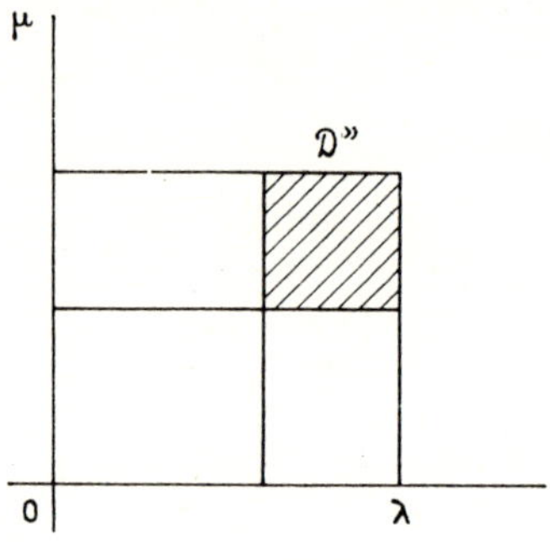

FIGURE 17

The domain $\mathscr{D}''$ corresponding to $\mathscr{D}$ in the $\lambda\mu$-plane is therefore defined by

$$3 < \lambda < \frac{33}{8}, \qquad 1 < \mu < 2.$$

If we differentiate relations (6), we obtain

$$d\lambda + d\mu = 2x\,dx + 2y\,dy, \qquad \lambda\,d\mu + \mu\,d\lambda = 6x\,dx + 2y\,dy,$$

from which we get

$$4x\,dx = (\mu - 1)\,d\lambda + (\lambda - 1)\,d\mu, \qquad 4y\,dy = (3 - \mu)\,d\lambda + (3 - \lambda)\,d\mu$$

and

$$\frac{D(x, y)}{D(\lambda, \mu)} = \frac{1}{16xy}[(\mu - 1)(3 - \lambda) - (\lambda - 1)(3 - \mu)] = \frac{\mu - \lambda}{8xy},$$

$$\left|\frac{D(x, y)}{D(\lambda, \mu)}\right| = \frac{\lambda - \mu}{8xy} = \frac{1}{4}(\lambda - \mu)(\lambda - 1)^{-1/2}(\mu - 1)^{-1/2}(\lambda - 3)^{-1/2}(3 - \mu)^{-1/2}.$$

Therefore, we may write

$$\mathscr{A} = \frac{1}{4}\iint_{\mathscr{D}'} (\lambda - \mu)(\lambda - 1)^{-1/2}(\mu - 1)^{-1/2}(\lambda - 3)^{-1/2}(3 - \mu)^{-1/2}\, d\lambda\, d\mu$$

or

$$\mathscr{A} = \frac{I - J}{4},$$

where

$$I = \iint_{\mathscr{D}''} (\lambda - 1)^{1/2}(\mu - 1)^{-1/2}(\lambda - 3)^{-1/2}(3 - \mu)^{-1/2}\, d\lambda\, d\mu,$$

$$J = \iint_{\mathscr{D}''} (\mu - 1)^{1/2}(\lambda - 1)^{-1/2}(\lambda - 3)^{-1/2}(3 - \mu)^{-1/2}\, d\lambda\, d\mu.$$

Here I can be expressed as the product of two integrals $I = I_1 I_2$, where

$$I_1 = \int_3^{33/8} (\lambda - 1)^{1/2}(\lambda - 3)^{-1/2}\, d\lambda, \qquad I_2 = \int_1^2 (\mu - 1)^{-1/2}(3 - \mu)^{-1/2}\, d\mu.$$

These two integrals can be calculated by a classical method by making the change of variable

$$t = \left(\frac{\lambda - 1}{\lambda - 3}\right)^{1/2}$$

in the first and

$$t = \left(\frac{3 - \mu}{\mu - 1}\right)^{1/2}$$

in the second. We then have

$$I_1 = 4\int_{5/3}^{\infty} \frac{t^2\, dt}{(t^2 - 1)^2} = 2\int_{5/3}^{\infty} \frac{dt}{t^2 - 1} - 2\left(\frac{t}{t^2 - 1}\right)_{5/3}^{\infty}$$
$$= \left(\log\left|\frac{1 - t}{1 + t}\right|\right)_{5/3}^{\infty} + 2 \cdot \frac{15}{16} = 2 \log 2 + \frac{15}{8},$$

$$I_2 = 2\int_1^{\infty} \frac{dt}{1 + t^2} = \frac{\pi}{2},$$

from which we get

$$I = \frac{\pi}{2}\left(2 \log 2 + \frac{15}{8}\right).$$

We calculate J in the same way by setting $J = J_1 J_2$, where

$$J_1 = \int_3^{33/8} (\lambda - 1)^{-1/2}(\lambda - 3)^{-1/2}\, d\lambda = 2\int_0^{3/5} \frac{dt}{1 - t^2} = 2 \log 2,$$

$$J_2 = \int_1^2 (\mu - 1)^{1/2}(3 - \mu)^{-1/2}\, d\mu = 4\int_0^1 \frac{t^2\, dt}{(1 + t^2)^2} = \frac{\pi}{2} - 1,$$

from which we get

$$J = 2 \log 2 \left(\frac{\pi}{2} - 1\right)$$

and, finally,

$$\mathscr{A} = \frac{1}{4}(I - J) = \frac{15\pi}{64} + \frac{\log 2}{2}.$$

11. Functions that are equal to their mean values

If $F(x, y)$ is a harmonic function, that is, if its Laplacian ΔF vanishes throughout some region, then F satisfies the mean-value theorem, which states that the mean value of F in the interior of every circle is equal to the value of F at the center of that circle and that the mean value of F on the edge of the circle is also equal to the value at the center. Here, we shall prove the converses of these properties.

Problems. A. Suppose that $F(x, y)$ is a function that is integrable in a region Ω. Let Γ be a circle of radius a with center at an arbitrary point (x, y) and suppose that Γ is contained in the interior of Ω. Suppose, finally, that

$$F(x, y) = \frac{1}{\pi a^2} \iint_{\mathscr{D}} F(u, v)\, du\, dv. \tag{1}$$

where $\mathscr{D}$ is the interior of Γ. Show that $F(x, y)$ is continuous and possesses partial derivatives of all orders. (*Hint*: Set $F = \partial G/\partial x$ and use Riemann's formula.)

B. Suppose that (1) holds for every value of the radius a. Show that $F(x, y)$ is equal to its mean value on the circle Γ and from this show that $F(x, y)$ is a harmonic function.

Solutions. **A.** The function

$$F(x, y) = \frac{1}{\pi a^2} \iint_{\mathscr{D}} F(u, v)\, du\, dv$$

is, by definition, bounded on every region over which it is integrable. We have

$$F(x + h, y + k) = \frac{1}{\pi a^2} \iint_{\mathscr{D}'} F(u, v)\, du\, dv,$$

where $\mathscr{D}'$ denotes the interior of the circle of radius a with center at the point $(x + h, y + k)$. Therefore,

$$F(x + h, y + k) - F(x, y) = \frac{1}{\pi a^2} \left[\iint_{\Delta_1} F(u, v)\, du\, dv - \iint_{\Delta} F(u, v)\, du\, dv \right]$$

where Δ and Δ_1 are the shaded regions shown in Figure 18. In these

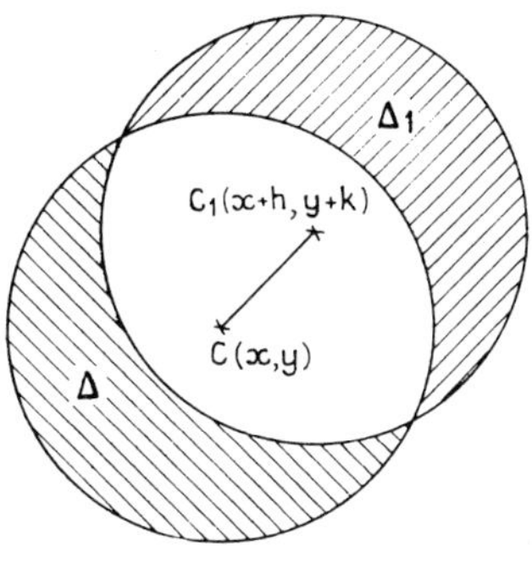

FIGURE 18

two regions, we have $|F(u, v)| < M$, where M is a constant. Furthermore, the areas of Δ_1 and Δ_2 tend to 0 as

$$CC_1 = (k^2 + h^2)^{1/2} \to 0,$$

so that

$$|F(x + h, y + k) - F(x, y)| \leqslant \frac{M}{\pi a^2} (\Delta_1 + \Delta)$$

also tends to 0, which proves the continuity of $F(x, y)$.

Since F is continuous, it is integrable with respect to x and its primitive G is continuous. Therefore, there exists a function $G(x, y)$ that is continuous in x and in y such that $F = \partial G/\partial x$.

We may write

$$F(x, y) = \frac{1}{\pi a^2} \iint_{\mathscr{D}} \frac{\partial G}{\partial x}(u, v)\, du\, dv.$$

According to Riemann's formula,

$$F(x, y) = \frac{1}{\pi a^2} \int_{\Gamma} G(u, v)\, dv,$$

where Γ is the circle bounding Δ. Therefore,

$$F(x, y) = \frac{1}{\pi a^2} \int_0^{2\pi} G(x + a\cos\theta, y + a\sin\theta) a\cos\theta\, d\theta.$$

Since $\partial G/\partial x$ exists by hypothesis and is continuous, we may differentiate under the integral sign with respect to x and write

$$\frac{\partial F}{\partial x} = \frac{1}{\pi a^2} \int_0^{2\pi} \frac{\partial G}{\partial x}(x + a\cos\theta, y + a\sin\theta) a\cos\theta\, d\theta$$

$$= \frac{1}{\pi a^2} \int_0^{2\pi} F(u, v)\, dv = \frac{1}{\pi a^2} \iint_{\mathscr{D}} \frac{\partial F}{\partial u}\, du\, dv.$$

Analogously, we can verify that

$$\frac{\partial F}{\partial y} = \frac{1}{\pi a^2} \iint_{\mathscr{D}} \frac{\partial F}{\partial v}\, du\, dv.$$

The functions $\partial F/\partial x$ and $\partial F/\partial y$ are integrable, they satisfy the same relation as does F, and they are continuous. Therefore, they have partial derivatives with the same properties.

We shall see later that $F(x, y)$ has partial derivatives of all orders.

B.

$$F(x, y) = \frac{1}{\pi a^2} \iint_{\mathscr{D}} F(u, v)\, du\, dv$$

is, by hypothesis, independent of a. Therefore,

$$0 = \frac{\partial F(x, y)}{\partial a} = -\frac{2}{\pi a^3} \iint_{\mathscr{D}} F(u, v)\, du\, dv + \frac{1}{\pi a^2} \left[\frac{\partial}{\partial a} \iint_{\mathscr{D}} F(u, v)\, du\, dv\right].$$

To evaluate the derivative of the double integral, we shift to polar coordinates:

$$\iint_{\mathscr{D}} F(u, v)\, du\, dv = \int_0^a \rho\, d\rho \int_0^{2\pi} F(x + \rho \cos \theta, y + \rho \sin \theta)\, d\theta$$

and

$$\frac{\partial}{\partial a}\left[\iint_{\mathscr{D}} F(u, v)\, du\, dv\right] = a \int_0^{2\pi} F(x + a \cos \theta, y + a \sin \theta)\, d\theta.$$

We thus have

$$0 = -\frac{2}{a} F(x, y) + \frac{1}{\pi a} \int^{2\pi} F(x + a \cos \theta, y + a \sin \theta)\, d\theta,$$

so that

$$F(x, y) = \frac{1}{2\pi} \int_0^{2\pi} F(x + a \cos \theta, y + a \sin \theta)\, d\theta.$$

Let us again write $\partial F/\partial a = 0$:

$$\begin{aligned} 0 &= \frac{1}{2\pi} \int_0^{2\pi} \left(\frac{\partial F}{\partial u} \cos \theta + \frac{F}{\partial v} \sin \theta\right) d\theta \\ &= \frac{1}{2\pi} \int_\Gamma \left(\frac{\partial F}{\partial u}\, dv - \frac{\partial F}{\partial v}\, du\right). \end{aligned}$$

Let us apply Riemann's formula:

$$0 = \frac{1}{2\pi} \iint_{\mathscr{D}} \Delta F\, du\, dv.$$

But, by virtue of the relations established above between functions and their derivatives, we have in general,

$$\Delta F(x, y) = \frac{1}{\pi a^2} \iint_{\mathscr{D}} \Delta F\, du\, dv.$$

Thus $\Delta F = 0$.

12. Spherical mean values and the Laplacian

In part this exercise supplements the preceding one. It gives a local definition of the Laplacian of a sufficiently regular function, and we show that if a regular function is such that in a neighborhood of a point its mean on a sphere of radius R (in the three-dimensional case) is independent of R, then it is locally harmonic.

Problem. Suppose that $U(x, y, z)$ is a continuous function and that it possesses continuous second derivatives at a point $M_0(x_0, y_0, z_0)$. We denote by $\mathfrak{M}(R)$ the mean value of $U(x, y, z)$ on the sphere S of radius R with center at M_0, so that

$$\mathfrak{M}(R) = \frac{1}{4\pi R^2} \iint_S U(x, y, z)\, dS.$$

Show that $\mathfrak{M}(R)$ tends to $U(x_0, y_0, z_0)$ as $R \to 0$ and find the principle part of the infinitesimal

$$\mathfrak{M}(R) - U(x_0, y_0, z_0).$$

Solutions. **A.** Suppose that $U(x, y, z)$ is a continuous function possessing continuous second derivatives at the point $M_0(x_0, y_0, z_0)$. Suppose that

$$\mathfrak{M}(R) = \frac{1}{4\pi R^2} \iint_S U(x, y, z)\, dS$$

is the mean value of U on the sphere of radius R with center at M_0.

As M tends to M_0, this mean value $\mathfrak{M}(R)$ tends to the value U_0 of U at M_0. Specifically,

$$\mathfrak{M}(R) - U_0 = \frac{1}{4\pi R^2} \iint_S [U(x, y, z) - U_0]\, dS$$

and

$$| \mathfrak{M}(R) - U_0 | \leqslant \frac{1}{4\pi R^2} \iint_S | U(x, y, z) - U_0 |\, dS.$$

Since $U(x, y, z)$ is continuous, there exists a number η such that, for $R < \eta$, we have

$$| U(x, y, z) - U_0 | < \epsilon,$$

so that

$$| \mathfrak{M}(R) - U_0 | < \frac{\epsilon}{4\pi R^2} \iint_S dS = \epsilon.$$

B. Since $U(x, y, z)$ has continuous second derivatives at M_0, we can apply Taylor's formula to it:

$$U(x, y, z) = U(x_0, y_0, z_0) + (x - x_0)\frac{\partial U}{\partial x_0} + (y - y_0)\frac{\partial U}{\partial y_0} + (z - z_0)\frac{\partial U}{\partial z_0}$$
$$+ \frac{1}{2}\left[(x - x_0)^2 \frac{\partial^2 U}{\partial x'^2} + \cdots + 2(y - y_0)(z - z_0)\frac{\partial^2 U}{\partial y' \partial z'} + \cdots\right],$$

where the bracketed expression is a quadratic form in $x - x_0$, $y - y_0$, and $z - z_0$, and where the second derivatives are taken at a point on the segment M_0M between M_0 and M.

Let us integrate this equation over S:

$$\mathfrak{M}(R) = U_0 + \frac{1}{4\pi R^2}\frac{\partial U}{\partial x_0}\iint_S (x - x_0)\,dS + \cdots$$
$$+ \frac{1}{2}\frac{1}{4\pi R^2}\iint_S (x - x_0)^2 \frac{\partial^2 U}{\partial x'^2}\,dS + \cdots$$
$$+ \frac{1}{4\pi R^2}\iint_S (y - y_0)(z - z_0)\frac{\partial^2 U}{\partial y'\partial z'}\,dS + \cdots.$$

The integral

$$\frac{1}{4\pi R^2}\iint_S (x - x_0)\,dS$$

vanishes, since it represents the abscissa of the center of gravity of the homogeneous sphere S relative to its geometric center. The same holds for the two analogous integrals.

The second integral may be written

$$\iint_S (x - x_0)^2 \frac{\partial^2 U}{\partial x'^2}\,dS = \iint_S (x - x_0)^2\left(\frac{\partial^2 U}{\partial x'^2} - \frac{\partial^2 U}{\partial {x_0}^2}\right)dS$$
$$+ \frac{\partial^2 U}{\partial {x_0}^2}\iint_S (x - {x_0}^2)\,dS.$$

The last term is equal to the moment of inertia of the sphere divided by its diameter. It is equal to one third of the integral

$$\iint_S [(x - x_0)^2 + (y - y_0)^2 + (z - z_0)^2]\,dS = R^2 \iint_S dS = 4\pi R^4.$$

The first integral tends to 0 as $R \to 0$. This is true because $\partial^2 U/\partial x'^2$ is, by hypothesis, continuous and it tends to $\partial^2 U/\partial x_0$. Since $(x - x_0)^2$ is bounded (less than R^2), we may follow a reasoning analogous to that for problem A. We may also remark that the integral is a proper integral depending on a parameter. The limit of the integral is the integral of the limit.

Therefore,

$$\lim_{R\to 0} \iint_S (x - x_0)^2 \frac{\partial^2 U}{\partial x'^2}\, dS = \tfrac{4}{3}\pi R^4.$$

The situation is the same for the two analogous integrals. Integrals of the type

$$\iint_S (y - y_0)(z - z_0) \frac{\partial^2 U}{\partial y'\, \partial z'}\, dS$$

tend to

$$\frac{\partial^2 U}{\partial y_0\, \partial z_0} \iint_S (y - y_0)(z - z_0)\, dS.$$

These limits vanish since they represent the moments of inertia of the sphere with respect to its center. Finally,

$$\mathfrak{M}(R) = U_0 + \frac{R^2}{6} \Delta U_0 + R^2 \varphi(R),$$

where $\varphi(R)$ tends to 0 as $R \to 0$. In other words,

$$\Delta U_0 = \lim_{R\to 0} \frac{6}{R^2} [\mathfrak{M}(R) - U_0].$$

A consequence. If, in a neighborhood of the point M_0, $\mathfrak{M}(R)$ is independent of R, then $\Delta U_0 = 0$. We again find a familiar property of functions that are harmonic at a point. In this case, $\mathfrak{M}(R) = U_0$ and $U(x, y, z)$ verifies the equation

$$U(x_0, y_0, z_0) = \frac{1}{4\pi R^2} \iint_S U(x, y, z)\, dS.$$

If this equation is verified not only at M_0 but throughout a neighborhood of M_0, the existence of continuous successive derivatives of U is proved and has not been taken as a hypothesis (see Exercise 10).

13. Improper double integrals. Fourier integrals. The gamma function

The object of this exercise is to evaluate a Fourier integral not satisfying the usual conditions for the validity of Fourier's reciprocity formulas. The evaluation uses indirectly a divergent improper double integral that results by passage to the limit from a convergent, and hence usable, double integral.

Problems. Consider the double integral

$$I = \iint_{\mathscr{D}} \cos tx\, e^{-\lambda t} s^{\alpha-1} e^{-ts}\, dt\, ds$$

over the region $t > 0$, $s > 0$. Here, λ is a positive parameter, x any real number, and α a number between 0 and 1.

A. Show that I is convergent. Find an expression for I by changing it to an iterated integral and evaluating the inner integral (with respect to t). Then reverse the order of integration and evaluate the then inner integral (with respect to s), and show that

$$\Gamma(\alpha) \int_0^\infty \frac{\cos tx\, e^{-\lambda t}}{t^\alpha}\, dt = \int_0^\infty \frac{s^{\alpha-1}(s+\lambda)}{(s+\lambda)^2 + x^2}\, ds. \tag{1}$$

B. Show that (1) is valid even for $\lambda = 0$, although the procedure just followed for proving it is not applicable. Then derive the value of the integral

$$\int_0^\infty \frac{\cos tx}{t^\alpha}\, dt.$$

C. May we apply Fourier's reciprocity formula to this integral?

Solutions. **A.** It is well known that, for a general (improper) integral to converge, it suffices that it converge absolutely. For a simple integral, this condition is not necessary. By using the fact that there is an order relation from the right, one can make the simple integral "semi-convergent" but not absolutely convergent. If, on the other hand, one considers a multiple integral over an unbounded region, for the integral to exist,† it is necessary that the integral of the absolute value of the function exist. For the double integral

$$I = \iint_{\mathscr{D}} \cos tx\, e^{-\lambda t} s^{\alpha-1} e^{-ts}\, dt\, ds$$

to be convergent, it is necessary and sufficient that it converge absolutely, which it will do if the integral

$$J = \iint_{\mathscr{D}} e^{-\lambda t} s^{\alpha-1} e^{-ts}\, dt\, ds$$

† The word "exist" is used here for the Lebesgue integral. For the special case of the Riemann integral, one may replace it with "converge."

converges. Now,

$$J = \int_0^\infty e^{-\lambda t} \int_0^\infty s^{\alpha-1} e^{-ts}\, ds\, dt.$$

But

$$\int_0^\infty s^{\alpha-1} e^{-ts}\, ds = \frac{1}{t^\alpha} \int_0^\infty u^{\alpha-1} e^{-u}\, du = \frac{\Gamma(\alpha)}{t^\alpha}.$$

Therefore,

$$J = \Gamma(\alpha) \int_0^\infty \frac{e^{-\lambda t}}{t^\alpha}\, dt.$$

Since λ is positive, J converges in a neighborhood of infinity. Since $0 < \alpha < 1$, the integral J converges in the neighborhood of the origin. Therefore, J converges and hence so does I. Therefore, we may write

$$I = \int_0^\infty \cos tx\, e^{-\lambda t}\, dt \int_0^\infty s^{\alpha-1} e^{-ts}\, ds = \Gamma(\alpha) \int_0^\infty \frac{\cos tx\, e^{-\lambda t}}{t^\alpha}\, dt.$$

But we can also get an expression for I by integrating first with respect to t:

$$I = \int_0^\infty s^{\alpha-1} \int_0^\infty \cos tx\, e^{-\lambda t} e^{-ts}\, dt\, ds.$$

The inner integral is the real part of

$$\int_0^\infty e^{itx} e^{-\lambda t} e^{-ts}\, dt = \frac{1}{s + \lambda - ix} = \frac{s + \lambda + ix}{(s + \lambda)^2 + x^2}.$$

Therefore,

$$I = \int_0^\infty \frac{s^{\alpha-1}(s + \lambda)}{(s + \lambda)^2 + x^2}\, ds.$$

Let us equate the two expressions that we have found for I:

$$\Gamma(\alpha) \int_0^\infty \frac{\cos tx\, e^{-\lambda t}}{t^\alpha}\, dt = \int_0^\infty \frac{s^{\alpha-1}(s + \lambda)}{(s + \lambda)^2 + x^2}\, ds. \tag{1}$$

B. This *result* holds for $\lambda > 0$. Let us show that it holds also for $\lambda = 0$ even though, for $\lambda = 0$, the original double integral diverges and the *method* employed is not valid for this case.

If the result is true, we must have

$$\Gamma(\alpha) \int_0^\infty \frac{\cos tx}{t^\alpha}\, dt = \int_0^\infty \frac{s^\alpha}{s^2 + x^2}\, ds. \tag{2}$$

The second integral can be evaluated by making the change of variable $s^2 = x^2u$. If $x > 0$, this integral becomes

$$\frac{x^{\alpha-1}}{2}\int_0^\infty \frac{u^{(\alpha-1)/2}}{1+u}\,du = \frac{x^{\alpha-1}}{2}\,\Gamma\left(\frac{1+\alpha}{2}\right)\Gamma\left(\frac{1-\alpha}{2}\right).$$

From the trigometrical formula for the sine of a sum, we have the identity

$$\frac{x^{\alpha-1}}{2}\,\frac{\pi}{\sin\,[(\alpha+1)\pi/2]} = \frac{x^{\alpha-1}}{2}\,\frac{\pi}{\cos\,(\alpha\pi/2)}.$$

Therefore, for $x > 0$,

$$\int_0^\infty \frac{\cos tx}{t^\alpha}\,dt = \frac{\pi}{2}\,\frac{x^{\alpha-1}}{\Gamma(\alpha)\cos(\alpha\pi/2)}.$$

The expression on the left is an even function of x, so that, for all $x \neq 0$, we have

$$\int_0^\infty \frac{\cos tx}{t^\alpha}\,dt = \frac{\pi}{2}\,\frac{|\,x\,|^{\alpha-1}}{\Gamma(\alpha)\cos\,(\alpha\pi/2)}.$$

It remains for us to justify this result, that is, to show that we can let λ approach 0 in (1). For $\lambda = 0$, the two members of (1) are meaningful. Are they continuous at $\lambda = 0$?

For this, it will be sufficient to show that the two integrals converge uniformly for $0 \leqslant \lambda \leqslant \lambda_0$, where λ_0 is a given positive number. Now, for every fixed $x > 0$,

$$\frac{s^{\alpha-1}(s+\lambda)}{(s+\lambda)^2+x^2} \leqslant \frac{s^{\alpha-1}(s+\lambda_0)}{s^2+x^2},$$

and the function on the right is integrable from 0 to $+\infty$.

The integral

$$\int_0^\infty \frac{s^{\alpha-1}(s+\lambda)}{(s+\lambda)^2+x^2}\,ds$$

therefore converges uniformly. It tends to

$$\int_0^\infty \frac{s^\alpha}{s^2+x^2}\,ds$$

as $\lambda \to 0$.

The first integral converges absolutely for $\lambda > 0$ but not for $\lambda = 0$. Let us apply the second mean-value theorem to it. The ratio $e^{-\lambda t}/t^\alpha$ is a positive decreasing function. Therefore, for $x > 0$,

$$\int_p^q \frac{\cos xt\, e^{-\lambda t}}{t^\alpha}\, dt = \frac{e^{-\lambda p}}{p^\alpha} \int_p^\xi \cos xt\, dt \qquad (p < \xi < q)$$

$$= \frac{1}{x} \frac{e^{-\lambda p}}{p^\alpha} (\sin x\xi - \sin xp),$$

$$\left| \int_p^q \cos xt \frac{e^{-\lambda t}}{t^\alpha}\, dt \right| < \frac{2}{x} \frac{e^{-\lambda p}}{p^\alpha}.$$

Consequently, whether λ be positive or zero,

$$\left| \int_p^q \cos xt \frac{e^{-\lambda t}}{t^\alpha}\, dt \right| < \frac{2}{x} \frac{1}{p^\alpha} \qquad \text{and} \qquad \left| \int_p^\infty \cos xt \frac{e^{-\lambda t}}{t^\alpha}\, dt \right| < \frac{2}{x} \frac{1}{p^\alpha}.$$

This inequality brings out the fact that the "remainder" $\int_p^\infty$ of the integral tends to 0 as $p \to \infty$. But this remainder is majorized by $2x^{-1}p^{-\alpha}$, which is independent of λ. Therefore, the convergence is uniform. The integral is a continuous function of λ. Its value for $\lambda = 0$ is the limit of its value as $\lambda \to 0$. Thus we justify the passage to the limit and the final formula.

Remark. Similarly, we can show that, if $x > 0$,

$$\int_0^\infty \frac{\sin tx}{t^\alpha}\, dt = \frac{\pi}{2} \frac{x^{\alpha-1}}{\Gamma(\alpha) \sin(\alpha\pi/2)}. \tag{3}$$

The first member is an odd function of x.

C. Under certain hypotheses, the formulas on Fourier integrals may be written as follows:

$$\varphi(x) = \int_0^\infty f(t) \cos xt\, dt, \qquad f(x) = \frac{2}{\pi} \int_0^\infty \varphi(t) \cos xt\, dt.$$

Here

$$f(t) = \frac{1}{t^\alpha}, \qquad \varphi(x) = \frac{\pi}{2} \frac{1}{x^\alpha \Gamma(\alpha) \cos(\alpha\pi/2)}.$$

Therefore,

$$\frac{1}{x^\alpha} = \frac{2}{\pi} \frac{\pi}{2} \frac{1}{\Gamma(\alpha) \cos(\alpha\pi/2)} \int_0^\infty \frac{\cos xt}{t^{1-\alpha}}\, dt.$$

Now, $1 - \alpha$, like α, lies between 0 and 1. Therefore,

$$\int_0^\infty \frac{\cos xt}{t^{1-\alpha}}\,dt = \frac{\pi}{2}\,\frac{1}{\Gamma(1-\alpha)\cos[(1-\alpha)\pi/2]\,x^\alpha}\,.$$

The formula is indeed verified if

$$\frac{\pi}{2}\,\frac{1}{\Gamma(\alpha)\cos(\alpha\pi/2)}\,\frac{1}{\Gamma(1-\alpha)\cos[(1-\alpha)\pi/2]} = 1$$

or

$$\Gamma(\alpha)\Gamma(1-\alpha) = \frac{\pi}{2\cos(\alpha\pi/2)\cos[(1-\alpha)\pi/2]}\,.$$

Hence, by the trigonometrical formula for the cosine of a difference,

$$2\cos\frac{\alpha\pi}{2}\cos\frac{(1-\alpha)\pi}{2} = 2\cos\frac{\alpha\pi}{2}\sin\frac{\alpha\pi}{2} = \sin\alpha\pi,$$

we have our conclusion.

However, this conclusion, the correctness of which has just been proved, cannot be considered a consequence of Fourier's formula unless these formulas are applicable.

The function $f(t) = 1/t^\alpha$ is a locally integrable function of bounded variation and even has a continuous derivative. However, it is neither summable from 0 to $+\infty$ nor square-summable. Therefore, the usual theories do not justify the preceding result, although this result is true.

It can be shown that Fourier's formulas are valid if the function $f(t)$ is nonincreasing and tends to 0 as $t \to +\infty$. This is the case with $1/t^\alpha$.

14. Evaluation of Fresnel's integrals. Cesaro-mean integration

We know that Cayley's integral

$$A = \iint_{\mathscr{D}} \sin(x^2 + y^2)\,dx\,dy$$

over the first quadrant $\mathscr{D}$ is divergent. This follows from the fact that the integral

$$\iint_{\mathscr{D}} |\sin(x^2 + y^2)|\,dx\,dy$$

is (properly) divergent. This can also be seen if we evaluate A by passing to the limit. We might try to write in Cartesian coordinates

$$A = \lim_{t\to\infty} \int_0^t dx \int_0^t \sin(x^2 + y^2)\, dy = 2 \lim_{t\to\infty} \int_0^t \sin x^2\, dx \times \lim_{t\to\infty} \int_0^t \cos y^2\, dy.$$

These two integrals have well-defined limits J_1 and J_2, which would encourage us to believe that

$$A = 2J_1 J_2 .$$

But, in polar coordinates, we would have

$$A = \lim_{t\to\infty} \int_0^{\pi/2} d\theta \int_0^t \sin r^2\, r\, dr = \lim_{t\to\infty} \frac{\pi}{4}(1 - \cos t^2).$$

Now, the right-hand member oscillates about $\pi/4$ but does not tend to any limit. Thus the symbol A has no meaning.

However, we might wonder if the "mean value" of the oscillations, namely $\pi/4$, is not equal to the product $2J_1J_2$. The expressions J_1 and J_2 are two of Fresnel's integrals. If we assume their values known, we can verify that this supposition is true.

The purpose of this exercise is to show by direct calculation that $2J_1J_2$ is equal to the mean value of the oscillations of $\pi(1 - \cos t^2)/4$, so that we may then derive the values of J_1 and J_2. The procedure used is an adaptation to integral calculus of the Cesaro-mean summation method that we have already expounded for the case of sequences in connection with Exercise 7 of Chapter 1.

Problems. Consider an arc C lying in the first quadrant of the xy-plane whose equation in polar coordinates is

$$r = f(\theta), \qquad 0 \leqslant \theta \leqslant \frac{\pi}{2}.$$

Suppose that $f(\theta)$ is a continuous function between two positive values of θ. We denote by C_t the homothetic curve of C with respect to the origin with a positive ratio t and we denote by $\mathscr{D}_t$ the region defined by

$$0 \leqslant r \leqslant tf(\theta), \qquad 0 \leqslant \theta \leqslant \frac{\pi}{2}.$$

A. Consider the iterated integral

$$I_T = \frac{1}{T} \int_0^T dt \iint_{\mathscr{D}_t} \sin(x^2 + y^2)\, dx\, dy.$$

Show that I_T tends to $\pi/4$ as $T \to \infty$.

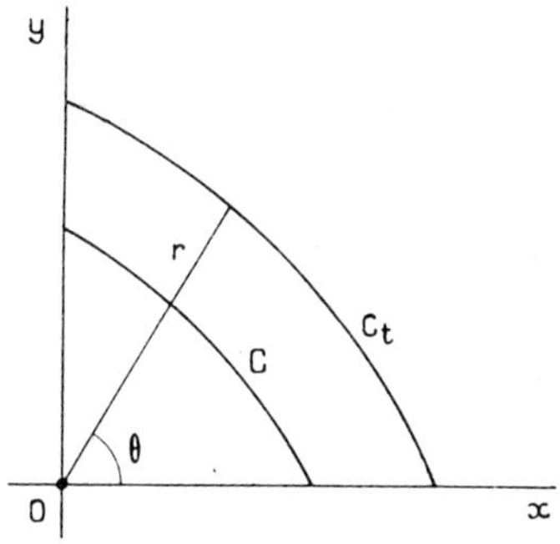

FIGURE 19

B. Suppose that $\mathscr{D}_t$ is the square defined by $x \leqslant t$, $y \leqslant t$. Show that the result of Problem A enables us to calculate the product of the two Fresnel integrals

$$J_1 = \int_0^\infty \sin x^2 \, dx, \qquad J_2 = \int_0^\infty \cos x^2 \, dx.$$

C. Redo the first two problems replacing I_T with

$$I_T' = \frac{1}{T} \int_0^T dt \iint_{\mathscr{D}_t} \cos(x^2 + y^2) \, dx \, dy.$$

From all these results, derive the values of J_1 and J_2. [You will find that $J_1 = J_2 = \pm\frac{1}{2}(\pi/2)^{1/2}$. Then show that $J_1 > 0$.]

Solutions. **A.** Consider a curve C in the xy-plane defined for $0 \leqslant \theta \leqslant \pi/2$ by the polar equation

$$r = f(\theta),$$

where $f(\theta)$ is a continuous strictly positive function (that is, it is, for all θ, greater than some fixed positive number).

Let C_t denote the homothetic arc of C with respect to the origin with positive ratio t. Let $\mathscr{D}_t$ denote the region bounded by C_t and the two coordinate axes

$$0 \leqslant \theta \leqslant \frac{\pi}{2}, \qquad 0 \leqslant r \leqslant tf(\theta).$$

Now consider the integral

$$I_T = \frac{1}{T} \int_0 dt \iint_{\mathscr{D}_t} \sin(x^2 + y^2) \, dx \, dy. \tag{1}$$

The double integral in this expression in polar coordinates becomes

$$\iint \sin r^2 r\, dr\, d\theta \qquad \text{for} \qquad 0 \leqslant \theta \leqslant \frac{\pi}{2}, \qquad 0 \leqslant r \leqslant tf(\theta)$$

or

$$\int_0^{\pi/2} d\theta \int_0^{tf(\theta)} \sin r^2 r\, dr = \frac{1}{2}\int_0^{\pi/2} (-\cos r^2)_0^{tf(\theta)}\, d\theta = \frac{1}{2}\int_0^{\pi/2} [1 - \cos t^2 f^2(\theta)]\, d\theta$$

$$= \frac{\pi}{4} - \frac{1}{2}\int_0^{\pi/2} \cos\, [t^2 f^2(\theta)]\, d\theta.$$

If we now integrate from 0 to T and then divide by T, we get

$$I_T = \frac{\pi}{4} - \frac{1}{2T}\int_0^T dt \int_0^{\pi/2} \cos\, [t^2 f^2(\theta)]\, d\theta.$$

The iterated integral contains only continuous functions. Therefore, we may reverse the order of the integrations, so that

$$I_T = \frac{\pi}{4} - \frac{1}{2T}\int_0^{\pi/2} d\theta \int_0^T \cos\, [t^2 f^2(\theta)]\, dt.$$

In the integral with respect to t, we set $tf(\theta) = s$. This gives us

$$\frac{1}{f(\theta)}\int_0^{Tf(\theta)} \cos s^2\, ds,$$

so that

$$\begin{aligned} I_T &= \frac{\pi}{4} - \frac{1}{2T}\int_0^{\pi/2} \frac{d\theta}{f(\theta)} \int_0^{Tf(\theta)} \cos s^2\, ds \\ &= \frac{\pi}{4} - \int_0^{\pi/2} \frac{d\theta}{f(\theta)}\, \frac{1}{2T} \int_0^{Tf(\theta)} \cos s^2\, ds. \end{aligned} \tag{2}$$

All these iterated integrals represent proper double integrals, since $1/f(\theta)$ is continuous over the region of integration.

Suppose now that T increases without bound. For every fixed value of θ, the integral

$$\int_0^{Tf(\theta)} \cos s^2\, ds$$

tends to a limit, which is the integral $\int_0^\infty \cos s^2\, ds$. We know that this integral converges since, when we set $s^2 = u$, it becomes

$$\int_0^\infty \frac{\cos u}{\sqrt{u}}\, du.$$

and $1/\sqrt{u}$ tends to 0, decreasing, as $u \to \infty$. Consequently, $(1/2T) \int_0^{Tf(\theta)} \cos s^2\, ds$ tends to 0. Does it follow that

$$\int_0^{\pi/2} \frac{d\theta}{f(\theta)} \left(\frac{1}{2T} \int_0^{Tf(\theta)} \cos s^2\, ds \right)$$

tends to 0?

The integral

$$g(z) = \int_0^z \cos s^2\, ds$$

is a continuous function of z and tends to a limit when $z \to \infty$. Therefore, the function $g[Tf(\theta)]$ is a continuous function of T and θ for $0 \leqslant \theta \leqslant \pi/2$ and $T \geqslant T_0$, where T_0 is an arbitrary fixed positive number. The same holds for $g[Tf(\theta)]/2Tf(\theta)$. It follows that the limit of the integral as $T \to \infty$ is equal to the integral of the limit, which is 0.

Although this is not necessary, we may prove the above result by a special reasoning. Since

$$\lim_{U \to \infty} \frac{1}{2U} \int_0^U \cos s^2\, ds = 0,$$

we may assign to every $\epsilon > 0$ a number $U_0(\epsilon)$ such that

$$\frac{1}{2U} \left| \int_0^U \cos s^2\, ds \right| < \epsilon \tag{3}$$

whenever $U > U_0(\epsilon)$.

Let us set $U = Tf(\theta)$. We need to choose $Tf(\theta) > U_0(\epsilon)$ or

$$T > \frac{U_0(\epsilon)}{f(\theta)}.$$

It is possible to satisfy this inequality with a value of T that does not depend on θ. Since $1/f$ is bounded, there exists a number A such that $1/f < A$. If we choose $T > AU_0(\epsilon)$, the condition

$$T > AU_0(\epsilon) \left(> \frac{U_0(\epsilon)}{f(\theta)} \right)$$

implies inequality (3). Then

$$\left| \int_0^{\pi/2} \frac{d\theta}{f(\theta)} \frac{1}{2T} \int_0^{Tf(\theta)} \cos s^2\, ds \right| < \epsilon \int_0^{\pi/2} d\theta = \frac{\pi}{2} \epsilon.$$

It follows from this that the left-hand member tends to 0 as $T \to \infty$. Thus

$$\lim_{T \to \infty} I_T = \frac{\pi}{4}.$$

B. If C consists of the two line segments $x = 1$ and $y = 1$ bounding the square $(0 \leqslant x \leqslant 1, 0 \leqslant y \leqslant 1)$, then $\mathscr{D}_t$ is the square $(0 \leqslant x \leqslant t, 0 \leqslant y \leqslant t)$.

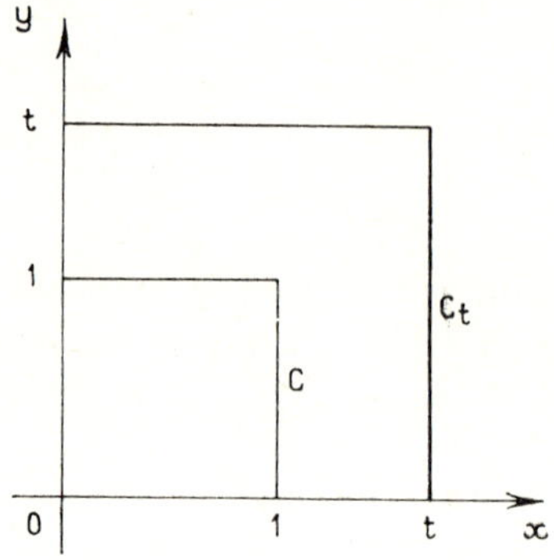

FIGURE 20

The function

$$\varphi(t) = \iint_{\mathscr{D}_t} \sin(x^2 + y^2)\, dx\, dy$$

is equal to

$$2 \int_0^t \sin x^2\, dx \int_0^t \cos x^2\, dx.$$

As $t \to \infty$, the integral $\int_0^t \sin x^2\, dx$ tends to a limit J_1, the integral $\int_0^t \cos x^2\, dx$ tends to a limit J_2, and, in the limit,

$$\lim_{t\to\infty} \iint_{\mathscr{D}_t} \sin(x^2 + y^2)\, dx\, dy = 2J_1 J_2 .$$

It follows immediately that the mean

$$I_T = \frac{1}{T} \int_0^T \varphi(t)\, dt$$

tends to $2J_1J_2$. It follows that

$$2J_1J_2 = \frac{\pi}{4}.$$

C. If we replace $\sin(x^2 + y^2)$ with $\cos(x^2 + y^2)$ in Problem A, we are led to a consideration of the integral

$$\frac{1}{2T} \int_0^{Tf(\theta)} \sin s^2\, ds,$$

which tends to 0 as $T \to \infty$.

In Problem B, we find first that

$$\lim_{t\to\infty} \iint_{\mathscr{D}_t} \cos(x^2 + y^2)\, dx\, dy = J_1^2 - J_2^2.$$

From this we conclude that

$$|J_1| = |J_2|.$$

Since $J_1 J_2$ is positive, we know that $J_1 = J_2$. Therefore,

$$J_1 = J_2 = \pm \frac{1}{2}\left(\frac{\pi}{2}\right)^{1/2}.$$

It remains for us to determine the sign. Now,

$$2J_1 = \int_0^\infty \frac{\sin s}{\sqrt{s}}\, ds.$$

Since $\sin s$ vanishes for $s = k\pi$ when k is an integer, we have

$$2J_1 = \int_0^\pi \frac{\sin s}{\sqrt{s}}\, ds + \cdots + \int_{k\pi}^{(k+1\pi)} \frac{\sin s}{\sqrt{s}}\, ds + \cdots .$$

Thus $2J_1$ is the limit of an *alternating* convergent series. According to the classical theorem on alternating series, this series converges. The

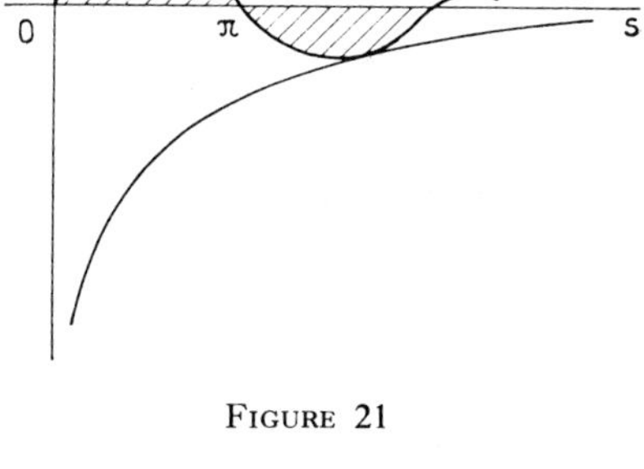

FIGURE 21

absolute value of the general term tends to 0 (as we already know) and it decreases. Specifically, it is equal to

$$\int_0^\pi \frac{\sin u}{(k\pi + u)^{1/2}}\, du,$$

which is the integral between constant limits of a function that decreases as k increases.

Therefore, the limit of the series has the same sign as does the first term, which is positive. Thus we know that J_1 is positive and hence

$$J_1 = J_2 = \frac{1}{2}\left(\frac{\pi}{2}\right)^{1/2}.$$

Thus we again find the values of Fresnel's integrals.

*15. Crofton's formula

In this exercise we define a measure on the set of straight lines $ux + vy + 1 = 0$ in a plane $\mathscr{P}$, that is, in the dual space $\mathscr{P}^*$ of $\mathscr{P}$. This measure is an additive positive set function assigned to all the quadrable subsets Ω of $\mathscr{P}^*$. It is represented by a double integral

$$\iint_\Omega F(u, v)\, du\, dv.$$

It has the property of being invariant under a change of coordinates.

We are particularly interested in the measure of the set of lines that intersect a convex closed curve C. Study of this measure leads to a general formula expressing the difference $L^2/2 - \pi s$, where L is the length of the curve C and S is the area of its interior, in the form of a remarkable integral.

We note that $L^2/2 - \pi s$ is positive. If L is given, we know that the maximum value of S is obtained when C is a circle. It is equal to $L^2/4\pi$, so that $L^2/2 - \pi s$ cannot be less than $L^2/4$.

Problems. We denote by D the straight line whose equation is

$$ux + vy + 1 = 0.$$

To this line D we assign the point whose coordinates are (u, v) in an auxiliary plane.

We shall study the integral

$$I = \iint\limits_{\Omega} \frac{du\, dv}{(u^2 + v^2)^{3/2}}$$

over a certain region Ω in the uv-plane.

A. We write the equation of the straight line D in the form

$$x \cos \alpha + y \sin \alpha = p \qquad (p \geqslant 0, \quad 0 \leqslant \alpha \leqslant 2\pi).$$

Show that

$$I = \iint\limits_{\Omega'} dp\, d\alpha,$$

where Ω' is the image of Ω in the $p\alpha$-plane. Show that I is invariant under a change of rectangular coordinates in the plane of the line D.

B. Consider the set of lines D that intersect a straight line segment of length a. Define the region Ω' and show that $I = 2a$.

C. Consider the set of lines D that intersect a closed convex curve C of length L enclosing an area S. Define the domain Ω' and show that $I = L$.

D. Under the hypotheses of Problem C, calculate the square of I in the form of the quadruple integral

$$J = I^2 = \iiiint dp\, dp_1\, d\alpha\, d\beta.$$

Transform J by making the change of variable

$$p = x \cos \alpha + y \sin \alpha, \qquad p_1 = x \cos \beta + y \sin \beta \qquad (\alpha = \alpha, \quad \beta = \beta),$$

which replaces the variables p, p_1, α, and β with x, y, α, and β. Define and interpret the new region of integration. Then show that

$$\iint\limits_{R} (\omega - \sin \omega)\, dx\, dy = \frac{L^2}{2} - \pi S,$$

where ω is the angle between the tangents drawn to C through an exterior point M whose coordinates are (x, y) and where R is that portion of the plane exterior to C.

E. Verify the preceding formula when C is a circle.

Solutions. **A.** We denote by $ux + vy + 1 = 0$ the equation of a straight line D in the plane $\mathscr{P}$. In the dual space $\mathscr{P}^*$ of $\mathscr{P}$, the image of this straight line is the point whose coordinates are (u, v). The line D does not pass through the origin. However, if the point (u, v) is moved in a certain direction infinitely far from the origin in the plane $\mathscr{P}^*$, the straight line D tends to a limiting position that does pass through the origin of the plane $\mathscr{P}$.

Consider the integral

$$I = \iint_{\Omega} \frac{du\, dv}{(u^2 + v^2)^{3/2}}$$

over a region Ω in $\mathscr{P}^*$, that is, over a certain set of straight lines.

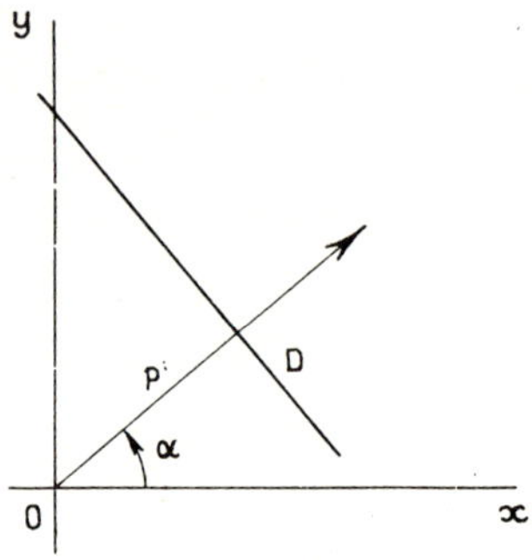

FIGURE 22

First, we shall transform I by representing D by the equation

$$x \cos \alpha + y \sin \alpha = p.$$

In this representation, two conventions are possible:

$$0 \leqslant \alpha \leqslant 2\pi \qquad (p \geqslant 0)$$

or

$$0 \leqslant \alpha < \pi \qquad (p \text{ arbitrary}).$$

In both cases, there is a one-to-one correspondence between $\mathscr{P}^*$ and a region in the $p\alpha$-plane. We shall choose the *first* of the above conventions (that is, $p \geqslant 0$). We have

$$u = -\frac{\cos \alpha}{p}, \qquad v = -\frac{\sin \alpha}{p},$$

so that

$$J = \frac{D(u, v)}{D(p, \alpha)} = \begin{vmatrix} \dfrac{\cos \alpha}{p^2} & \dfrac{\sin \alpha}{p^2} \\ \dfrac{\sin \alpha}{p} & -\dfrac{\cos \alpha}{p} \end{vmatrix} = -\frac{1}{p^3}, \qquad |J| = \frac{1}{p^3},$$

with

$$u^2 + v^2 = \frac{1}{p^2}, \qquad (u^2 + \nu^2)^{3/2} = \frac{1}{p^3}.$$

Consequently,

$$I = \iint_{\Omega'} dp\, d\alpha,$$

where Ω' is the image of Ω in the plane E of the variables p and α.

The integral I remains invariant under a change of rectangular coordinates in the plane $\mathscr{P}$. To show this, we set

$$x = x' \cos \theta - y' \sin \theta + x_0, \qquad y = x' \sin \theta + y' \cos \theta + y_0.$$

Then

$$x \cos \alpha + y \sin \alpha - p = x' \cos (\alpha - \theta) + y' \sin (\alpha - \theta) + x_0 \cos \alpha + y_0 \sin \alpha - p,$$

so that

$$\alpha' = \alpha - \theta, \qquad p' = p - x_0 \cos \alpha - y_0 \sin \alpha$$

and

$$\frac{D(p', \alpha')}{D(p, \alpha)} = 1.$$

Therefore,

$$\iint_{\Omega'} dp\, d\alpha = \iint_{\Omega_1'} dp'\, d\alpha',$$

where Ω_1' is the image of Ω' after the transformation.

B. Consider the set of straight lines that intersect a given segment of length a. Evaluate the integral I over this segment.

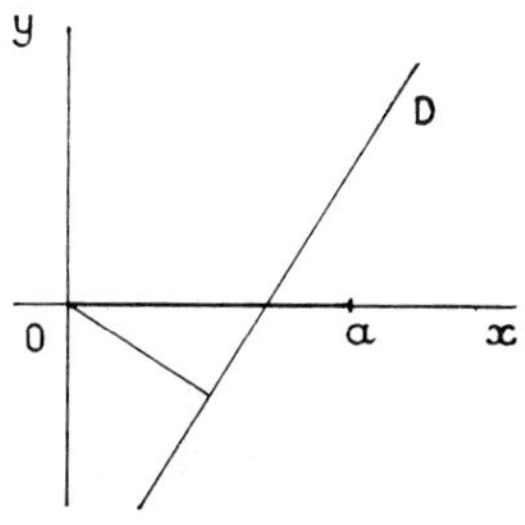

FIGURE 23

By virtue of the invariance property, we may place the axes around the segment as we find it convenient. Accordingly, we place one end of the segment at the origin and the other at the point $(a, 0)$.

A straight line D intersects the x-axis at the point

$$x = \frac{p}{\cos \alpha}.$$

The set Ω' is thus defined by

$$0 \leqslant \frac{p}{\cos \alpha} \leqslant a \qquad (p \geqslant 0, \quad 0 \leqslant \alpha \leqslant 2\pi)$$

or

$$0 \leqslant p \leqslant a \cos \alpha \qquad (\cos \alpha > 0).$$

Therefore,

$$I = \int_{\cos \alpha \geqslant 0} d\alpha \int_0^{a \cos \alpha} dp = 2 \int_0^{\pi/2} a \cos \alpha \, d\alpha = 2a.$$

Thus the value of I is *twice the length of the segment.*

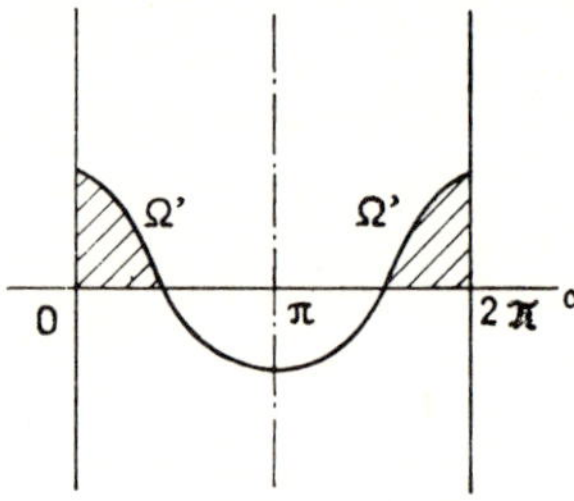

FIGURE 24

C. Let Z be a bounded region in the plane $\mathscr{P}$ and let C be the curve bounding it. Consider the set of straight lines D that intersect Z. We assume without proof that D intersects the boundary at at most two points, except possibly in the case when certain arcs of C are line segments, in which case D may have in common with Z (which we assume to be closed, that is, containing C) the set of points of such a segment.

We can always place the origin in the interior of Z. To describe the region Δ that is the image of the set of lines D, we need to fix α and let p vary from 0 to the value $p(\alpha)$ that is the least upper bound of the positive values of p for which D intersects Z and then let α vary from 0 to 2π. Then $p(\alpha)$ is a continuous function of α that is periodic with period 2π.

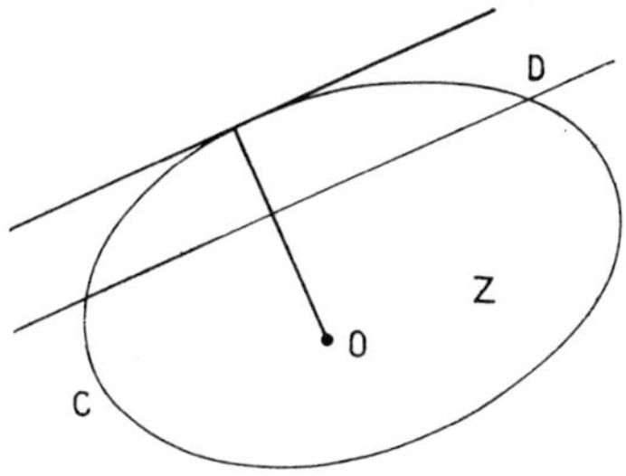

FIGURE 25

Let us suppose first that $p(\alpha)$ has continuous first and second derivatives. Then the curve C is the *envelope* of the family of straight lines D thus selected. The point of intersection of D with C is defined by the equations

$$x \cos \alpha + y \sin \alpha = p(\alpha), \qquad -x \sin \alpha + y \cos \alpha = p'(\alpha),$$

which may be rewritten

$$x = p \cos \alpha - p' \sin \alpha, \qquad y = p \sin \alpha + p' \cos \alpha.$$

From this we conclude that

$$x' = -(p + p'') \sin \alpha = (p + p'') \cos \left(\alpha + \frac{\pi}{2}\right),$$

$$y' = (p + p'') \cos \alpha = (p + p'') \sin \left(\alpha + \frac{\pi}{2}\right).$$

An element of arc of C is equal to

$$ds = | p + p'' | \, d\alpha.$$

But, since the curve C is convex, the measure of the projection of the speed (x', y') onto the tangent oriented in the sense of increasing α is nonnegative. This is true because, if the continuous function $p + p''$ changed sign, it would pass through the value 0 and then the curve C, possessing an inflection point, would not be convex. On the other hand, at the point at which the curve C crosses the positive half of the x-axis, we have $y' > 0$, $\cos \alpha > 0$, so that $p + p'' > 0$. Therefore, the function $p + p''$ cannot be negative. Thus

$$ds = (p + p'') \, d\alpha.$$

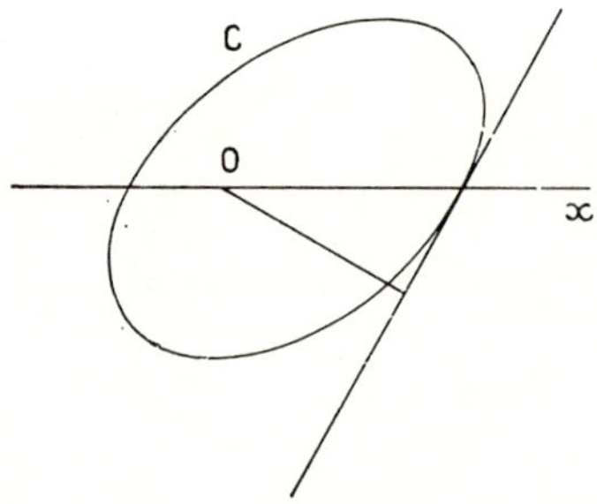

FIGURE 26

The length L of C is equal to

$$\int_0^{2\pi} (p + p'')\, d\alpha = \int_0^{2\pi} p\, d\alpha + (p')_0^{\pi}.$$

Since p' is continuous and periodic, the difference between its values at 0 and 2π is zero. Finally,

$$L = \int_0^{2\pi} p(\alpha)\, d\alpha.$$

Now,

$$I = \iint dp\, d\alpha = \int_0^{2\pi} d\alpha \int_0^{p(\alpha)} dp = \int_0^{2\pi} p(\alpha)\, d\alpha.$$

Thus *the integral I is equal to the length of the curve C.*

Suppose now that C has angular points or that it contains straight line segments but that there are only finitely many of each of these. At an angular point (x_0, y_0), the value of p varies in such a way that

$$x_0 \cos \alpha + y_0 \sin \alpha = p.$$

Therefore, between two values of α, p is a circular function verifying the relation

$$p + p'' = 0.$$

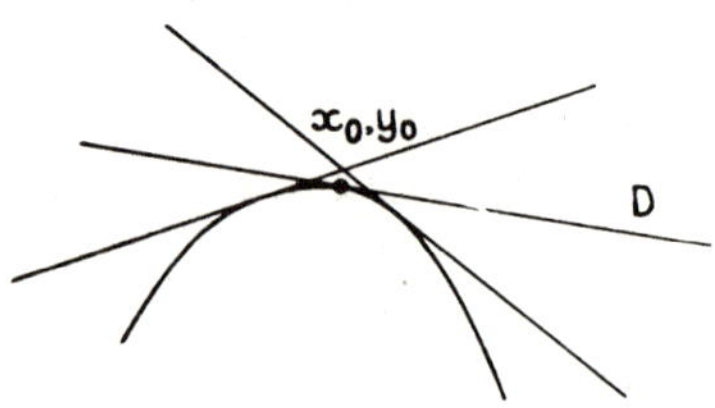

FIGURE 27

It follows from this that the angular point has no effect on the length L of C, although the representation of C as an envelope assigns to this point a finite arc in the $p\alpha$-plane. Thus the presence of angular points is not a singularity.

Now, if the curve C contains a segment AB, the continuous function $p(\alpha)$ remains constant on this segment, to which a definite value of α corresponds. It follows that the derivative $p'(\alpha)$ is discontinuous at α. It undergoes a saltus $\Delta p'$, which corresponds to a displacement of the point (x, y) on the straight line $x \cos \alpha + y \sin \alpha = p$. The components of the vector **AB**

$$\Delta x = -\sin \alpha \, \Delta p', \qquad \Delta y = \cos \alpha \Delta p'.$$

Its projection onto the straight line D, which is oriented in the sense of increasing α, is equal to $\Delta p'$. The latter quantity is necessarily positive and its value is the length of the segment AB.

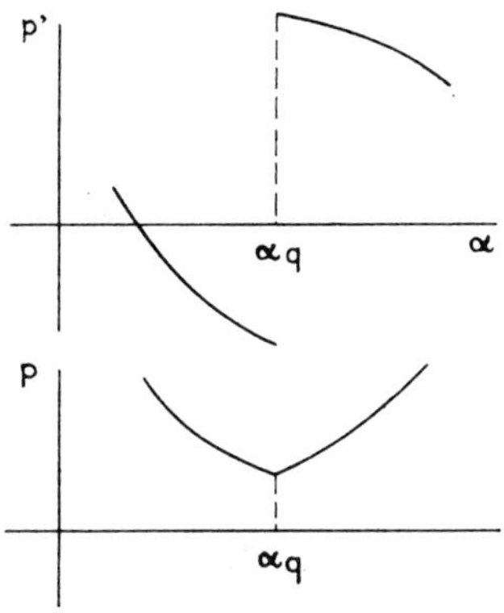

FIGURE 28

Let us suppose that the curve C contains n segments placed at the points whose abscissas in its representation are $\alpha_1, \ldots, \alpha_q, \ldots, \alpha_n$.

The length of C is equal to the sum of the arc lengths for which p'' is continuous, namely,

$$\sum \int_{\alpha_q}^{\alpha_{q+1}} (p + p'') \, d\alpha$$

and the lengths of the segments, namely,

$$\sum \Delta p'(\alpha_q).$$

Now,

$$\sum \int_{\alpha_q}^{\alpha_{q+1}} (p + p'')\, d\alpha = \int_0^{2\pi} p\, d\alpha + \sum [p'(\alpha_{q+1} - 0) - p'(\alpha_q + 0)].$$

By regrouping the terms in the summation Σ, we see that

$$-p'(\alpha_q + 0) + p'(\alpha_q - 0) = -\Delta p'(\alpha_q).$$

We finally have

$$L = \sum \int_{\alpha_q}^{\alpha_{q+1}} (p + p'')\, d\alpha + \sum \Delta p'(\alpha_q) = \int_0^{2\pi} p\, d\alpha - \sum \Delta p'(\alpha_q) + \sum \Delta p'(\alpha_q)$$

$$= \int_0^{2\pi} p\, d\alpha.$$

The formula

$$I = \iint dp\, d\alpha = \int_0^{2\pi} p\, d\alpha = L$$

thus remains valid if the boundary curve C of the region Z is the union of a finite number of segments and arcs on which p has continuous first and second derivatives.

Remark. The general case can be treated in the following manner: We inscribe in C a sequence of regular convex polygons Π_n of length L_n such that L_n tends to the length L of C as $n \to \infty$. For the polygon Π_n, the corresponding function $p_n(\alpha)$ consists of a succession of sinusoidal arcs separated by discontinuities of the first kind. We have just seen that

$$\int_0^{2\pi} p_n(\alpha)\, d\alpha = L_n .$$

If we assume that it has been proved that $p_n(\alpha)$ tends to $p(\alpha)$ for every α as $n \to \infty$, the formula $I = L$ follows. This is true because the functions $p_n(\alpha)$ are bounded by a constant and they tend to a limit for each α. According to Lebesgue's theorem, the integral $L = \int_0^{2\pi} p_n(\alpha)\, d\alpha$ must therefore tend to a limit, which is equal to $\int_0^{2\pi} p(\alpha)\, d\alpha$. Thus $\int_0^{2\pi} p(\alpha)\, d\alpha = L$.

However, this reasoning would require a precise study of the conditions under which the continuous function $p(\alpha)$ defines a rectifiable curve in its dual form.

D. We shall now evaluate the square of the integral $I = \iint dp\, d\alpha$ for the set of straight lines intersecting the convex region Z. Here, I^2 is a quadruple integral of the form

$$I^2 = \iiiint dp\, dp_1\, d\alpha\, d\alpha_1 .$$

We make the change of variables

$$p = x \cos \alpha + y \sin \alpha, \qquad p_1 = x \cos \beta + y \sin \beta,$$

but we keep α and β as they are. The Jacobian of this transformation is equal to

$$\frac{D(p, p_1, \alpha, \beta)}{D(x, y, \alpha, \beta)} = \begin{vmatrix} \cos \alpha & \sin \alpha \\ \cos \beta & \sin \beta \end{vmatrix} = \sin(\beta - \alpha).$$

Therefore,

$$I^2 = \iiiint |\sin(\beta - \alpha)| \, dx \, dy \, d\alpha \, d\beta.$$

The meaning of this change of variables is, in the plane $\mathscr{P}$, the following: Through the point $M(x, y)$, we pass two straight lines (p, α) and (p_1, β). Instead of defining these straight lines by p, p_1, α, and β, we define them by (x, y) and (α, β). Thus we consider a point $M(x, y)$ in the plane $\mathscr{P}$. With M fixed, we let (α, β) range over all the ordered pairs that pass through M and intersect the region Z. Then M describes the plane.

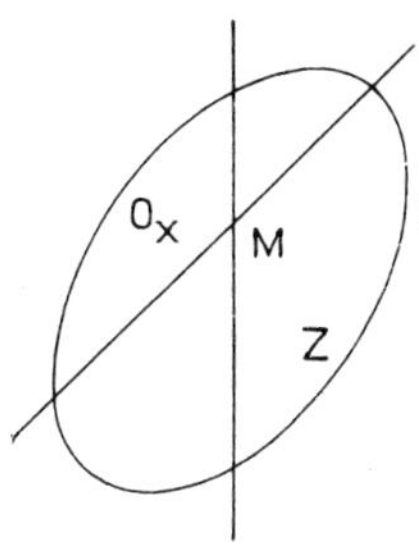

FIGURE 29

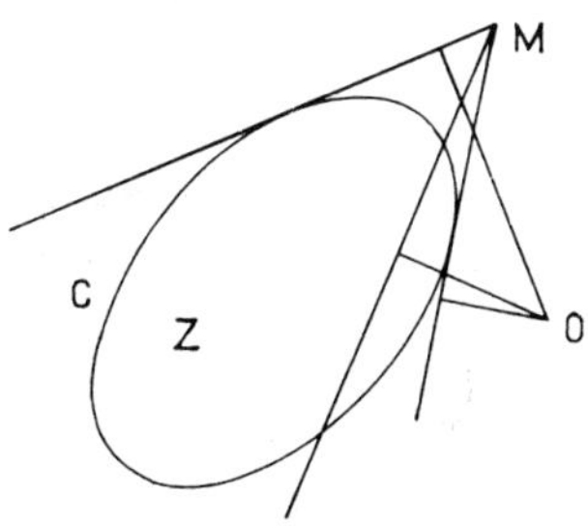

FIGURE 30

We distinguish two cases:

Case 1. *M is in the interior of Z.* To obtain all the straight lines D, we must let α vary from 0 to π. Thus, α and β vary independently from 0 to π. We have

$$\iint |\sin(\beta - \alpha)|\, d\alpha\, d\beta = \int_0^\pi d\beta \left[\int_0^\beta \sin(\beta - \alpha)\, d\alpha - \int_\beta^\pi \sin(\beta - \alpha)\, d\alpha\right]$$
$$= 2\int_0^\pi (1 - \cos\beta)\, d\beta$$
$$= 2\pi.$$

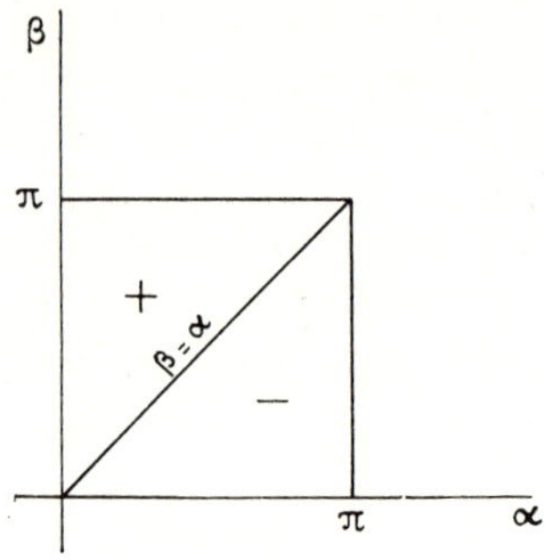

FIGURE 31

It remains to integrate with respect to x and y:

$$2\pi \iint dx\, dy = 2\pi S,$$

where S is the area enclosed by C.

Case 2. *M lies outside Z.* The line D now lies inside the acute angle between the tangents from M to the curve C. They may not be actual tangents but simply lines containing segments of C or passing through angular points.

It is possible to choose the origin outside C and the direction of the x-axis in such a way that α and β vary from 0 to ω, where ω is the acute angle between the two tangents. In this case, we have

$$\iint |\sin(\beta - \alpha)|\, d\alpha\, d\beta = \int_0^\omega d\beta \left[\int_0^\beta \sin(\beta - \alpha)\, d\alpha - \int_\beta^\omega \sin(\beta - \alpha)\, d\alpha\right]$$
$$= \int_0^\omega [2 - \cos\beta - \cos(\beta - \omega)]\, d\beta$$
$$= 2\omega - 2\sin\omega.$$

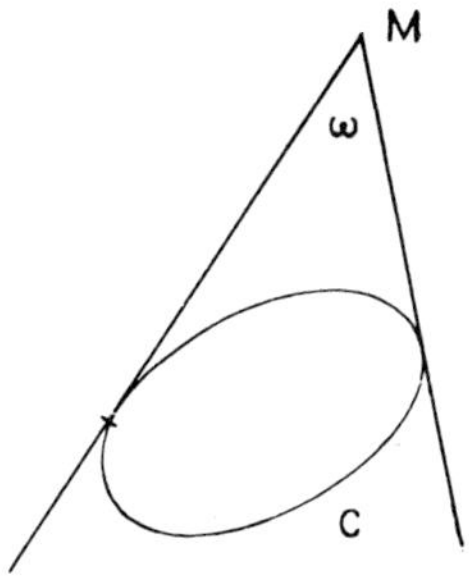

FIGURE 32

When we integrate over the region $\mathscr{R}$ of the portion of the plane exterior to Z, we obtain

$$2 \iint_{\mathscr{R}} (\omega - \sin \omega)\, dx\, dy.$$

This integral converges and has a positive value. Finally, we have

$$l^2 = 2\pi S + 2 \iint_{\mathscr{R}} (\omega - \sin \omega)\, dx\, dy$$

or

$$\iint_{\mathscr{R}} (\omega - \sin \omega)\, dx\, dy = \frac{L^2}{2} - \pi S.$$

We recall that $\mathscr{R}$ is the exterior of a convex bounded domain of area S whose boundary curve is of length L and that Z subtends the angle ω at a point $M(x, y)$ in the plane. Since $\Omega > \sin \Omega$, we note that

$$L^2 > 2\pi S.$$

This is a strict inequality.

E. *An example.* Suppose that C is the circle of radius R with center at the origin. We have

$$L = 2\pi R, \qquad S = \pi R^2, \qquad \frac{L^2}{2} - \pi S = \pi^2 R^2.$$

Let us evaluate directly

$$\iint_{Z} (\omega - \sin \omega)\, dx\, dy.$$

In polar coordinates (r, θ), we have

$$\sin\frac{\omega}{2} = \frac{R}{r}, \qquad r = \frac{R}{\sin(\omega/2)}.$$

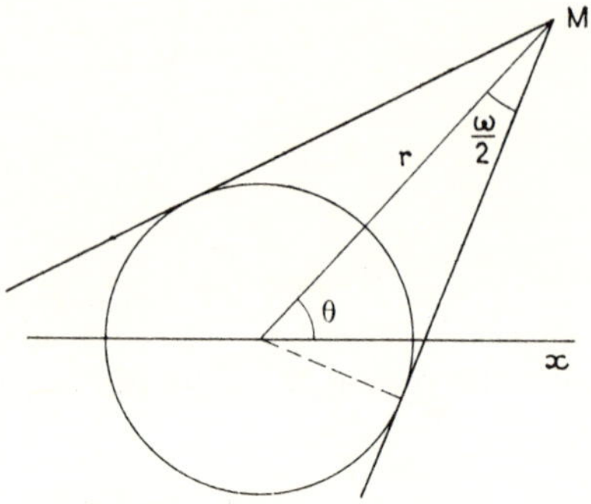

FIGURE 33

Letting r vary from R to ∞ is equivalent to letting ω vary from π to 0. In terms of the variables ω and θ,

$$dx\,dy = r\,dr\,d\theta = \frac{R^2}{2}\frac{\cos(\omega/2)}{\sin^3(\omega/2)}\,d\omega\,d\theta.$$

From this, let us evaluate the integral

$$\frac{R^2}{2}\iint(\omega - \sin\omega)\frac{\cos(\omega/2)}{\sin^3(\omega/2)}\,d\omega\,d\theta = \pi R^2\int_0^\pi(\omega - \sin\omega)\frac{\cos(\omega/2)}{\sin^3(\omega/2)}\,d\omega$$

over the region $0 < \omega < \pi$, $0 < \theta < 2\pi$. Integration by parts yields the result $\pi^2 R^2$.

16. Evaluation of the area of a closed surface

Problem. Find the area of the surface S defined by the equation

$$(x^2 + y^2 + z^2)^2 = x^2 - y^2.$$

Solution. In spherical coordinates, the equation of the surface S is

$$r^2 = \sin^2\theta\cos 2\varphi.$$

An element of length in this space is given by

$$ds^2 = dr^2 + r^2 d\theta^2 + r^2 \sin^2\theta \, d\varphi^2.$$

On S, r is a function of θ and φ, and we have

$$ds^2 = E \, d\theta^2 + 2F \, d\theta \, d\varphi + G \, d\varphi^2,$$

where E, F, and G are functions of θ and φ to be determined. First of all, we have

$$r \, dr = \sin\theta \cos\theta \cos 2\varphi \, d\theta - \sin^2\theta \sin 2\varphi \, d\varphi,$$

so that

$$dr^2 = \cos^2\theta \cos 2\varphi \, d\theta^2 - 2 \sin\theta \cos\theta \sin 2\varphi \, d\theta \, d\varphi + \frac{\sin^2\theta \sin^2 2\varphi}{\cos 2\varphi} \, d\varphi^2.$$

We must add dr^2 to

$$\sin^2\theta \cos 2\varphi \, (d\theta^2 + \sin^2\theta \, d\varphi^2).$$

We find

$$\begin{aligned} E &= \cos 2\varphi, \\ F &= -\sin\theta \cos\theta \sin 2\varphi, \\ G &= \frac{\sin^2\theta}{\cos 2\varphi} (\sin^2 2\varphi + \sin^2\theta \cos^2 2\varphi), \end{aligned}$$

from which we conclude that

$$EG - F^2 = \sin^4\theta.$$

The surface S has the three coordinate planes for planes of symmetry. The desired area $\mathscr{A}$ is eight times the area of that portion of the surface corresponding to positive values of x, y, and z, that is, corresponding to $0 < \theta < \pi/2$, $0 < \varphi < \pi/2$.

But, for r^2 to be positive, it is necessary that $\cos 2\varphi$ be positive. Therefore, $0 < \varphi < \pi/4$.

The area is calculated from the formula

$$\mathscr{A} = 8 \iint (EG - F^2)^{1/2} \, d\theta \, d\varphi,$$

where

$$0 < \theta < \frac{\pi}{2}, \qquad 0 < \varphi < \frac{\pi}{4}.$$

Therefore, we have

$$\mathcal{A} = 8 \iint \sin^2\theta \, d\theta \, d\varphi = 8 \int_0^{\pi/4} d\varphi \int_0^{\pi/2} \sin^2\theta \, d\theta$$
$$= 8 \cdot \frac{\pi}{4} \frac{\pi}{4} = \frac{\pi^2}{2}.$$

Thus

$$\mathcal{A} = \frac{\pi^2}{2}.$$

*17. The mapping of surfaces of revolution onto helicoids

It is said of surfaces S and Σ that they can be mapped 1 : 1 onto each other if locally they have the same linear element. This means that if we use one system of curvilinear coordinates u and v on the first surface and another system u' and v' on the second, there exists a continuous transformation mapping (u', v') into (u, v) such that the linear element of Σ expressed in terms of (u, v) is identical to that of S.

In this problem we show that surfaces whose element of length is of the form $E\,du^2 + 2F\,du\,dv + G\,dv^2$, where E, F, and G are functions of u only, can be mapped 1 : 1 onto surfaces of revolution whose element of length is of the form

$$f(u')\,du'^2 + u'^2\,dv'^2 \qquad \text{with} \qquad u' = r,\ v' = \theta$$

(polar coordinates in the projection onto a plane perpendicular to the axis of revolution). We shall apply this result to certain helicoids that are applicable onto hyperboloids of revolution.

Problems. A. Show that the element of length of a surface of revolution S about the z-axis is given in cylindrical coordinates r, θ, z by

$$ds^2 = f(r)\,dr^2 + r^2\,d\theta^2. \tag{1}$$

Find the equation of the meridian of S.

B. Denote by Σ a surface whose element of length is given in curvilinear coordinates u and v by

$$ds^2 = E\,du^2 + 2F\,du\,dv + G\,dv^2, \tag{2}$$

where E, F, and G are functions of u only (not v).

Find a change of variables

$$\theta = \theta(u, v), \qquad r = r(u),$$

that will put (2) in the form (1). Note that $\partial\theta/\partial v$ is a constant $1/k$.

C. Show that the equations

$$x = \rho \cos \varphi, \qquad y = \rho \sin \varphi, \qquad z = g(\rho) + a\varphi,$$

where g is a given function and a is a given number, represent a helicoid Σ in cylindrical coordinates. Show that the element of length of Σ is of the form (2) with $u = \rho$ and $v = \varphi$.

D. Suppose that Σ is the surface of a screw with triangular thread (a helicoid generated by a straight line that intersects the z-axis). Treat in detail the change of variables of Problem C. Show that for a suitable value of the constant k, the surface S is a hyperboloid of one sheet.

Solutions. **A.** We know that, in general, an element of length in space is given in cylindrical coordinates by

$$ds^2 = dr^2 + r^2\, d\theta^2 + dz^2.$$

If the equation of a surface S is given in the form

$$z = z(r, \theta),$$

we have

$$dz = \frac{\partial z}{\partial r}\, dr + \frac{\partial z}{\partial \theta}\, d\theta.$$

On the surface, ds^2 reduces to

$$ds^2 = \left[1 + \left(\frac{\partial z}{\partial r}\right)^2\right] dr^2 + 2\,\frac{\partial z}{\partial r}\frac{\partial z}{\partial \theta}\, dr\, d\theta + \left[r^2 + \left(\frac{\partial z}{\partial \theta}\right)^2\right] d\theta^2.$$

If the surface S is a surface of revolution about the z-axis, then $\partial z/\partial \theta = 0$, and the above expression simplifies to

$$ds^2 = \left[1 + \left(\frac{\partial z}{\partial r}\right)^2\right] dr^2 + r^2\, d\theta^2$$

or

$$ds^2 = f(r)\, dr^2 + r^2\, d\theta^2 \tag{1}$$

with $f(r) = 1 + (\partial z/\partial r)^2$. The meridian of S is defined in its rz-plane by an equation of the form $z = z(r)$, where

$$\frac{\partial z}{\partial r} = \pm [f(r) - 1]^{1/2}.$$

By integrating, we obtain

$$z = z_0 \pm \int_{r_0}^{r} [f(r') - 1]^{1/2}\, dr'.$$

B. We know that an element of length on a surface Σ can be written

$$ds^2 = E\, du^2 + 2F\, du\, dv + G\, dv^2. \tag{2}$$

Our hypothesis is that E, F, and G are functions of u only. We wish to convert (2) to the form (1) by a change of variables. We set

$$\theta = \theta(u, v), \qquad r = r(u).$$

By differentiating, we get

$$d\theta = \frac{\partial\theta}{\partial u}\, du + \frac{\partial\theta}{\partial v}\, dv, \qquad dr = \frac{dr}{du}\, du,$$

so that

$$du = \frac{du}{dr}\, dr, \qquad dv = -\frac{du}{dr}\frac{(\partial\theta/\partial u)}{(\partial\theta/\partial v)}\, dr + \frac{1}{\partial\theta/\partial v}\, d\theta.$$

Thus (2) may be written

$$ds^2 = \left(\frac{du}{dr}\right)^2 \left[E - 2F\frac{\partial\theta/\partial u}{\partial\theta/\partial v} + G\frac{(\partial\theta/\partial u)^2}{(\partial\theta/\partial v)^2}\right] dr^2$$
$$+ 2\frac{du}{dr}\left[\frac{F}{\partial\theta/\partial v} - G\frac{\partial\theta/\partial u}{(\partial\theta/\partial v)^2}\right] dr\, d\theta + \frac{G}{(\partial\theta/\partial v)^2}\, d\theta^2.$$

If we wish to put ds^2 in the form (1), we must have

$$F\frac{\partial\theta}{\partial v} - G\frac{\partial\theta}{\partial u} = 0, \qquad \frac{G}{(\partial\theta/\partial v)^2} = r^2.$$

The last condition may be written

$$\frac{\partial\theta}{\partial v} = \frac{[G(u)]^{1/2}}{r(u)}.$$

We then have

$$\frac{\partial\theta}{\partial u} = \frac{F(u)}{G(u)}\frac{\partial\theta}{\partial v} = \frac{F(u)}{[G(u)]^{1/2}}\frac{1}{r(u)}.$$

Thus $\partial\theta/\partial u$ and $\partial\theta/\partial v$ are functions of u alone. Therefore, $(\partial/\partial v)(\partial\theta/\partial u)$ and, hence, $(\partial/\partial u)(\partial\theta/\partial v)$ vanish and $\partial\theta/\partial v$ is a constant $1/k$.

Finally,

$$\frac{\partial\theta}{\partial u} = \frac{1}{k}\frac{F(u)}{G(u)}.$$

By integrating, we get

$$\theta = \frac{1}{k}\left[\int \frac{F(u)}{G(u)}\,du + v + \text{const}\right]$$
$$r = k\,[G(u)]^{1/2}$$

and

$$f(r) = \left(\frac{du}{dr}\right)^2\left[E - 2F\frac{\partial\theta/\partial u}{\partial\theta/\partial v} + G\frac{(\partial\theta/\partial u)^2}{(\partial\theta/\partial v)^2}\right] = \left(\frac{du}{dr}\right)^2\frac{EG - F^2}{G}.$$

C. A helicoid about the z-axis is generated by a curve that undergoes a helicoidal motion with pitch $2\pi a$. Suppose that the equation of the intersection of the helicoid and a plane passing through the z-axis is

$$z = g(\rho).$$

This helicoid has the parametric representation

$$x = \rho\cos\varphi, \qquad y = \rho\sin\varphi, \qquad z = g(\rho) + a\varphi,$$

which means that as φ varies, every point on the curve $z = g(\rho)$ describes a helix of pitch $2\pi a$.

For the helicoid, we therefore have

$$ds^2 = d\rho^2 + \rho^2\,d\varphi^2 + dz^2 = \left[1 + \left(\frac{dg}{d\rho}\right)\right]d\rho^2 + 2a\frac{dg}{d\rho}\,d\varphi\,d\rho + (\rho^2 + a^2)\,d\varphi^2.$$

If we replace ρ with u and φ with v, we have

$$ds^2 = E\,du^2 + 2F\,du\,dv + G\,dv^2,$$

where

$$E(u) = 1 + \left(\frac{dg}{du}\right)^2, \qquad F(u) = a\frac{dg}{du}, \qquad G(u) = u^2 + a^2.$$

An element of length on the helicoid is of the form (2).

D. If the surface Σ is the surface of a screw with triangular thread, we may write

$$g(\rho) = \alpha\rho + \beta$$

or

$$g(u) = \alpha u + \beta,$$

where α and β are constants. Therefore,

$$E(u) = 1 + \alpha^2, \qquad F(u) = a\alpha, \qquad G(u) = u^2 + a^2$$

and

$$\begin{aligned} \theta &= \frac{1}{k}\left(\alpha \int \frac{a\, du}{u^2 + a^2} + v + \text{const}\right) = \frac{1}{k}\left(\alpha \arctan \frac{u}{a} + v + \text{const}\right), \\ r &= k\,(u^2 + a^2)^{1/2}, \\ f(r) &= \frac{a^2 + u^2 + \alpha^2 u^2}{a^2 + u^2}\left(\frac{du}{dr}\right)^2. \end{aligned}$$

Since

$$\left(\frac{du}{dr}\right)^2 = \frac{r^2}{k^2(r^2 - k^2a^2)},$$

we have

$$f(r) = \frac{r^2}{k^2(r^2 - k^2a^2)} + \frac{\alpha^2}{k^2}.$$

The corresponding equation of the meridian of the surface of revolution S is given by

$$z - z_0 = \pm \int [f(r) - 1]^{1/2}\, dr = \pm \int \left[\frac{r^2}{k^2(r^2 - k^2a^2)} + \frac{\alpha^2 - k^2}{k^2}\right]^{1/2} dr.$$

Evaluation of z as a function of r in general involves an elliptic integral. However, if

$$\alpha^2 - k^2 = 0,$$

that is, if

$$\alpha = \pm k,$$

then

$$z - z_0 = \pm \int \frac{r\, dr}{k(r^2 - k^2a^2)^{1/2}} = \pm \frac{1}{k}(r^2 - k^2a^2)^{1/2}$$

or

$$k(z - z_0) = \pm (r^2 - k^2a^2)^{1/2}.$$

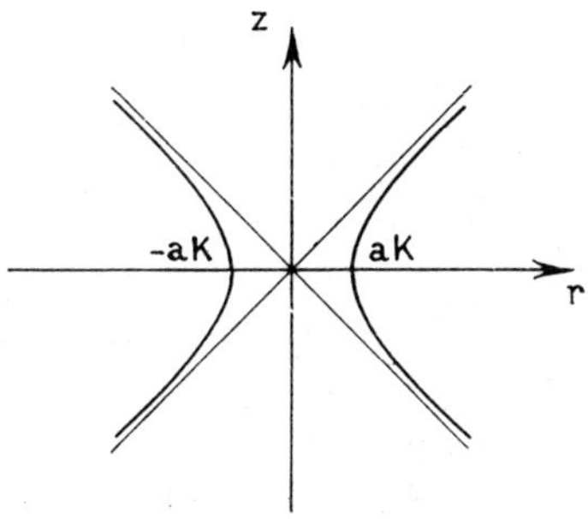

FIGURE 34

We recognize the equation in z and r of a hyperbola that intersects the r-axis at the points $\pm ak$. Therefore, in this case, the surface of revolution is a hyperboloid of one sheet.

18. Conic fields

Problems. A. Find the vector fields whose flux is zero through every portion of the surface of any cone whose apex is a given point A.

B. Suppose that Σ is a portion of a surface bounded by a closed contour Γ. We assume that Σ is intersected at at most one point by a ray originating at the point A. We denote by $\mathscr{D}_A$ the region bounded by Σ and the cone with apex at A whose directrix is the curve Γ. The region $\mathscr{D}_A$ is filled with material of constant density 1. Let I_A be the moment of inertia with respect to A of the mass contained in $\mathscr{D}_A$.

If we take the coordinate origin at A, show that I_A can be expressed by the flux through Σ of a certain vector field. If the components of this field are homogeneous functions of the same degree, they are necessarily given by

$$\tfrac{1}{5}\,x(x^2+y^2+z^2), \qquad \tfrac{1}{5}\,y(x^2+y^2+z^2), \qquad \tfrac{1}{5}\,z(x^2+y^2+z^2).$$

C. Let A, B, C, and D be the points whose coordinates in the xy-plane are $(\pm a, \pm b)$. We denote by I_A, I_B, I_C, and I_D the four moments of inertia analogous to I_A that are defined in terms of the same curve Γ. Show that there exist three numbers ϵ_1, ϵ_2, and ϵ_3 each equal to 1 or -1 such that

$$I_A + \epsilon_1 I_B + \epsilon_2 I_C + \epsilon_3 I_D$$

can be expressed by a line integral of the form

$$\int_{\Gamma} P(x, y, z)\, dz$$

and then determine the function $P(x, y, z)$.

Begin by assuming that the four regions $\mathscr{D}_A$, $\mathscr{D}_B$, $\mathscr{D}_C$, and $\mathscr{D}_D$ all lie entirely on the same side of the plane containing the rectangle $ABCD$.

Solutions. **A.** Since S is a portion of the surface of a cone with apex at A, the flux of the vector $\mathbf{V}(x, y, z)$ through S is

$$\Phi = \iint_S (\mathbf{V}, \mathbf{N})\, dS,$$

where N denotes the outer normal to S.

For Φ to vanish for every surface S, it is necessary and sufficient, if $\mathbf{V}$ is continuous, that

$$(\mathbf{V}, \mathbf{N}) = 0$$

at every point $M(x, y, z)$ of S; that is, it is necessary and sufficient that $\mathbf{V}$ be in the tangent plane to the cone at M.

Since this must be true for *any* cone with apex at A that passes through M, $\mathbf{V}$ must be collinear with the generator AM that is common to all the tangent planes.

Conversely, if $\mathbf{V}$ passes through A, we have $\Phi = 0$ for every portion of a cone with apex at A. Therefore, the desired vector fields are the fields of radial vectors that are concurrent at A.

B. Since Σ is intersected by at most one ray originating at A, it defines a definite region $\mathscr{D}_A$ whose moment of inertia about A is

$$I_A = \iiint_{\mathscr{D}_a} (x^2 + y^2 + z^2)\, dx\, dy\, dz.$$

Suppose that a set of vectors that are concurrent at A constitutes a field $\mathbf{V}(X, Y, Z)$, where X, Y, and Z are the components in the x, y, and z directions. If this field is continuous and has continuous first derivatives in $\mathscr{D}_A$ and on its boundary and if

$$\operatorname{div} \mathbf{V} = x^2 + y^2 + z^2, \tag{1}$$

we may write, in accordance with Ostrogradskiy's formula,

$$\iiint_{\mathscr{D}_A} (x^2 + y^2 + z^2)\, dx\, dy\, dz = \iiint_{\mathscr{D}_A} \operatorname{div} \mathbf{V}\, dx\, dy\, dz = \iint_{\Sigma} (\mathbf{V}, \mathbf{N})\, d\Sigma.$$

The last integral represents the flux of $\mathbf{V}$ through Σ since the flux of $\mathbf{V}$ through the surface of the cone is zero. Thus $\mathbf{V}$ is determined by (1) and by

$$\frac{X}{x} = \frac{Y}{y} = \frac{Z}{z}. \tag{2}$$

Let us denote by $\lambda(x, y, z)$ the common value of these three ratios. Then

$$\frac{\partial X}{\partial x} = \lambda + x\frac{\partial \lambda}{\partial x}, \qquad \frac{\partial Y}{\partial y} = \lambda + y\frac{\partial \lambda}{\partial y}, \qquad \frac{\partial Z}{\partial z} = \lambda + z\frac{\partial \lambda}{\partial z},$$

so that

$$\operatorname{div} \mathbf{V} = 3\lambda + x\frac{\partial \lambda}{\partial x} + y\frac{\partial \lambda}{\partial y} + z\frac{\partial \lambda}{\partial z} = x^2 + y^2 + z^2.$$

If we assume in addition that X, Y, and Z are homogeneous functions of x, y, and z, they are, by virtue of (2), of the same degree. Let n denote their common degree. Then λ is of degree $n - 1$ and, by virtue of Euler's identity,

$$x\frac{\partial \lambda}{\partial x} + y\frac{\partial \lambda}{\partial y} + z\frac{\partial \lambda}{\partial z} = (n-1)\lambda.$$

Equation (1) may therefore be written

$$(n+2)\lambda = x^2 + y^2 + z^2,$$

which shows that λ is of degree 2. Therefore,

$$n = 3 \qquad \text{and} \qquad \lambda = \frac{x^2 + y^2 + z^2}{5}.$$

Finally, the components of $\mathbf{V}$ are

$$\begin{aligned} X &= \lambda x = \tfrac{1}{5}\, x(x^2 + y^2 + z^2), \\ Y &= \lambda y = \tfrac{1}{5}\, y(x^2 + y^2 + z^2), \\ Z &= \lambda z = \tfrac{1}{5}\, z(x^2 + y^2 + z^2). \end{aligned}$$

C. Let $\mathbf{V}_A(X_A, Y_A, Z_A)$ be the vector corresponding to the point $A(x = a, y = b)$ and in the region $\mathscr{D}_A$ defined as above and let

$$\begin{aligned} X_A &= \tfrac{1}{5}\,(x-a)[(x-a)^2 + (y-b)^2 + z^2], \\ Y_A &= \tfrac{1}{5}\,(y-b)[(x-a)^2 + (y-b)^2 + z^2], \\ Z_A &= \tfrac{1}{5}\, z[(x-a)^2 + (y-b)^2 + z^2] \end{aligned}$$

in the region $\mathscr{D}_A$ defined above.

We define similarly the vector $\mathbf{V}_B$, $\mathbf{V}_C$, and $\mathbf{V}_D$ corresponding to the other three corners of the rectangle $ABCD$,

$$B(-a, b), \qquad C(-a, -b), \qquad D(a, -b).$$

If the normal to Σ is oriented in the same sense for all four regions $\mathscr{D}_A$, $\mathscr{D}_B$, $\mathscr{D}_C$, and $\mathscr{D}_D$, that is, if these four regions are on the same side of Σ, we may write

$$I_A + \epsilon_1 I_B + \epsilon_2 I_C + \epsilon_3 I_D = \iint_\Sigma (\mathbf{V}_A + \epsilon_1 \mathbf{V}_B + \epsilon_2 \mathbf{V}_C + \epsilon_3 \mathbf{V}_D , \mathbf{N})\, d\Sigma.$$

For the double integral in this expression to be expressible as a line integral

$$\int_\Gamma P(x, y, z)\, dz,$$

it is necessary that

$$\operatorname{div} (\mathbf{V}_A + \epsilon_1 \mathbf{V}_B + \epsilon_2 \mathbf{V}_C + \epsilon_3 \mathbf{V}_D) = 0.$$

Now,

$$\begin{aligned}
\operatorname{div} \mathbf{V} &= x^2 + y^2 + z^2, \\
\operatorname{div} \mathbf{V}_A &= (x - a)^2 + (y - b)^2 + z^2, \\
\operatorname{div} \mathbf{V}_B &= (x + a)^2 + (y - b)^2 + z^2, \\
\operatorname{div} \mathbf{V}_C &= (x + a)^2 + (y + b)^2 + z^2, \\
\operatorname{div} \mathbf{V}_D &= (x - a)^2 + (y + b)^2 + z^2
\end{aligned}$$

and

$$\begin{aligned}
\operatorname{div} (\mathbf{V}_A &+ \epsilon_1 \mathbf{V}_B + \epsilon_2 \mathbf{V}_C + \epsilon_3 \mathbf{V}_D) \\
&= (1 + \epsilon_1 + \epsilon_2 + \epsilon_3)(x^2 + y^2 + z^2 + a^2 + b^2) \\
&\quad + (-1 + \epsilon_1 + \epsilon_2 - \epsilon_3) \cdot 2ax \\
&\quad + (-1 - \epsilon_1 + \epsilon_2 + \epsilon_3) \cdot 2by.
\end{aligned}$$

From this we conclude that

$$\epsilon_2 = 1 \qquad \text{and} \qquad \epsilon_1 = \epsilon_3 = -1.$$

Therefore, there exists a vector U such that

$$\mathbf{V}_A + \epsilon_1 \mathbf{V}_B + \epsilon_2 \mathbf{V}_C + \epsilon_3 \mathbf{V}_D = \operatorname{curl} \mathbf{U}.$$

Let us try to get the first two components of U equal to 0. If P is the third component of $\mathbf{U}$, the components of curl $\mathbf{U}$ will be

$$+\frac{\partial P}{\partial y}, \qquad -\frac{\partial P}{\partial x}, \qquad 0,$$

so that

$$\frac{\partial P}{\partial y} = X_A - X_B + X_C - X_D = \frac{8}{5}\, aby,$$
$$-\frac{\partial P}{\partial x} = Y_A - Y_B + Y_C - Y_D = \frac{8}{5}\, abx,$$

and it is easily verified that $0 = Z_A - Z_B + Z_C - Z_D$. From this we immediately obtain

$$dP = \frac{8}{5}\, ab(y\, dy - x\, dx)$$

and

$$P = \frac{4}{5}\, ab(y^2 - x^2) + \text{const.}$$

Remark 1. If the four regions $\mathscr{D}_A$, $\mathscr{D}_B$, $\mathscr{D}_C$, and $\mathscr{D}_D$ do not all lie on the same side of Σ, we should write

$$I_A + \epsilon_1 I_B + \epsilon_2 I_C + \epsilon_3 I_D = \iint_\Sigma (\mathbf{V}_A + \epsilon_1\epsilon_1'\mathbf{V}_B + \epsilon_2\epsilon_2'\mathbf{V}_C + \epsilon_3\epsilon_3'\mathbf{V}_D\, , \mathbf{N})\, d\Sigma,$$

where ϵ_1' is equal to $+1$ if $\mathscr{D}_B$ is on the same side of Σ as $\mathscr{D}_A$ but equal to -1 if it is not (and similarly for ϵ_2' and ϵ_3'). We then find

$$\epsilon_2\epsilon_2' = 1, \qquad \epsilon_1\epsilon_1' = \epsilon_3\epsilon_3' = -1.$$

The result remains unchanged for P.

Remark 2 (*geometric interpretation*). By hypothesis,

$$\text{div}\, \mathbf{V} = \mathbf{OM}^2, \qquad \text{div}\, \mathbf{V}_A = \mathbf{AM}^2, \qquad \text{etc.}$$

The condition

$$\text{div}(\mathbf{V}_A + \epsilon_1\mathbf{V}_B + \epsilon_2\mathbf{V}_C + \epsilon_3\mathbf{V}_D) = 0$$

is represented geometrically by

$$\mathbf{AM}^2 + \epsilon_1\mathbf{BM}^2 + \epsilon_2\mathbf{CM}^2 + \epsilon_3\mathbf{DM}^2 = 0.$$

If we drop perpendiculars MH_1 and MH_2 from M to the segments AB and CD, we obviously have

$$AH_1 = DH_2$$

and

$$BH_1 = CH_2$$

(since the plane MH_1H_2 is perpendicular to AB and CD).

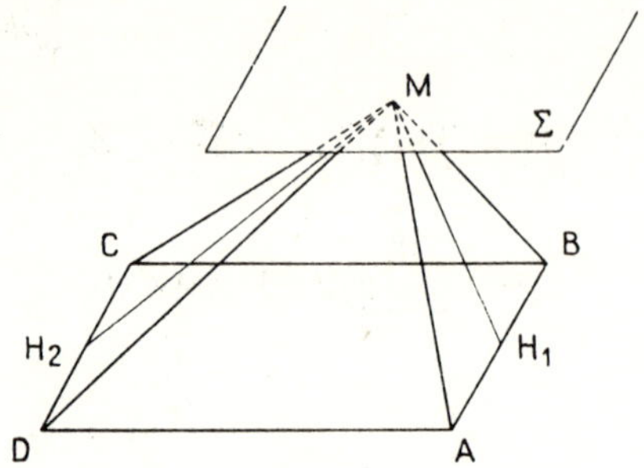

FIGURE 35

But we also have

$$\mathbf{AM}^2 - \mathbf{BM}^2 = \overline{AH_1}^2 - \overline{BH_1}^2 = \overline{DH_2}^2 - \overline{CH_2}^2 = \overline{DM}^2 - \overline{CM}^2,$$

so that, for any M,

$$AM^2 - BM^2 + CM^2 - DM^2 = 0.$$

19. Another example of a conic field

Problems. Suppose that:
Γ is a simple closed curve;
S is a surface region bounded by Γ;
dS is the element of area on S;
G is a straight line passing through the origin that intersects S at a single point $M(x, y, z)$;
α, β, and γ are the direction cosines of the normal to S at the point M directed away from the origin;
Q is an algebraic surface of degree n that touches the line G at the points M_1, M_2, ..., M_n.

We write the equation for Q in the form

$$\varphi_0 + \varphi_1(x, y, z) + \cdots + \varphi_n(x, y, z) = 0,$$

where $\varphi_k(x, y, z)$ is a homogeneous polynomial of degree k.

Finally, we set $OM = r$ and $OM_k = r_k$.

As M describes S, the segments OM_k sweep conic regions of algebraic volumes V_k.

A. Show that the sum $V = V_1 + V_2 + \cdots + V_n$ can be represented by an integral of the form

$$V = \iint_S L(x, y, z)(\alpha x + \beta y + \gamma z)\, dS,$$

where L is a homogeneous function of degree -3, which can be expressed in terms of the φ_k.

B. Evaluate V when Q is a sphere of radius R with center at the point on the z-axis whose z-coordinate is a and S is a portion of the sphere of radius 1 with center at the origin.

C. Show that there exist three functions A, B, and C that are homogeneous of degree -2 such that

$$V = \int_\Gamma \begin{vmatrix} dx & dy & dz \\ x & y & z \\ A & B & C \end{vmatrix}$$

Solutions. **A.** The point M describes a surface region S bounded by a closed curve Γ that is intersected at a single point M by a straight line G passing through the origin. We also have an algebraic surface Q of degree n. The straight line G intersects Q at n points $M_1, M_2, \ldots, M_n$.

We set $OM = r$ and $OM_k = r_k$. We denote by x, y, and z the coordinates of M and we denote by $\lambda_k x$, $\lambda_k y$, and $\lambda_k z$ the coordinates of M_k.

If $\varphi(x, y, z) = 0$ is the equation of the surface Q, the numbers λ_k are roots of the equation

$$\varphi(\lambda x, \lambda y, \lambda z) = 0.$$

When M describes the surface S, each segment OM_k sweeps a conical region of volume V_k.

Consider an element of area dS around M on the surface S. When M

ranges over this element, the segment OM_k sweeps an element of volume equal to

$$\frac{1}{3}\frac{r_k^3\, dS \cos\theta}{r^2},$$

where θ is the acute angle between the straight line OG and the normal N to the surface S at M.

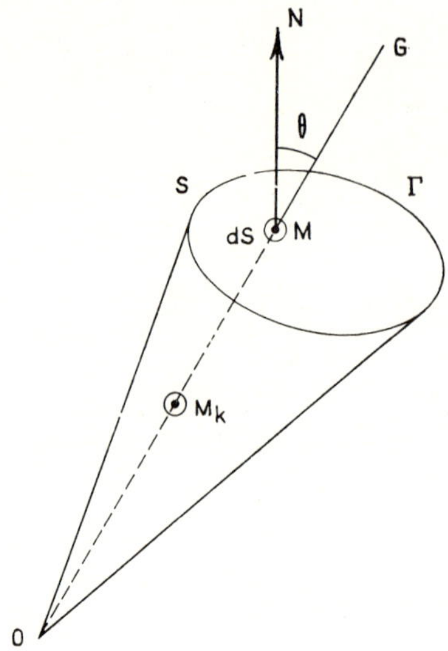

FIGURE 36

If α, β, and γ are the direction cosines of the normal, then

$$\cos\theta = \frac{\alpha x + \beta y + \gamma z}{r},$$

and the element of volume becomes

$$\frac{1}{3}\left(\frac{r_k}{r}\right)^3 (\alpha x + \beta y + \gamma z)\, dS.$$

Here $\alpha x + \beta y + \gamma z$ and r^3 are positive by hypothesis. We assign r_k the sign $+$ if M_k and M are on the same side of the origin, and we assign it the sign $-$ if they are not. We assign to the elements of volume the same sign as r_k. With the volume V_k thus defined as regards both magnitude and sign, we set

$$V = \sum_k V_k = \frac{1}{3}\iint_S (\alpha x + \beta y + \gamma z)\left[\sum\left(\frac{r_k}{r}\right)^3\right] dS.$$

The ratio r_k/r is equal to the number λ_k which is a root of the equation

$$\varphi(\lambda x, \lambda y, \lambda z) = 0.$$

Let us group the terms in the polynomial φ in such a way that φ will be the sum of $n + 1$ terms each being a homogeneous polynomial of degree k for $k = 0, \ldots, n$:

$$\varphi = \varphi_0 + \varphi_1(x, y, z) + \cdots + \varphi_n(x, y, z).$$

The equation in λ is written explicitly

$$\varphi_0 + \lambda\varphi_1(x, y, z) + \cdots + \lambda^n\varphi_n(x, y, z) = 0,$$

where x, y, and z are the coordinates of the point M that describes S. Then

$$V = \tfrac{1}{3} \iint_S (\alpha x + \beta y + \gamma z) \left(\sum \lambda_k^3\right) dS.$$

Let us set

$$\begin{aligned} S_1 &= \lambda_1 + \cdots + \lambda_n, \\ S_2 &= \lambda_1\lambda_2 + \lambda_1\lambda_3 + \cdots + \lambda_{n-1}\lambda_n, \\ S_3 &= \lambda_1\lambda_2\lambda_3 + \cdots + \lambda_{n-2}\lambda_{n-1}\lambda_n. \end{aligned}$$

It is easily verified that

$$\sum \lambda_k^3 = S_1^3 - 3S_1S_2 + 3S_3.$$

Now,

$$S_1 = -\frac{\varphi_{n-1}}{\varphi_n}, \qquad S_2 = \frac{\varphi_{n-2}}{\varphi_n}, \qquad S_3 = -\frac{\varphi_{n-3}}{\varphi_n}.$$

Therefore,

$$\sum \lambda_k^3 = -\frac{\varphi_{n-1}^3}{\varphi_n^3} + 3\frac{\varphi_{n-1}\varphi_{n-2}}{\varphi_n^2} - 3\frac{\varphi_{n-3}}{\varphi_n}$$

and

$$V = \iint_S L(x, y, z)(\alpha x + \beta y + \gamma z)\, dS,$$

where

$$L(x, y, z) = -\frac{1}{3}\frac{\varphi_{n-1}^3}{\varphi_n^3} + \frac{\varphi_{n-1}\varphi_{n-2}}{\varphi_n^2} - \frac{\varphi_{n-3}}{\varphi_n}$$

is a homogeneous function of degree -3.

If $n = 2$, Q is a quadric surface, and this expression is simply

$$L(x, y, z) = -\frac{1}{3}\frac{{\varphi_1}^3}{{\varphi_2}^3} + \frac{\varphi_1\varphi_0}{{\varphi_2}^2}.$$

B. *Example.* Suppose that Q is the sphere of radius R with center at the point on the z-axis whose abscissa is some positive number a. Suppose that S is a portion of the sphere of radius 1 with center at the origin:

$$\varphi = x^2 + y^2 + (z - a)^2 - R^2, \qquad \varphi_0 = a^2 - R^2,$$
$$\varphi_1 = -2az, \qquad \varphi_2 = x^2 + y^2 + z^2 = r^2,$$
$$L = \frac{8}{3} a^3 \frac{z^3}{r^6} - 2a \frac{(a^2 - R^2)z}{r^4}, \qquad \alpha = x, \qquad \beta = y, \qquad \gamma = z,$$
$$V = \iint_S \left[\frac{8}{3} a^3 \frac{z^3}{r^4} - 2a \frac{(a^2 - R^2)z}{r^2}\right] dS.$$

On S we have $r = 1$, $z = \cos\theta$, and $dS = \sin\theta\, d\theta\, d\varphi$, so that

$$V = \iint_S [\tfrac{8}{3} a^3 \cos^3\theta - 2a(a^2 - R^2)\cos\theta] \sin\theta\, d\theta\, d\varphi.$$

Particular cases. (1) $a = R$, $0 < \varphi < 2\pi$, $0 < \theta < \pi/2$. Here, the sphere Q passes through the origin and V is the volume of the sphere, since

$$V = \frac{8}{3} R^3 \times 2\pi \times \int_0^{\pi/2} \sin\theta \cos^3\theta\, d\theta = \frac{16}{3}\pi R^3 \left(\frac{\cos^4\theta}{4}\right)_{\pi/2}^{0} = \frac{4}{3}\pi R^3.$$

(2) Suppose that S is bounded by a small circle with pole on the z-axis, so that φ varies from 0 to 2π and θ from 0 to a given value ω lying between 0 and $\pi/2$ (conical volume of revolution around the z-axis). We then have

$$V = 2\pi a \int_0^{\omega} (\tfrac{8}{3} a^2 \cos^3\theta - 2(a^2 - R^2)\cos\theta) \sin\theta\, d\theta$$
$$= \frac{4}{3}\pi a^3 (1 - \cos^4\omega) - 2\pi a(a^2 - R^2)\sin^2\omega.$$

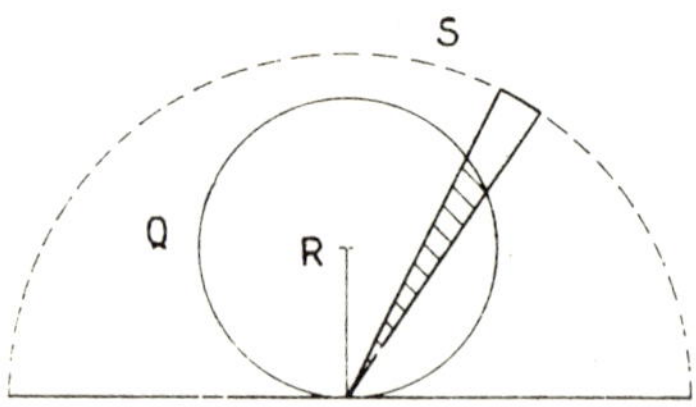

FIGURE 37

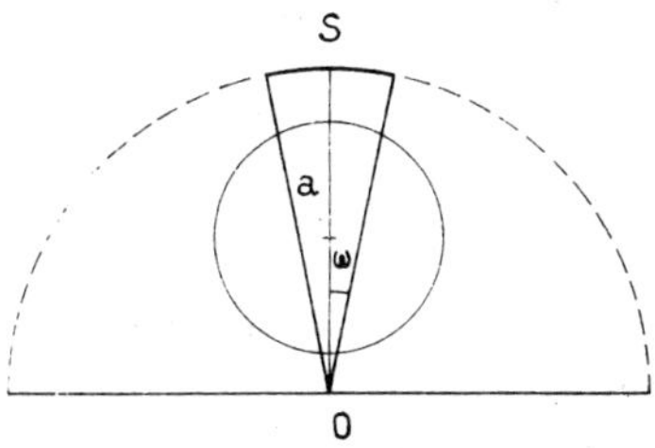

FIGURE 38

For $a = R$ and $\omega = \pi/2$, we again obtain the volume of the sphere of radius R. We can verify that V vanishes for $a = 0$ (compensation of opposite volumes). If we set

$$\sin^2 \omega = s,$$

we have

$$V = \frac{2\pi a}{3} s(a^2 + 3R^2 - 2a^2 s).$$

It might be noted that if $a > R$, then V is real and positive whenever all the lines OG intersect the sphere Q at real points, that is, when

$$\sin \omega < \frac{R}{a}.$$

However, V remains real, although it becomes negative, when ω exceeds the limiting value arcsin R/a.

If $a < R\sqrt{3}$, and in particular if $a < R$, then V is always positive.

C. The volume V is given by

$$V = \iint_S L(x, y, z)(\alpha x + \beta y + \gamma z)\, dS,$$

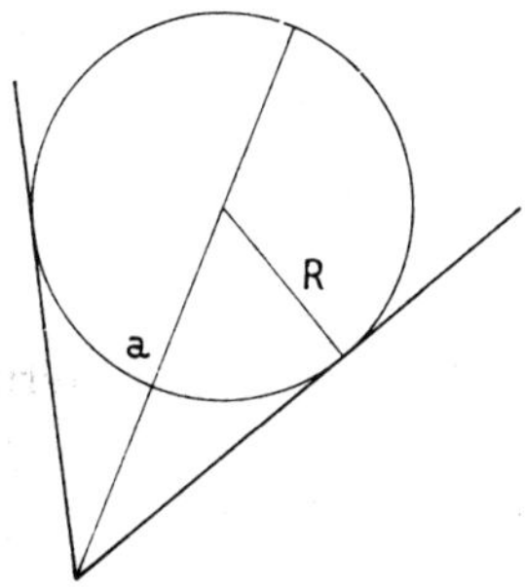

FIGURE 39

where α, β, and γ are the direction cosines of the normal to S at the point (x, y, z).

Let us show that if $L(x, y, z)$ is a homogeneous function of degree -3, V can be represented as a line integral over the boundary Γ of S. This result is valid without any other hypotheses, and it is not necessary for L to have the particular form given in Problem A.

For that, it is sufficient for the divergence of the vector (xL, yL, zL) to be zero or, in other words, for the exterior differential form

$$p = L(x\, dy \wedge dz + y\, dz \wedge dx + z\, dx \wedge dy)$$

have a zero exterior differential. Now,

$$\frac{\partial}{\partial x}(Lx) + \frac{\partial}{\partial y}(Ly) + \frac{\partial}{\partial z}(Lz) = 3L + x\frac{\partial L}{\partial x} + y\frac{\partial L}{\partial y} + z\frac{\partial L}{\partial z}.$$

The left-hand member will be recognized as Euler's equation. For L to be homogeneous of degree -3 is equivalent to the vanishing of the divergence of the vector (Lx, Ly, Lz).

Thus there exists a linear form

$$q = A\, dx + B\, dy + C\, dz$$

whose outer differential dq is equal to p. In other words, there exists a vector $U(X, Y, Z)$ whose curl is the vector (Lx, Ly, Lz). Let us show that there exists a vector field $V(A, B, C)$ whose components are homogeneous functions of degree -2 such that

$$U = OM \wedge V.$$

Let us evaluate curl U. Its components are of the form

$$\frac{\partial}{\partial y}(Bx - Ay) - \frac{\partial}{\partial z}(Az - Cx) = -2A + x\frac{\partial B}{\partial y} + x\frac{\partial C}{\partial z} - y\frac{\partial A}{\partial y} - z\frac{\partial A}{\partial z}$$

$$= -\left(2A + x\frac{\partial A}{\partial x} + y\frac{\partial A}{\partial y} + z\frac{\partial A}{\partial z}\right) + x\left(\frac{\partial A}{\partial x} + \frac{\partial B}{\partial y} + \frac{\partial C}{\partial z}\right).$$

According to Euler's identity, the first term in the second member is zero. Therefore, the vector curl U is radial. For it to coincide with the vector (Lx, Ly, Lz), it is sufficient that

$$\frac{\partial A}{\partial x} + \frac{\partial B}{\partial y} + \frac{\partial C}{\partial z} = L,$$

which is possible since both members are homogeneous of degree -3. Any vector field that is homogeneous of degree -2 whose divergence is L thus enables us to put V in the form of a line integral

$$V = \int_{\Gamma} \begin{vmatrix} dx & dy & dz \\ x & y & z \\ A & B & C \end{vmatrix}$$

20. Stokes' formula

Problems. Let S denote an oriented surface region bounded by a closed curve Γ. Let α, β, and γ denote the direction cosines of the oriented normal to S.

A. Transform the surface integral

$$I = \iint_S [\alpha x(z^2 - y^2) + \beta y(x^2 - z^2) + \gamma z(y^2 - x^2)]\, dS$$

into a line integral.

B. Evaluate I when S is the portion of the plane $x = 1$ defined by

$$y^2 + z^2 \leqslant 1, \qquad z + y \geqslant 0, \qquad z - y \geqslant 0.$$

Solutions. **A.** We have the integral

$$I = \iint_S [\alpha x(z^2 - y^2) + \beta y(x^2 - z^2) + \gamma z(y^2 - x^2)]\, dS$$

over an oriented surface region S bounded by a closed curve Γ, where α, β, and γ are the direction cosines of the oriented normal to S.

We may also write

$$I = \iint_S \omega,$$

where

$$\varpi = x(z^2 - y^2)\, dy \wedge dz + y(x^2 - z^2)\, dz \wedge dx + z(y^2 - x^2)\, dx \wedge dy$$

is an outer differential form of order 2.

We can verify that the outer differential of ω is zero, that is, that the divergence of the vector $x(z^2 - y^2)$, $y(x^2 - z^2)$, $z(y^2 - x^2)$ is zero. Therefore, there exists a linear form

$$\omega = X\, dx + Y\, dy + Z\, dz,$$

of which ω is the outer differential. We have

$$\iint_S d\omega = \int_\Gamma \varpi.$$

In other words, there exists a vector (X, Y, Z) whose curl is equal to the given vector:

$$\frac{\partial Z}{\partial y} - \frac{\partial Y}{\partial z} = x(z^2 - y^2) = xz^2 - xy^2,$$

$$\frac{\partial X}{\partial z} - \frac{\partial Z}{\partial x} = y(x^2 - z^2) = yx^2 - yz^2,$$

$$\frac{\partial Y}{\partial x} - \frac{\partial X}{\partial y} = z(y^2 - x^2) = zy^2 - zx^2.$$

To evaluate X, Y, and Z, we may, for example, seek a particular solution such that $Z = 0$. But we spot an obvious system of solutions obtained by writing

$$\frac{\partial Z}{\partial y} = xz^2, \qquad \frac{\partial Y}{\partial z} = xy^2,$$

$$\frac{\partial X}{\partial z} = yx^2, \qquad \frac{\partial Z}{\partial x} = yz^2,$$

$$\frac{\partial Y}{\partial x} = zy^2, \qquad \frac{\partial X}{\partial y} = zx^2.$$

From this we find

$$X = x^2yz, \qquad Y = y^2zx, \qquad Z = z^2xy,$$

so that

$$I = \int_\Gamma xyz(x\,dx + y\,dy + z\,dz) = \int_\Gamma xyz\,\frac{d\rho^2}{2} \qquad (\rho^2 = x^2 + y^2 + z^2).$$

B. Let us orient the normal to S in the direction of increasing values of x. The curve Γ bounding S will then be oriented in the sense of the arrow shown in Figure 40. We then have

$$\alpha = 1, \qquad \beta = 0, \qquad \gamma = 0,$$

so that

$$I = \iint_S (z^2 - y^2)\,dS.$$

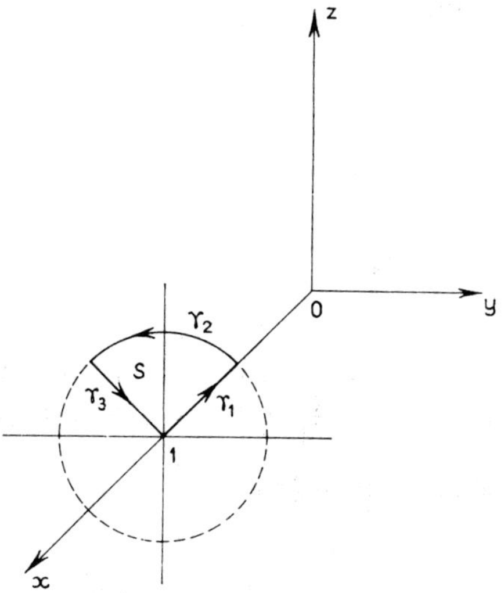

FIGURE 40

This integral is easily evaluated by changing to polar coordinates:

$$y = r\cos\theta, \qquad z = r\sin\theta,$$

$$\begin{aligned} I &= \iint_S r^2(\sin^2\theta - \cos^2\theta)\, r\,dr\,d\theta \\ &= \int_0^1 r^3\,dr \int_{\pi/4}^{3\pi/4} (-\cos 2\theta)\,d\theta \\ &= \left(\frac{r^4}{4}\right)_0^1 \left(-\frac{\sin 2\theta}{2}\right)_{\pi/4}^{3\pi/4} = \frac{1}{4}. \end{aligned}$$

By using a line integral, we have three successive integrals to evaluate:

$$\int_\Gamma = \int_{\gamma_1} + \int_{\gamma_2} + \int_{\gamma_3}$$

On the arc γ_2,

$$dr^2 = 0 \qquad \text{and} \qquad \int_{\gamma_2} = 0.$$

On the radius γ_1, we have $y = z$, so that

$$\int_{\gamma_1} 2y^3\,dy = \int_0^{1/\sqrt{2}} 2y^3\,dy = 2\left(\frac{y^4}{4}\right)_0^{1/\sqrt{2}} = \frac{1}{8}.$$

Similarly, on the radius γ_3,

$$y = -z$$

and

$$\int_{\gamma_3} -2y^2\,dy = -\int_{-1/\sqrt{2}}^{0} 2y^3\,dy = \tfrac{1}{8}.$$

Thus we again find $I = \frac{1}{4}$.

21. Evaluation of an improper multiple integral. The gamma and beta functions

Problem. Evaluate the integral

$$I = \iiint_{\mathscr{D}} \frac{dx\,dy\,dz}{1 - \cos x \cos y \cos z}$$

over the region $0 < x < \pi$, $0 < y < \pi$, $0 < z < \pi$.

Solution. The integrand of the integral

$$I = \iiint_{\mathscr{D}} \frac{dx\,dy\,dz}{1 - \cos x \cos y \cos z}$$

over the cube $\mathscr{D}$

$$0 < x < \pi, \qquad 0 < y < \pi, \qquad 0 < z < \pi$$

is always positive. Hence, it has a value if we can find any particular method of evaluating it that is convergent.

Let us make the change of variables

$$\tan\frac{x}{2} = u, \qquad \tan\frac{y}{2} = v, \qquad \tan\frac{z}{2} = w.$$

We have

$$\frac{dx}{2} = \frac{du}{1 + u^2}, \qquad \cos x = \frac{1 - u^2}{1 + u^2}$$

and analogous equations for v and w. Evaluating the Jacobian of the coordinate transformation, we see that

$$\left|\frac{D(x, y, z)}{D(u, v, w)}\right| = \frac{8}{(1 + u^2)(1 + v^2)(1 + w^2)}.$$

Therefore,

$$I = \iiint\limits_{\mathscr{D}'} \frac{8\,du\,dv\,dw}{(1+u^2)(1+v^2)(1+w^2)-(1-u^2)(1-v^2)(1-w^2)},$$

where $\mathscr{D}'$ is defined by

$$0 < u < \infty, \qquad 0 < v < \infty, \qquad 0 < w < \infty.$$

Let us make a new change of variables and shift to spherical coordinates:

$$u = r\sin\theta\cos\varphi, \qquad v = r\sin\theta\sin\varphi, \qquad w = r\cos\theta.$$

Calculation shows that

$$\left|\frac{D(u, v, w)}{D(r, \theta, \varphi)}\right| = r^2\sin\theta, \qquad 0 < \theta < \frac{\pi}{2}.$$

We can also verify that

$$(1+u^2)(1+v^2)(1+w^2)-(1-u^2)(1-v^2)(1-w^2) = 2(u^2+v^2+w^2)$$
$$+ 2u^2v^2w^2 = 2r^2 + 2r^6\cos^2\varphi\sin^2\varphi\sin^4\theta\cos^2\theta.$$

Therefore,

$$I = 4\iiint\limits_{\mathscr{D}''} \frac{\sin\theta\,dr\,d\theta\,d\varphi}{1 + r^4\cos^2\varphi\sin^2\varphi\cos^2\theta\sin^4\theta},$$

where $\mathscr{D}''$ is the parallelepiped

$$0 < \varphi < \frac{\pi}{2}, \qquad 0 < \theta < \frac{\pi}{2}, \qquad 0 < r < \infty.$$

We have

$$I = 4\int_0^{\pi/2} \sin\theta\,d\theta \int_0^{\pi/2} d\varphi \int_0^{\infty} \frac{dr}{1+\lambda r^4} \quad \text{where } \lambda = \cos^2\varphi\sin^2\varphi\cos^2\theta\sin^4\theta.$$

The change of variable $\lambda r^4 = s$ shows that

$$\int_0^{\infty} \frac{dr}{1+\lambda r^4} = \frac{1}{4\lambda^{1/4}} B\left(\frac{1}{4}, \frac{3}{4}\right) = \frac{1}{4\lambda^{1/4}} \Gamma\left(\frac{1}{4}\right)\Gamma\left(\frac{3}{4}\right).$$

Therefore,

$$I = \Gamma\left(\frac{1}{4}\right)\Gamma\left(\frac{3}{4}\right)\int_0^{\pi/2} \frac{d\theta}{(\cos\theta)^{1/2}} \int_0^{\pi/2} \frac{d\varphi}{(\sin\varphi\cos\varphi)^{1/2}}.$$

The integral

$$I_1 = \int_0^{\pi/2} \frac{d\theta}{(\cos\theta)^{1/2}}$$

can be calculated by setting $\cos^2\theta = t$. We obtain

$$I_1 = \frac{1}{2}\int_0^1 t^{-3/4}(1-t)^{-1/2}\,dt = \frac{1}{2} B\left(\frac{1}{4}, \frac{1}{2}\right) = \frac{1}{2}\frac{\Gamma(\frac{1}{4})\,\Gamma(\frac{1}{2})}{\Gamma(\frac{3}{4})}.$$

By taking 2φ as the variable of integration in $I_2 = \int_0^{\pi/2} [d\varphi/(\sin\varphi\cos\varphi)^{1/2}]$, we see that $I_2 = \sqrt{2}I_1$.

Finally,

$$I = \frac{1}{4}\frac{\Gamma(\frac{1}{4})\Gamma(\frac{3}{4})\Gamma(\frac{1}{4})\Gamma(\frac{1}{2})\Gamma(\frac{1}{4})^2}{\Gamma(\frac{3}{4})\Gamma(\frac{1}{2})} = \frac{1}{4}[\Gamma(\tfrac{1}{4})]^4.$$

We know that $\Gamma(\frac{1}{4}) = 3.6256\ldots$.

22. Evaluation of a second improper multiple integral

Problem. Consider the integral

$$I = \iiint_\Omega \frac{dx\,dy\,dz}{1 + x^\alpha + y^\beta + z^\gamma}$$

over the region $x > 0$, $y > 0$, $z > 0$.

For what values of α, β, and γ does this integral converge? Use the gamma and beta functions.

Solution. The integrand is positive. Therefore, we shall try to evaluate the integral, and the validity of the calculation will serve to reveal the conditions of convergence. We know that every convergent multiple integral is absolutely convergent and that, for the integral of a positive function to converge, it is sufficient for a calculating procedure to yield a convergent result. Then all other procedures will necessarily yield the same result.

Let us make the change of variables

$$x^\alpha = u^2, \qquad y^\beta = v^2, \qquad z^\gamma = w^2.$$

By an easy calculation, we find that

$$\left|\frac{D(x, y, z)}{D(u, v, w)}\right| = 8\,\frac{u^{(2/\alpha)-1}v^{(2/\beta)-1}w^{(2/\gamma)-1}}{\alpha\beta\gamma},$$

so that

$$I = \frac{8}{\alpha\beta\gamma} \iiint_{\Omega} \frac{u^{(2/\alpha)-1} v^{(2/\beta)-1} w^{(2/\gamma)-1}}{1 + u^2 + v^2 + w^2} \, du \, dv \, dw,$$

where Ω now denotes the region $u > 0$, $v > 0$, $w > 0$.

Let us now shift to spherical coordinates r, θ, φ in uvw-space:

$$u = r \sin\theta \cos\varphi, \qquad v = r \sin\theta \sin\varphi, \qquad w = r\cos\theta.$$

Then

$$I = \frac{8}{\alpha\beta\gamma} \iiint_{\Omega} \frac{r^{2[(1/\alpha)+(1/\beta)+(1/\gamma)]-3}}{1 + r^2} (\sin\theta)^{(2/\alpha)-1} (\cos\varphi)^{(2/\alpha)-1} (\sin\theta)^{(2/\beta)-1}$$

$$\times (\sin\varphi)^{(2/\beta)-1} (\cos\theta)^{(2/\gamma)-1} r^2 \sin\theta \, dr \, d\theta \, d\varphi$$

$$= \frac{8}{\alpha\beta\gamma} \int_0^{\infty} \frac{r^{2[(1/\alpha)+(1/\beta)+(1/\gamma)]-1}}{1 + r^2} \, dr \int_0^{\pi/2} (\sin\theta)^{(2/\alpha)+(2/\beta)-1} (\cos\theta)^{(2/\gamma)-1} \, d\theta$$

$$\times \int_0^{\pi/2} (\cos\varphi)^{(2/\alpha)-1} (\sin\varphi)^{(2/\beta)-1} \, d\varphi.$$

Now, we know that, if $p > 0$, $q > 0$, then

$$B(p, q) = \int_0^{\infty} \frac{s^{p-1}}{(1 + s)^{p+q}} \, ds = \int_0^1 t^{p-1}(1 - t)^{q-1} \, dt$$

$$= 2 \int_0^{\pi/2} \cos^{2p-1}\theta \sin^{2q-1}\theta \, d\theta.$$

If we set $r^2 = s$, we obtain

$$\int_0^{\infty} \frac{r^{2[(1/\alpha)+(1/\beta)+(1/\gamma)]-1}}{1 + r^2} \, dr = \frac{1}{2} B\left[\frac{1}{\alpha} + \frac{1}{\beta} + \frac{1}{\gamma}, 1 - \left(\frac{1}{\alpha} + \frac{1}{\beta} + \frac{1}{\gamma}\right)\right]$$

$$= \frac{1}{2} \Gamma\left(\frac{1}{\alpha} + \frac{1}{\beta} + \frac{1}{\gamma}\right) \cdot \Gamma\left[1 - \left(\frac{1}{\alpha} + \frac{1}{\beta} + \frac{1}{\gamma}\right)\right]$$

when

$$0 < \frac{1}{\alpha} + \frac{1}{\beta} + \frac{1}{\gamma} < 1;$$

also,

$$\int_0^{\pi/2} (\sin\theta)^{(2/\alpha)+(2/\beta)-1} \cos^{(2/\gamma)-1}\theta \, d\theta = \frac{1}{2} B\left(\frac{1}{\alpha} + \frac{1}{\beta}, \frac{1}{\gamma}\right)$$

$$= \frac{1}{2}\left[\Gamma\left(\frac{1}{\alpha} + \frac{1}{\beta}\right) \Gamma\left(\frac{1}{\gamma}\right) \Big/ \Gamma\left(\frac{1}{\alpha} + \frac{1}{\beta} + \frac{1}{\gamma}\right)\right]$$

when $1/\gamma > 0$ and $1/\alpha + 1/\beta > 0$; finally,

$$\int_0^{\pi/2} (\cos\varphi)^{(2/\alpha)-1} (\sin\varphi)^{(2/\beta)-1}\, d\varphi = \frac{1}{2} B\left(\frac{1}{\alpha}, \frac{1}{\beta}\right)$$

when α and β are positive.

Finally,

$$I = \frac{1}{\alpha\beta\gamma} \Gamma\left(\frac{1}{\alpha}\right) \Gamma\left(\frac{1}{\beta}\right) \Gamma\left(\frac{1}{\gamma}\right) \Gamma\left[1 - \left(\frac{1}{\alpha} + \frac{1}{\beta} + \frac{1}{\gamma}\right)\right].$$

The conditions under which this calculation will be valid are the following:

$$0 < \frac{1}{\alpha} + \frac{1}{\beta} + \frac{1}{\gamma} < 1, \qquad \gamma > 0, \qquad \frac{1}{\alpha} + \frac{1}{\beta} > 0, \qquad \alpha > 0, \qquad \beta > 0.$$

These reduce to

$$\alpha > 0, \qquad \beta > 0, \qquad \gamma > 0, \qquad \frac{1}{\alpha} + \frac{1}{\beta} + \frac{1}{\gamma} < 1.$$

In particular, it follows that $\alpha > 1$, $\beta > 1$, and $\gamma > 1$.

Another method. We integrate first with respect to z by setting

$$z^\gamma = t(1 + x^\alpha + y^\beta).$$

We find

$$\int_0^\infty \frac{dz}{1 + x^\alpha + y^\beta + z^\gamma} = \frac{1}{\gamma}(1 + x^\alpha + y^\beta)^{(1/\gamma)-1}\, \Gamma\left(\frac{1}{\gamma}\right) \Gamma\left(1 - \frac{1}{\gamma}\right)$$

if $\gamma > 0$.

We then set

$$y^\beta = (1 + x^\alpha)u$$

and find

$$\frac{1}{\beta\gamma} \Gamma\left(\frac{1}{\gamma}\right) \Gamma\left(1 - \frac{1}{\gamma}\right) B\left(\frac{1}{\beta}, 1 - \frac{1}{\beta} - \frac{1}{\gamma}\right) (1 + x^\alpha)^{(1/\beta)+(1/\gamma)-1}$$

if $\beta > 0$ and $1/\beta + 1/\gamma < 1$.

It remains to integrate with respect to x. We set $x^\alpha = v$ and find

$$\frac{1}{\beta\gamma} \Gamma\left(\frac{1}{\gamma}\right) \Gamma\left(1 - \frac{1}{\gamma}\right) B\left(\frac{1}{\beta}, 1 - \frac{1}{\beta} - \frac{1}{\gamma}\right) \frac{1}{\alpha} B\left(\frac{1}{\alpha}, 1 - \frac{1}{\alpha} - \frac{1}{\beta} - \frac{1}{\gamma}\right),$$

$$\alpha > 0, \qquad \frac{1}{\alpha} + \frac{1}{\beta} + \frac{1}{\gamma} < 1.$$

Since

$$B\left(\frac{1}{\beta}, 1 - \frac{1}{\beta} - \frac{1}{\gamma}\right) = \Gamma\left(\frac{1}{\beta}\right) \Gamma\left(1 - \frac{1}{\beta} - \frac{1}{\gamma}\right) \Big/ \Gamma\left(1 - \frac{1}{\gamma}\right)$$

and

$$B\left(\frac{1}{\alpha}, 1 - \frac{1}{\alpha} - \frac{1}{\beta} - \frac{1}{\gamma}\right) = \Gamma\left(\frac{1}{\alpha}\right) \Gamma\left(1 - \frac{1}{\alpha} - \frac{1}{\beta} - \frac{1}{\gamma}\right) \Big/ \Gamma\left(1 - \frac{1}{\beta} - \frac{1}{\gamma}\right)$$

we again find the value of I that we obtained above by the first method.

CHAPTER 4

Analytic Functions

The exercises in this chapter deal with the elements of the theory of analytic functions:

The Cauchy conditions. Harmonic conjugate functions. Conformal representation. Homographic functions. Elementary functions. Singular points. Multiple-valued functions. Cauchy's theorem. Cauchy's formula. Taylor and Laurent series. Integration of analytic functions. The residue theorem.

We shall invoke only classical concepts requiring a minimum familiarity with functions of a complex variable. The general results on conformal mapping are not included in the program, nor are the Schwarz transformation, the general properties of Riemann surfaces, Mittag-Leffler's theorem, or the properties of entire functions.

We shall encounter various applications of analytic functions in the following chapters. These applications will constitute a supplement to the exercises in this chapter.

1. The Cauchy conditions in polar coordinates

Problem. Show that the Cauchy conditions in polar coordinates are

$$\frac{\partial P}{\partial r} = \frac{1}{r}\frac{\partial Q}{\partial \theta}, \qquad \frac{\partial Q}{\partial r} = -\frac{1}{r}\frac{\partial P}{\partial \theta}.$$

As an application, find the holomorphic functions of the complex variable $z = x + iy$ whose real part does not depend on $r = | z |$.

Solutions. **A.** We have, in polar coordinates,

$$x = r\cos\theta, \qquad dx = \cos\theta\,dr - r\sin\theta\,d\theta,$$
$$y = r\sin\theta, \qquad dy = \sin\theta\,dr + r\cos\theta\,d\theta,$$

from which we get

$$dr = \cos\theta\,dx + \sin\theta\,dy, \qquad r\,d\theta = -\sin\theta\,dx + \cos\theta\,dy.$$

The Cauchy conditions expressing the fact that

$$f(z) = P(x, y) + iQ(x, y)$$

is a holomorphic function of $z = x + iy$ are written

$$\frac{\partial P}{\partial x} = \frac{\partial Q}{\partial y}, \qquad \frac{\partial P}{\partial y} = -\frac{\partial Q}{\partial x}$$

or, in polar coordinates,

$$\frac{\partial P}{\partial r}\frac{\partial r}{\partial x} + \frac{\partial P}{\partial \theta}\frac{\partial \theta}{\partial x} = \frac{\partial Q}{\partial r}\frac{\partial r}{\partial y} + \frac{\partial Q}{\partial \theta}\frac{\partial \theta}{\partial y},$$
$$\frac{\partial P}{\partial r}\frac{\partial r}{\partial y} + \frac{\partial P}{\partial \theta}\frac{\partial \theta}{\partial y} = -\frac{\partial Q}{\partial r}\frac{\partial r}{\partial x} - \frac{\partial Q}{\partial \theta}\frac{\partial \theta}{\partial x},$$

from which we get

$$\frac{\partial P}{\partial r}\cos\theta - \frac{\partial P}{\partial \theta}\frac{\sin\theta}{r} = \frac{\partial Q}{\partial r}\sin\theta + \frac{\partial Q}{\partial \theta}\frac{\cos\theta}{r},$$
$$\frac{\partial P}{\partial r}\sin\theta + \frac{\partial P}{\partial \theta}\frac{\cos\theta}{r} = -\frac{\partial Q}{\partial r}\cos\theta + \frac{\partial Q}{\partial \theta}\frac{\sin\theta}{r},$$

which, by means of obvious linear combinations, is equivalent to

$$\frac{\partial P}{\partial r} = \frac{1}{r}\frac{\partial Q}{\partial \theta}, \qquad -\frac{1}{r}\frac{\partial P}{\partial \theta} = \frac{\partial Q}{\partial r}.$$

An application. If P depends only on r, the second condition expresses the fact that Q depends only on θ. Therefore,

$$r\frac{\partial P}{\partial r} \quad \text{and} \quad \frac{\partial Q}{\partial \theta}$$

have a common constant value k. Then

$$P = k \log r + a, \qquad Q = k\theta + b,$$

where a, b, and k are arbitrary real constants.

Finally,

$$\begin{aligned} Z = P + iQ &= k(\log r + i\theta) + a + bi \\ &= k \log z + z_0 , \end{aligned}$$

where k is a real number and z_0 is a complex number, both arbitrary.

B. *A variant proof.* As we move along the radius vector, we have

$$dZ = (P_r{}' + iQ_r{}')\, dr, \qquad dz = e^{i\theta}\, dr,$$

so that

$$\frac{dZ}{dz} = \frac{P_r{}' + iQ_r{}'}{e^{i\theta}} .$$

If we move perpendicularly to the radius vector, we have

$$dZ = (P_\theta{}' + iQ_\theta{}')\, d\theta, \qquad dz = ire^{i\theta}\, d\theta,$$

so that

$$\frac{dZ}{dz} = \frac{P_\theta{}' + iQ_\theta{}'}{ire^{i\theta}} .$$

Equality of the two derivatives is written

$$P_r{}' + iQ_r{}' = \frac{P_\theta{}' + iQ_\theta{}'}{ir} ,$$

so that

$$P_r{}' = \frac{Q_\theta{}'}{r} , \qquad Q_r{}' = -\frac{P_\theta{}'}{r} .$$

2. The orthogonal trajectories of a family of curves

If $P(x, y)$ is a harmonic function, the equation for the orthogonal trajectories of the family of curves

$$P(x, y) = \text{const}$$

is

$$Q(x, y) = \text{const},$$

where Q is the harmonic conjugate of P such that $P + iQ$ is a holomorphic function of $z = x + iy$.

If P is not harmonic, we can find the orthogonal trajectories by writing the differential equation of the given curves

$$\frac{\partial P}{\partial x} + y' \frac{\partial P}{\partial y} = 0$$

and then replacing y by y' in this equation. The resulting equation

$$y' \frac{\partial P}{\partial x} - \frac{\partial P}{\partial y} = 0$$

is the differential equation of the orthogonal trajectories. Thus the problem requires solution of a differential equation.

However, if P is not harmonic, it may happen that there exists a function F of a single variable such that $F(P)$ is harmonic. Since the curves $P =$ const are also described by the equation $F(P) =$ const, our search for the orthogonal trajectories again leads to a problem concerning holomorphic functions. The present exercise consists in finding out under what conditions on P will there exist a function F such that $F(P)$ is harmonic.

Problems. A. Suppose that $P(x, y)$ is a real function of two real variables that is three times continuously differentiable. For there to exist a function F such that $F(P)$ is harmonic, it is necessary and sufficient that $\arctan(P_y'/P_x')$ be harmonic. How do we then find the function F?

Note that $F(P)$ must be the real part of a holomorphic function $Z(z)$ of $z = x + iy$ and examine $Z'(z)$.

B. Study the following example: $P = (x^2 + y^2 - 1)/y$ (a pencil of circles). Construct the function $Z(z)$.

C. Consider the family of straight lines

$$x \cos \lambda + y \sin \lambda = \varphi(\lambda).$$

The equation for this family, when solved for λ, takes the form $P(x, y) = \lambda$. Show that the condition of Problem A is solved if and only if the straight lines in question are all either concurrent or parallel.

Solutions. **A.** The function P is not assumed to be harmonic. But, by hypothesis, there exists a function F such that $F(P)$ is harmonic. Then $F(P)$ is the real part of a holomorphic function

$$Z = F(P) + iQ.$$

Since Z is holomorphic, Z' exists and is holomorphic. Now, if we move in the direction of the positive x-axis, we have

$$Z' = F'(P)\frac{\partial P}{\partial x} + i\frac{\partial Q}{\partial x}.$$

From the Cauchy conditions, $\partial Q/\partial x = -F'(P)\,\partial P/\partial y$, so that

$$Z' = F'(P)\left(\frac{\partial P}{\partial x} - i\frac{\partial P}{\partial y}\right).$$

Let us set

$$Z' = Ae^{i\Phi}.$$

Then

$$\log Z' = \log A + i\Phi,$$

and Φ, being the imaginary part of the holomorphic function $\log Z'$, is harmonic. Now,

$$\tan \Phi = -\frac{P_y'}{P_x'}.$$

Thus we conclude that, *for there to exist a function F such that F(P) is harmonic, it is necessary that* $\arctan(P_y'/P_x')$ *be harmonic.*

Conversely, suppose that $\arctan(P_y'/P_x')$ is harmonic. There exists then a holomorphic function such that $\arctan(P_y'/P_x')$ is its imaginary part. Thus there exists a function $H(x, y)$ such that

$$H\left(\frac{\partial P}{\partial x} - i\frac{\partial P}{\partial y}\right)$$

is holomorphic. We wish to show that H is a function of P. The condition for this to be the case is that the Jacobian

$$J = \frac{D(H, P)}{D(x, y)}$$

vanish identically. Now, HP_x' and $-HP_y'$ satisfy the Cauchy conditions

$$\frac{\partial}{\partial x}(HP_x') = -\frac{\partial}{\partial y}(HP_y'), \qquad \frac{\partial}{\partial x}(HP_y') = \frac{\partial}{\partial y}(HP_x').$$

By performing the differentiation in the second of these, we obtain

$$J = \frac{\partial H}{\partial x}\frac{\partial P}{\partial y} - \frac{\partial H}{\partial y}\frac{\partial P}{\partial x} = 0.$$

Therefore, there exists a function, which we denote by F', such that $H = F'(P)$ and

$$F'(P)\left(\frac{\partial P}{\partial x} - i\frac{\partial P}{\partial y}\right)$$

is holomorphic. But this is simply the derivative of $F(P) + iQ$. Therefore, $F(P)$ is harmonic and the existence of F is established.

Under these conditions, the family of curves $P =$ const is also given by the equation $F(P) =$ const. The equation for the orthogonal trajectories is $Q =$ const, where Q is the harmonic conjugate of $F(P)$.

We derive the condition for $F(P)$ to be harmonic. Observe that $F(P)$ is a function of two variables.

$$\frac{\partial}{\partial x}F(P) = P_x'F', \qquad \frac{\partial}{\partial y}F(P) = P_y'F',$$

$$\frac{\partial^2}{\partial x^2}F(P) = P''_{x^2}F' + P'^2_xF'', \qquad \frac{\partial^2}{\partial y^2}F(P) = P''_{y^2}F' + P'^2_yF'',$$

so that

$$\Delta F = F'\Delta P + F''(P'^2_x + P'^2_y).$$

Since $\Delta F = 0$, we have

$$F'(P)\Delta P + F''(P)(P'^2_x + P'^2_y) = 0.$$

Thanks to the possibility condition, this is a differential equation in $F(P)$. It is easily verified that the condition

$$\arctan\frac{P_y'}{P_x'} \text{ is harmonic}$$

is equivalent to

$$\frac{P'^2_x + P'^2_y}{\Delta P} \text{ is a function of } P.$$

B. *An example.* The curves $P =$ const, where

$$P = \frac{x^2 + y^2 - 1}{y}$$

are circles passing through the points $y = 0$, $x = \pm 1$. Their orthogonal trajectories constitute a pencil of circles conjugate to the preceding pencil. Let us find this result again by calculation. The function P is not harmonic. However, it is easily verified that $\arctan(P_y'/P_x')$ is harmonic. Let us therefore seek, without carrying out this verification, a function F such that $F(P)$ is harmonic. We have seen that

$$F'(P)\Delta P + F^2(P)(P_x'^2 + P_y'^2) = 0.$$

If the function F exists, this equation must be an ordinary differential equation in F and P; that is, $(P_x'^2 + P_y'^2)/\Delta P$ must be a function of P. Now,

$$P_x' = \frac{2x}{y}, \quad P_y' = \frac{-x^2 + y^2 + 1}{y^2}, \quad P_x'^2 + P_y'^2 = \frac{(x^2 - y^2 - 1)^2 + 4x^2y^2}{y^4},$$

$$P''_{x^2} = \frac{2}{y}, \qquad P''_{y^2} = \frac{2(x^2 - 1)}{y^3}, \qquad \Delta P = \frac{2(x^2 + y^2 - 1)}{y^3}.$$

We see that $\Delta P = 2P/y^2$, and we can verify that

$$\frac{P^2 + 4}{y^2} = P_x'^2 + P_y'^2.$$

Therefore,

$$2PF' + (P^2 + 4)F'' = 0,$$

which is indeed a differential equation in F and P. Its solution is immediate:

$$\frac{F''}{F'} = -\frac{2P}{P^2 + 4}, \qquad F' = \frac{1}{P^2 + 4} \text{ (up to a constant factor)}.$$

Therefore, $F = (\frac{1}{2})\arctan(P/2)$ up to an additive constant.

We can verify that $F'(P)(P_x' - iP_y')$ is a holomorphic function that can serve as Z'. Specifically,

$$P_x' - iP_y' = \frac{2xy + i(x^2 - y^2 - 1)}{y^2} = i\,\frac{\bar{Z}^2 - 1}{y^2},$$

$$P^2 + 4 = \frac{(Z\bar{Z} - 1)^2 - (Z - \bar{Z})^2}{y^2} = \frac{(Z^2 - 1)(\bar{Z}^2 - 1)}{y^2}.$$

Therefore,

$$F'(P)(P_x' - iP_y') = \frac{y^2}{(Z^2 - 1)(\bar{Z}^2 - 1)}\, i\,\frac{\bar{Z}^2 - 1}{y^2} = \frac{i}{Z^2 - 1}.$$

The function Z' is equal to $i/(Z^2 - 1)$ and the function Z is equal to $(i/2)\log(Z - 1)/(Z + 1)$.

The orthogonal trajectories of the circles $(x^2 + y^2 - 1)/y = \text{const}$ are defined by

$$\left|\frac{Z-1}{Z+1}\right| = \text{const} \qquad \text{or} \qquad \frac{(x-1)^2 + y^2}{(x+1)^2 + y^2} = \text{const}.$$

These do indeed constitute a pencil of circles conjugate to the given pencil.

C. Consider the family of straight lines

$$x \cos \lambda + y \sin \lambda = \varphi(\lambda),$$

each defined by a value of the parameter λ. Let us suppose that this equation has been solved for λ in the form

$$P(x, y) = \lambda.$$

Let us see under what condition $\arctan(P_y'/P_x')$ is harmonic.

We may consider P as a solution of the implicit equation

$$x \cos P + y \sin P = \varphi(P).$$

If we differentiate with respect to x, we get

$$\cos P - (x \sin P - y \cos P)P_x' = P_x'\varphi',$$

so that

$$P_x' = \frac{\cos P}{x \sin P - y \cos P + \varphi'}\,.$$

Similarly,

$$P_y' = \frac{\sin P}{x \sin P - y \cos P + \varphi'}\,.$$

We see that

$$\arctan \frac{P_y'}{P_x'} = P.$$

This problem has a solution only in the trivial case in which P is itself a harmonic function, that is, when F is a linear function.

Let us seek the corresponding form of the function φ. If we calculate P''_{x^2} and P''_{y^2}, we see, after some brief computation, that the condition $P''_{x^2} + P''_{y^2} = 0$ reduces to

$$\varphi + \varphi'' = 0.$$

Therefore,

$$\varphi(\lambda) = A \cos \lambda + B \sin \lambda,$$

where A and B are constants. Thus the corresponding straight lines pass through the fixed point whose coordinates are (A, B) and their orthogonal trajectories are circles with centers at that point.

Remark. The form of the equation

$$x \cos \lambda + y \sin \lambda = \varphi(\lambda)$$

excludes the case in which the straight lines have a fixed direction. The equation for these lines is of the form

$$ax + by = \lambda,$$

and $P = ax + by$ is obviously a harmonic function.

In sum, the method of harmonic conjugate functions does not enable us to find the orthogonal trajectories of a family of straight lines except when this family is a pencil of lines.

3. Directional derivatives of nonholomorphic functions

Problem. Suppose that Z is a function of a complex variable z. Consider a given point z and denote by Z' the derivative of Z in the direction of the slope λ. Evaluate Z' and show that, in general, as λ varies, the image of the complex number Z' describes a circle γ. Find the diameter of γ. Consider the case in which Z is a holomorphic function. For convenience, assume that the point z is at the origin.

Solution. Let $Z = P + iQ$ be a function of a complex variable $z = x + iy$. Suppose that P and Q have continuous first derivatives.

We choose the origin as our point of reference. If the derivative of Z exists, it is the limit of $\Delta Z/\Delta z$ as $|\Delta z|$ tends to 0 in any manner whatsoever. If the derivative does not exist, there is still, in general, a derivative in every direction; that is, as Δz approaches 0 along a line determined by the given parameter λ. This derivative is equal to

$$Z' = \lim_{\substack{h \to 0 \\ k \to 0}} \frac{hP_x' + kP_y' + i(hQ_x' + kQ_y')}{h + ik},$$

with h and k approaching 0 in such a way that the ratio k/h tends to λ. Thus

$$Z' = \frac{P_x' + iQ_x' + \lambda(P_y' + iQ_y')}{1 + i\lambda}.$$

In general, Z' depends on λ and is, in fact, a homographic function of λ. Therefore, as the real number λ varies, the complex number Z' describes a circle γ, which is the image, under the homographic mapping, of the real axis. The circle γ passes through the points

$$A = P_x' + iQ_x' \qquad \text{image of} \quad \lambda = 0,$$
$$B = \frac{P_y' + iQ_y'}{i} \qquad \text{image of} \quad \lambda = \infty.$$

Let us find the exact position of γ. We may write

$$Z' = \frac{A - B}{1 + i\lambda} + B.$$

As λ ranges over the real axis, the quantity $t = (1 + i\lambda)^{-1}$ describes the circle γ' one diameter of which is the interval [0, 1]. Therefore,

$$Z' = (A - B)t + B.$$

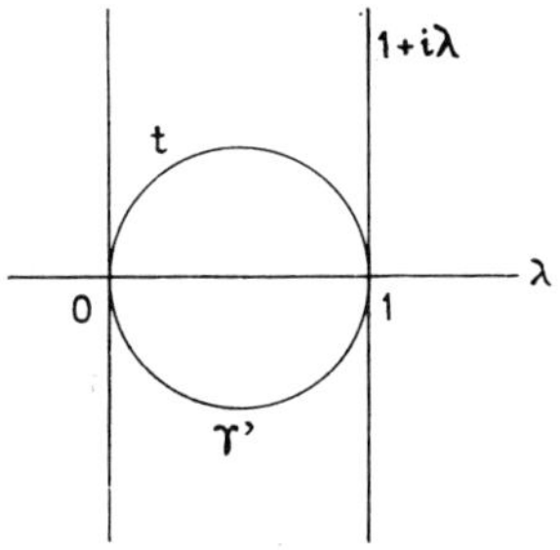

FIGURE 1

The point $(A - B)t$ describes a circle γ'' of diameter $|A - B|$, which is obtained from γ' by a similarity mapping with center at the origin. The circle γ is obtained from γ'' by a translation by an amount B.

For Z to have a derivative, it is necessary that Z' have the same value in all directions, that is, for it to be independent of λ. The circle γ must be reduced to a point. Its diameter must vanish: $|A - B| = 0$. In other words, the two complex numbers A and B are equal. If we equate the real and imaginary parts of these numbers, we obtain the Cauchy conditions.

Example. $Z = \bar{z}$ is not a holomorphic function of z. At the origin, $Z' = (1 - i\lambda)/(1 + i\lambda)$, but Z' describes the unit circle about the origin.

4. Determination of holomorphic functions

Problems. Suppose that $Z = f(z) = P + iQ$ is a holomorphic function of the complex variable $z = x + iy$. Determine Z so that

A. P depends only on $x \cos \varphi + y \sin \varphi$, where φ is a given angle.

B. $P^2 + Q^2$ depends only on x.

C. P/Q depends only on x.

D.

$$P = \frac{x(1 + x) + y^2}{(1 + x)^2 + y^2}.$$

E. $P = (x/2) \log(x^2 + y^2) - y \arctan(y/x)$.

F.

$$P = \frac{\sin 2x}{\cosh 2y - \cos 2x}.$$

G. P is the product of a function of x and a function of y.
(*N.B.*: These seven questions are independent.)

Solutions. **A.** Let us find a real function F of a single real variable such that

$$P(x, y) = F(x \cos \varphi + y \sin \varphi).$$

We have

$$\frac{\partial^2 P}{\partial x^2} = F''(x \cos \varphi + y \sin \varphi) \cos^2 \varphi,$$
$$\frac{\partial^2 P}{\partial y^2} = F''(x \cos \varphi + y \sin \varphi) \sin^2 \varphi.$$

The condition $\Delta P = 0$ implies $F'' = 0$, so that $F(u) = au + b$, where a and b are real constants. Thus

$$P(x, y) = a(x \cos \varphi + y \sin \varphi) + b$$

and, from the Cauchy conditions,

$$Q(x, y) = a(-x \sin \varphi + y \cos \varphi) + c.$$

Therefore, the holomorphic function of which P is the real part is

$$ae^{-i\varphi} z + b + ic,$$

where a, b, and c are arbitrary real constants.

B. Since $P^2 + Q^2$ depends only on x, the same is true of $\log|f(z)|$, which is the real part of $\log f(z)$. Therefore,

$$\log f(z) = az + b + ic,$$

where a, b, and c are real numbers, and

$$f(z) = z_0 e^{az},$$

where a is an arbitrary real number and z_0 is an arbitrary complex number.

C. Let us set

$$Z = P + iQ = Ae^{iB}.$$

Then

$$\log Z = \log A + iB, \qquad \tan B = \frac{Q}{P}.$$

If P/Q depends only on x, the same is true of B and the real part of $-i \log Z$ depends only on x. From Problem A (with $\varphi = 0$),

$$-i \log Z = az + b + ic,$$

so that

$$\log Z = iaz + ib - c$$

and

$$Z = z_0 e^{iaz},$$

where z_0 is an arbitrary complex number.

D. Consider $z = x + iy$ and its conjugate $\bar{z} = x - iy$. We have

$$P = \frac{z\bar{z} + [(z + \bar{z})/2]}{(1 + z)(1 + \bar{z})} = \frac{1}{2}\left(\frac{z}{1 + z} + \frac{\bar{z}}{1 + \bar{z}}\right).$$

Thus a solution of the problem posed is

$$Z = \frac{z}{1 + z},$$

since

$$P = \mathscr{R}Z = \tfrac{1}{2}(Z + \bar{Z}).$$

Therefore, the general solution is

$$Z = f(z) = \frac{z}{1+z} + ia \qquad (a \text{ real}).$$

E. Let us set $z = re^{i\theta}$. We have

$$P = x \log r - y\theta.$$

If

$$\zeta = \log z = \log r + i\theta,$$

we see that

$$\zeta z = x \log r - y\theta + i(x\theta + y \log r).$$

Therefore,

$$f(z) = z \log z + ia,$$

where a is real.

F. We have

$$P = \frac{1}{i} \frac{e^{2ix} - e^{-2ix}}{e^{2y} + e^{-2y} - (e^{2ix} + e^{-2ix})}.$$

We set

$$\alpha = e^{iz} \qquad \bar{\alpha} = e^{-i\bar{z}}$$

$$\beta = e^{-iz} = \frac{1}{\alpha}, \qquad \bar{\beta} = e^{i\bar{z}}.$$

We have

$$e^{2ix} = \alpha\bar{\beta}, \qquad e^{-2ix} = \beta\bar{\alpha},$$
$$e^{2y} = \beta\bar{\beta}, \qquad e^{-2y} = \alpha\bar{\alpha},$$

so that

$$P = \frac{1}{i} \frac{\alpha\bar{\beta} - \beta\bar{\alpha}}{\beta\bar{\beta} + \alpha\bar{\alpha} - (\alpha\bar{\beta} + \beta\bar{\alpha})} = \frac{1}{i} \frac{\alpha\bar{\beta} - \beta\bar{\alpha}}{(\alpha - \beta)(\bar{\alpha} - \bar{\beta})} = \frac{1}{i}\left(\frac{\bar{\alpha}}{\bar{\alpha} - \bar{\beta}} - \frac{\alpha}{\alpha - \beta}\right)$$
$$= \left(\frac{i\alpha}{\alpha - \beta}\right) + \overline{\left(\frac{i\alpha}{\alpha - \beta}\right)}$$

If we set

$$Z = P + iQ,$$

we have

$$P = \tfrac{1}{2}(Z + \bar{Z}).$$

Therefore, one solution of the problem is

$$Z = \frac{2i\alpha}{\alpha - \beta} = \frac{2ie^{iz}}{e^{iz} - e^{-iz}} = \frac{e^{iz}}{\sin z} = \frac{\cos z}{\sin z} + i.$$

The general solution is obtained from this by adding a purely imaginary constant. Thus

$$Z = \cot z + ia \qquad (a \text{ real}).$$

G. If P is of the form $\varphi(x)\psi(y)$, the condition $\Delta P = 0$ may be written

$$\frac{\varphi''(x)}{\varphi(x)} = -\frac{\psi''(y)}{\psi(y)}.$$

Thus the two members of this equation must both be equal to some real constant. If, for example, this constant is positive, we have

$$\varphi(x) = Ae^{\omega x} + Be^{-\omega x}, \qquad \psi(y) = A'e^{i\omega y} + B'e^{-i\omega y},$$

where A, B, A', B' are constants and A', B' are complex conjugates. For φ to be real, A and B must be real; for ψ to be real, A' and B' must be complex conjugates. From this we conclude that

$$P = AA'e^{\omega z} + BB'e^{-\omega z} + AB'e^{\omega \bar{z}} + BA'e^{-\omega \bar{z}}.$$

Since $B' = \bar{A}'$, we may set

$$AA' = \frac{\alpha}{2}, \qquad BB' = \frac{\beta}{2}, \qquad AB' = \frac{\bar{\alpha}}{2}, \qquad BA' = \frac{\bar{\beta}}{2},$$

so that

$$P = \tfrac{1}{2}(\alpha e^{\omega z} + \beta e^{-\omega z}) + \tfrac{1}{2}(\bar{\alpha} e^{\omega \bar{z}} + \bar{\beta} e^{-\omega \bar{z}}).$$

One solution of the problem posed is

$$Z = \alpha e^{\omega z} + \beta e^{-\omega z},$$

where ω is real and α and β are complex constants. The general solution is obtained from this by the addition of a purely imaginary constant.

We obtain a second class of solutions by replacing the real constant ω with a purely imaginary one $i\omega$. Examples of particular solutions are

$$Z = \cosh z, \qquad Z = \sinh z, \qquad Z = \cos z, \qquad Z = \sin z.$$

5. Poincaré's geometry

Plane geometry can be thought of as the study of the properties of plane figures that are invariant under transformations belonging to a certain group of transformations. If this group is the group of displacements of the plane (translations, rotations, possibly including symmetries), we are interested in the congruence properties of figures, and two homologous arcs have the same Euclidean length. Length is an invariant of the group of displacements.

One can conceive of other groups of transformations of the plane that have an invariant and that may serve as the foundation of a (non-Euclidean) plane geometry. If we think of the plane as the plane of a complex variable z, the set of homographic transformations constitutes a group. This group has various subgroups (in particular, the group of displacements $z' = az + b$, for $|a| = 1$). One can easily verify that the transformations

$$z' = \frac{az + b}{cz + d},$$

where a, b, c, and d are real numbers such that $ad - bc = 1$, constitute a group and that this group preserves the upper half-plane. A geometry that does not satisfy the same axioms as Euclidean geometry and is modeled after Lobachevskian geometry has been constructed with this group as its foundation.

In this geometry, straight lines are replaced with semicircles lying in the upper half-plane with their centers on the x-axis. The "length" of an arc γ is no longer the usual length $\int_\gamma ds$ but the integral $\int_\gamma y^{-1}\, ds$. The homographic transformation has an invariant, which is the birapport of four points. This birapport is associated with the length of a "non-Euclidean line segment." Specifically, consider a semicircle whose diameter is the segment of the x-axis between two points α and β. If z_1 and z_2 are two points on this semicircle and γ is the arc between them, we have

$$\int_\gamma \frac{ds}{y} = |\log(z_1, z_2, \alpha, \beta)|$$

and the arc length of such non-Euclidean lines is additive on that arc.

The double integral $\int\int_\Omega dx\, dy/y^2$ plays the role of non-Euclidean area. The area of a triangle (formed by three arcs of circles with centers on the x-axis) is equal to $\pi - (A + B + C)$, where A, B, and C are the angles of such a triangle; thus the sum of the angles of one of these triangles is less than π.

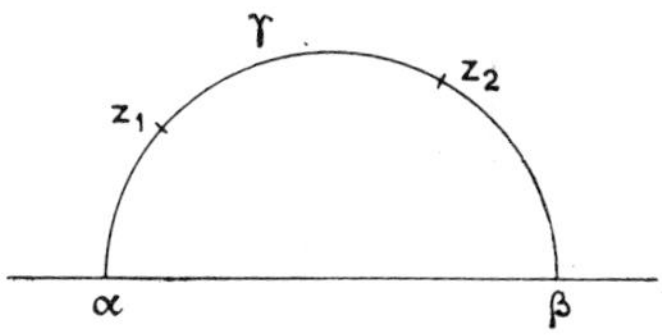

FIGURE 2

It can be verified with the aid of elementary methods of the calculus of variations that the "line" segment joining two points is the curve of minimum "length" joining these two points.

The points at infinity of such a "line" are the end points α and β of the semicircle representing it. Through any point there pass two "lines" parallel to a given "line" δ, that is, two lines having a point at infinity in common with δ. However, infinitely many of these "lines" pass through the given point that do not intersect δ and yet are not parallel to δ.

The purpose of the following exercise is to prove those of the above-mentioned properties that are not obvious.

Problems. Let

$$z' = \frac{az + b}{cz + d},$$

be a homographic transformation whose coefficients are real and related by

$$ad - bc = 1.$$

A. Show that, if the imaginary part y of z is positive, the imaginary part y' of z' is also positive. The point z describes an arc γ lying in the half-plane $y > 0$. Let γ' denote the arc described by z'. Denote by $ds = |\, dz \,|$ and $ds' = |\, dz' \,|$ the elements of arc length. Show that

$$\int_\gamma \frac{ds}{y} = \int_{\gamma'} \frac{ds'}{y'}.$$

B. Let γ be a circular arc with end points z_1 and z_2 (located in the half-plane $y > 0$) and with center on the real axis. Let α and β denote the points at which the circle containing this arc intersects the real axis. Show that

$$\int_\gamma \frac{ds}{y} = |\log(z_1, z_2, \alpha, \beta)|,$$

where the parenthesized expression is the cross ratio of z_1, z_2, α, and β.

C. Let Ω and Ω' be two regions in the xy-plane that are mapped one into the other by the homographic transformation. Show that

$$\iint_{\Omega} \frac{dx\,dy}{y^2} = \iint_{\Omega'} \frac{dx'\,dy'}{y'^2}.$$

D. Let z_1, z_2, and z_3 be three points in the half-plane $y > 0$. These three points define three circular arcs with centers on the x-axis. These three arcs form a curvilinear triangle with angles A, B, and C. Let Ω denote the area inside this triangle. Show that

$$\iint_{\Omega} \frac{dx\,dy}{y^2} = \pi - (A + B + C).$$

Solutions. **A.** The homographic transformation

$$z' = \frac{az + b}{cz + d},$$

where a, b, c, and d are real numbers such that

$$ad - bc = 1,$$

maps z into z'. We have

$$z' = \frac{(az + b)(c\bar{z} + d)}{|\,cz + d\,|^2} = \frac{acz\bar{z} + bd + adz + bc\bar{z}}{|\,cz + d\,|^2}.$$

From this we conclude that the imaginary part y' of z' is obtained from the imaginary part of z by the formula

$$y' = \frac{y}{|\,cz + d\,|^2}.$$

Thus y and y' have the same sign. This homographic transformation preserves the two half-planes $y > 0$ and $y < 0$ globally.

We have

$$\frac{dz'}{dz} = \frac{1}{(cz + d)^2},$$

so that

$$\left|\frac{dz'}{dz}\right| = \frac{y'}{y}.$$

If γ and γ' are two homologous arcs, then $|dz|$ and $|dz'|$ are curvilinear elements of abscissas on these arcs measured positively in the counterclockwise sense on γ and in the corresponding sense on γ'. Thus we have

$$\frac{ds'}{ds} = \frac{y'}{y} \quad \text{or} \quad \frac{ds'}{y'} = \frac{ds}{y},$$

so that

$$\int_\gamma \frac{ds}{y} = \int_{\gamma'} \frac{ds'}{y'}.$$

B. We take for γ a circular arc $z_1 z_2$ in the half-plane $y > 0$ with its center on the x-axis. We denote by α and β its points of intersection with the x-axis. Consider the homographic transformation

$$z' = a\,\frac{z - \alpha}{z - \beta}.$$

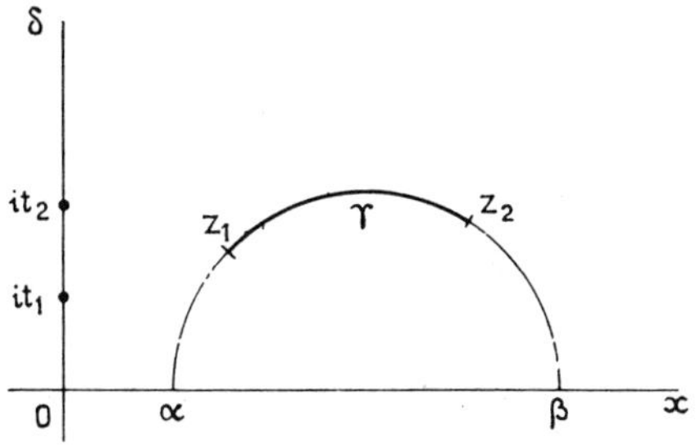

FIGURE 3

If $a(\alpha - \beta) = 1$, this transformation is of the type indicated. It maps the point α into the origin, the point β into the point at infinity of the complex plane, the circle γ into a straight line δ, and the real axis into itself. Since angles are conserved, the line δ is orthogonal to the real axis. Since it passes through the origin, it is the imaginary axis. Let us denote by it_1 and it_2 the images of z_1 and z_2, and let us denote by γ' the segment joining the points it_1 and it_2. Thus, if we assume, for example, that $t_1 < t_2$,

$$\int_{\gamma'} \frac{ds'}{y'} = \int_{t_1}^{t_2} \frac{|dy'|}{y'} = \left| \int_{t_1}^{t_2} \frac{dy'}{y'} \right| = \left| \log \frac{t_2}{t_1} \right|.$$

Now,

$$\frac{t_1}{t_2} = \frac{it_1}{it_2} = (it_1, it_2, 0, \infty),$$

which is the cross ratio of the four points it_1, it_2, 0, ∞.

The homographic transformation preserves the cross ratio. Now, it_1, it_2, 0, and ∞ are the images of z_1, z_2, α, and β, respectively. Therefore, in accordance with Problem A,

$$\int_\gamma \frac{ds}{y} = |\log(z_1, z_2, \alpha, \beta)|.$$

C. Let us set $z' = f(x, y) + ig(x, y)$. From the theory of change of variable in double integrals, we have

$$\iint_{\Omega'} \frac{dx'\,dy'}{y'^2} = \iint_{\Omega} \left| \frac{D(f, g)}{D(x, y)} \right| \frac{|cz + d|^4}{y^2}\, dx\,dy.$$

Now,

$$\frac{D(f, g)}{D(x, y)} = \left| \frac{dz'}{dz} \right|^2 = \frac{1}{|cz + d|^4}.$$

Thus we have

$$\iint_{\Omega'} \frac{dx'\,dy'}{y'^2} = \iint_{\Omega} \frac{dx\,dy}{y^2}.$$

D. According to Riemann's formula,

$$\iint_{\Omega} \frac{dx\,dy}{y^2} = \int_\gamma \frac{dx}{y},$$

where the contour γ is described in the positive sense.

Let us suppose that γ is a curvilinear triangle $z_1z_2z_3$, the sides of which are three circular arcs with centers on the real axis.

To evaluate $\int\int_\Omega dx\,dy/y^2$, we need only calculate $\int dx/y$ on the three sides of the triangle in succession going in the counterclockwise direction.

Let l denote the abscissa of the center of the circle that passes through z_1 and z_2 and let r denote its radius. Then, on this circle,

$$x = l + r\cos\theta, \qquad y = r\sin\theta,$$
$$dx = -r\sin\theta\,d\theta, \qquad \int \frac{dx}{y} = -\int d\theta.$$

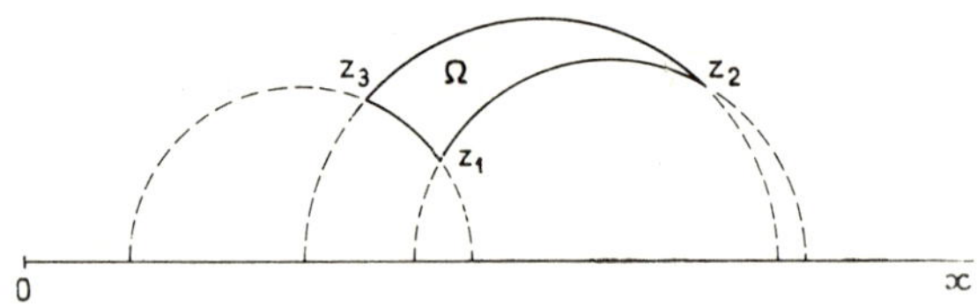

FIGURE 4

Thus the value of the integral on one side of the triangle is equal to the negative of the change in θ on that side. Instead of the radius vector, we may introduce the tangent to γ oriented counterclockwise. The change in θ on one side is equal to the change in the angle between this tangent and the x-axis. As we move along γ, the total change in this angle is 2π. But this change includes the changes on the sides plus the sudden changes at the vertices, these last being equal to $\pi - A$, $\pi - B$, and $\pi - C$, respectively, where A, B, and C are the angles of the triangle. Thus we have

$$\iint_{\Omega} \frac{dx\, dy}{y^2} = \int_{\gamma} \frac{dx}{y} = -[2\pi - (\pi - A) - (\pi - B) - (\pi - C)]$$

$$= \pi - (A + B + C).$$

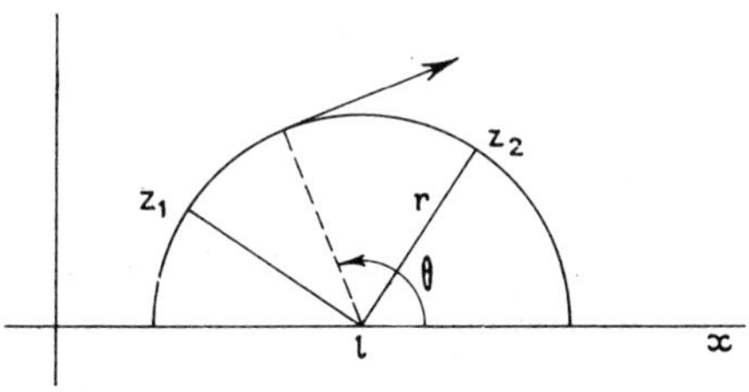

FIGURE 5

From this, it follows, in particular, that $A + B + C$ is at most equal to π.

Example. Consider the triangle shown in Figure 6, in which each semicircle is tangent to each of the other two at the ends of their diameters on the x-axis. Here, we have

$$A = B = C = 0, \qquad \text{so that} \qquad \iint_{\Omega} \frac{dx\, dy}{y^2} = \pi.$$

Let us verify this result by evaluating the double integral for each of the three semicircles separately. One difficulty appears because, if the

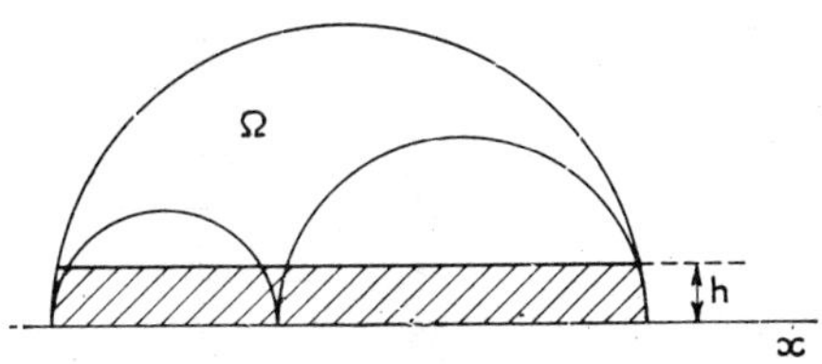

FIGURE 6

integral over the entire figure exists, each of the three integrals over the triangles diverges. Therefore, we shall consider the complement with respect to Ω of a strip of altitude h parallel to the x-axis and then let h tend to 0.

Thus it is a matter of evaluating $\int\int dx\,dy/y^2$, over an area consisting of a closed semicircle of radius R minus this strip of altitude h parallel to the diameter which is the limiting position of the upper edge of the strip. We can translate the center of the circle to the origin. Thus we need to evaluate

$$\int_h^R \frac{dy}{y^2} \int_{-(R^2-y^2)^{1/2}}^{(R^2-y^2)^{1/2}} dx = 2\int_h^R \frac{(R^2-y^2)^{1/2}}{y^2}\,dy.$$

If we set $y = R\sin\theta$, we get

$$2\int_{\theta_0}^{\pi/2} \frac{\cos^2\theta}{\sin^2\theta}\,d\theta = -\pi + 2(\operatorname{cotan}\theta_0 + \theta_0)$$

$$= \frac{2R}{h}\left(1 - \frac{h^2}{R^2}\right)^{1/2} - \pi + 2\arcsin\frac{h}{R} \qquad \text{with } \sin\theta_0 = \frac{h}{R}.$$

If h is infinitesimally small, this expression has the abbreviated expansion

$$\frac{2R}{h} - \pi + \cdots .$$

Thus if R_1 and R_2 denote the radii of the interior circles, we have

$$\iint_\Omega \frac{dx\,dy}{y^2} = \frac{2}{h}(R_1 + R_2) - \frac{2}{h}R_1 - \frac{2}{h}R_2 - \pi + \pi + \pi + \cdots = \pi + \cdots,$$

where the omitted terms tend to 0 as h tends to 0. The unwritten terms cancel and, in the limit, we have

$$\iint_\Omega \frac{dx\,dy}{y^2} = \pi.$$

Remark. In Euclidean geometry, for any rectilinear triangle, the difference $\pi - (A + B + C)$ is zero.

In Poincaré's geometry, the difference $\pi - (A + B + C)$ is a positive number equal to the area of the curvilinear triangle.

In the case of a spherical triangle, consisting of three arcs of great circles, the difference $\pi - (A + B + C)$ is negative and the corresponding positive quantity, that is,

$$A + B + C - \pi$$

is equal to the area of the triangle (on a unit sphere).

6. Examples of conformal transformations

In this exercise we shall map a sector S formed by a 60° arc of a unit circle conformally onto the half-plane $x > 0$. Now we might open the sector by means of the transformation $u = z^6$ to make it cover the entire circle and then map the circle onto the half-plane by means of a homographic transformation. But the transformation obtained in this way does not satisfy the problem posed because the image of the interior of the sector S is not the interior of the half-plane. The segment (0,1) of the real axis *interior* to the half-plane appears twice in the image of the *boundary* of S. Thus we would actually be mapping S onto the half-plane *cut* along the segment (0,1). A closed curve in the interior of the half-plane that touches the cut (for example, a circle with center at the origin and of radius less than 1) is not the image of a closed curve in the interior of S.

To map S onto the half-plane, we are led to a somewhat more complicated conformal transformation.

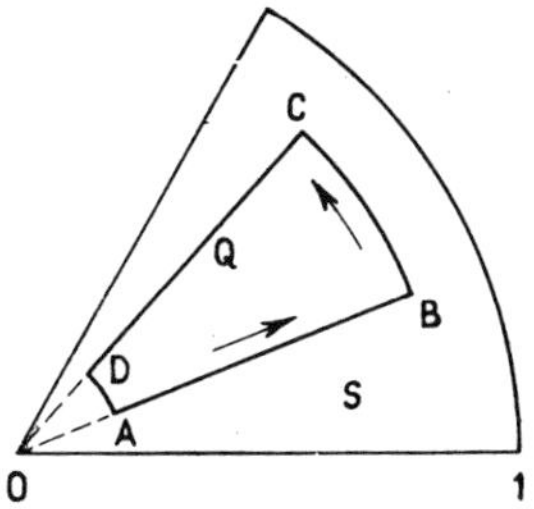

FIGURE 7

Problems. Consider a circular sector S in the complex z-plane that lies in the interior of the unit circle and whose boundary is defined in polar coordinates by

$$\theta = 0, \quad 0 \leqslant r \leqslant 1, \quad r = 1,$$

$$0 \leqslant \theta \leqslant \frac{\pi}{3}, \quad \theta = \frac{\pi}{3}, \quad 0 \leqslant r \leqslant 1.$$

Let Q denote a curvilinear quadrilateral in the interior of S whose sides are defined by

$$\begin{array}{ll} r_1 \leqslant r \leqslant r_2\,, & \theta = \theta_1\,, \\ r_1 \leqslant r \leqslant r_2\,, & \theta = \theta_2\,, \\ r = r_1\,, & \theta_1 \leqslant \theta \leqslant \theta_2\,, \\ r = r_2\,, & \theta_1 \leqslant \theta \leqslant \theta_2\,, \end{array}$$

with

$$0 < \theta_1 < \theta_2 < \frac{\pi}{3}\,, \qquad 0 < r_1 < r_2 < 1.$$

A. What is the image of Q under the transformation

$$u = z^6\,?$$

B. What is the image of Q under the transformation

$$(T_1) \quad Z = \frac{1-u}{1+u} = \frac{1-z^6}{1+z^6}\,?$$

What is the image of the boundary of the vector S?

C. What is the image of Q under the transformation

$$v = z^3\,?$$

D. What is the image of Q under the transformation

$$w = \frac{1-v}{1+v} = \frac{1-z^3}{1+z^3}\,?$$

E. What is the image of Q under the transformation

$$(T_2) \quad Z = iw^2 = i\left(\frac{1-z^3}{1+z^3}\right)^2\,?$$

What is the image of the boundary of the sector S?

F. Show that the image of S under (T_1) is the half-plane $\mathscr{R}(z) > 0$ cut by the segment $(0,1)$ and that the image of S under (T_2) is the uncut half-plane $\mathscr{R}(z) > 0$.

Solutions. (In all that follows, it will be convenient to say that θ_1 is small, θ_2 is close to $\pi/3$, that r_1 is small, and that r_2 is close to 1.)

A. The transformation $u = z^6$ maps Q into an analogous quadrilateral Q' defined by

$$\rho_1 \leqslant r \leqslant \rho_2 , \qquad \theta = \varphi_1 \quad \text{and} \quad \theta = \varphi_2 ,$$
$$\varphi_1 \leqslant \theta \leqslant \varphi_2 , \qquad r = \rho_1 \quad \text{and} \quad r = \rho_2 ,$$

where

$$\rho_1 = r_1^{\,6}, \qquad \rho_2 = r_2^{\,6}, \qquad \varphi_1 = 6\theta_1 , \qquad \varphi_2 = 6\theta_2 ,$$

ρ_1 is small, ρ_2 is close to 1, φ_1 is small, and φ_2 is close to 2π.

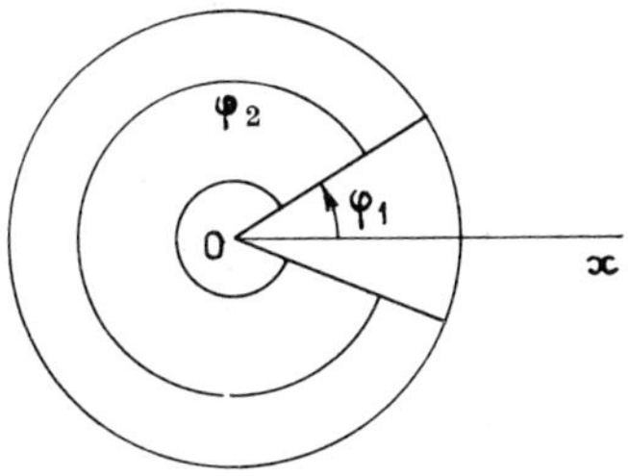

FIGURE 8

B. The transformation $Z = (1 - z^6)/(1 + z^6)$ is the product of the transformations

$$u = z^6 \qquad \text{and} \qquad Z = \frac{1 - u}{1 + u} .$$

This last is a homographic transformation. We have

$$Z = \frac{(1 - u)(1 + \bar{u})}{| 1 + u |^2} = \frac{1 - | u |^2}{| 1 + u |^2} + \frac{\bar{u} - u}{| 1 + u |^2} .$$

If $| u | < 1$, then $\mathscr{R}Z > 0$; if $| u | > 1$, then $\mathscr{R}Z < 0$.

Thus the image of the interior of the unit circle is the half-plane $\mathscr{R}(z) > 0$.

If u describes the real axis, Z describes the real axis. If u describes a circle γ of radius r with center at the origin, Z describes a circle Γ orthogonal to the real axis and passing through the points

$$Z = \frac{1 - r}{1 + r} \, (u = r) \qquad \text{and} \qquad Z = \frac{1 + r}{1 - r} \, (u = -r).$$

If u describes a ray issuing from the origin, Z describes a circular arc orthogonal to the circle Γ and passing through the points

$$Z = 1 \qquad (u = 0) \qquad \text{and} \qquad Z = -1 \qquad (u = \infty).$$

Figure 9 shows the image Q'' of the quadrilateral Q'.

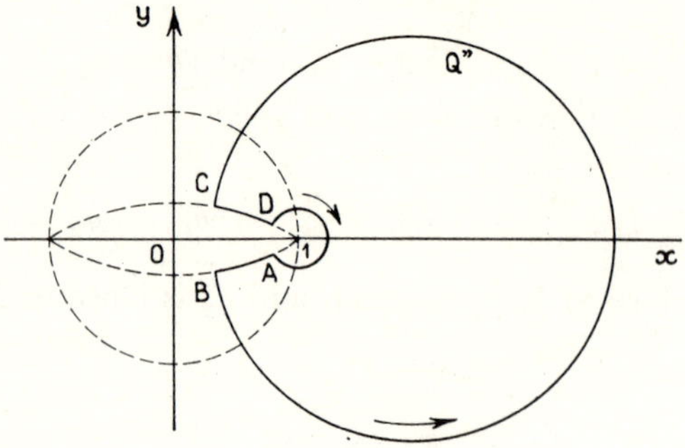

FIGURE 9

We can now define the image of the boundary of the circular sector S. The image of $z = 0$ is $Z = 1$. The image of the circle $|z| = 1$ is the imaginary axis. The straight lines

$$\theta = 0 \qquad \text{and} \qquad \theta = \frac{\pi}{3}$$

are both mapped onto the real axis. The image of S is therefore the half-plane

$$\mathscr{R}Z > 0$$

cut along the segment (0, 1).

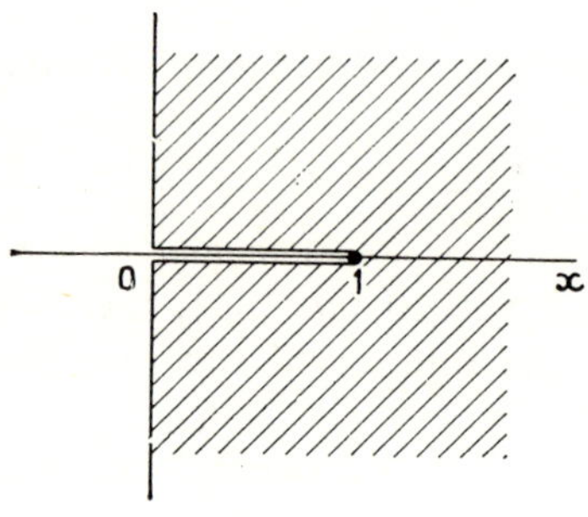

FIGURE 10

C. The transformation $v = z^3$ maps the given quadrilateral Q into a quadrilateral Q_1 shown in Figure 11.

D. The transformation

$$w = \frac{1 - v}{1 + v} = \frac{1 - z^3}{1 + z^3}$$

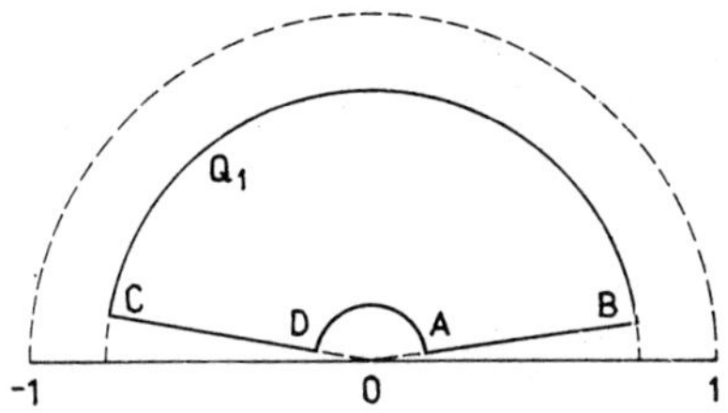

FIGURE 11

was studied in Question B. It maps Q_1 into a quadrilateral Q_2 lying in one quadrant of the plane, whereas Q'' was located in the right *half*-plane.

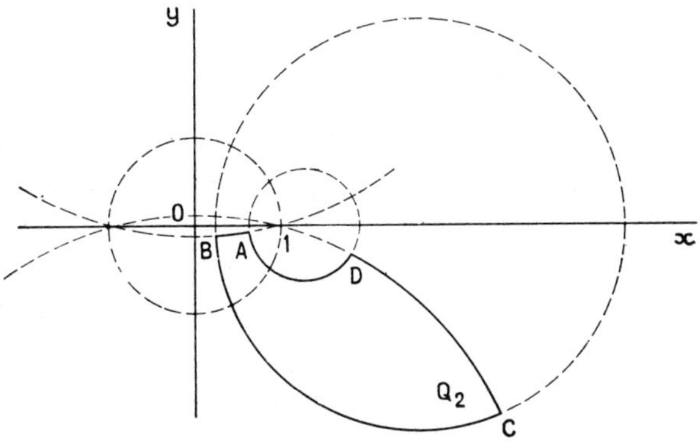

FIGURE 12

E. Finally, let us set

$$Z = iw^2 = i\left(\frac{1 - z^3}{1 + z^3}\right)^2.$$

If the argument of w lies between $-\pi/2$ and 0, the argument of w^2 lies between $-\pi$ and 0 and the argument of Z lies between $-\pi/2$ and $\pi/2$. The transformation $Z = iw^2$ maps the quadrant occupied by the quadrilaterals Q_2 into the half-plane $\mathscr{R}(z) > 0$.

Thus the transformation

$$Z = i\left(\frac{1 - z^3}{1 + z^3}\right)^2$$

maps the initial sector S onto the half-plane $\mathscr{R}(z) > 0$.

The boundary of S is transformed as follows: $z = 0$ is mapped into $Z = i$; $|z| = 1$ into the negative half of the imaginary axis; the segment (0, 1) of the real axis into the segment (0, 1) of the imaginary axis; the segment (0, 1) of the ray $\theta = \pi/3$ into the segment $(1, \infty)$ of the imaginary axis.

F. Let us compare the transformations

$$Z = \frac{1 - z^6}{1 + z^6} \qquad \text{and} \qquad Z = i\left(\frac{1 - z^3}{1 + z^3}\right)^2.$$

These two transformations map the sector S onto the half-plane $\mathscr{R}(z) > 0$. However, in the first case, the half-plane is cut by the segment (0, 1) of the real axis. The boundaries of the two images are quite different.

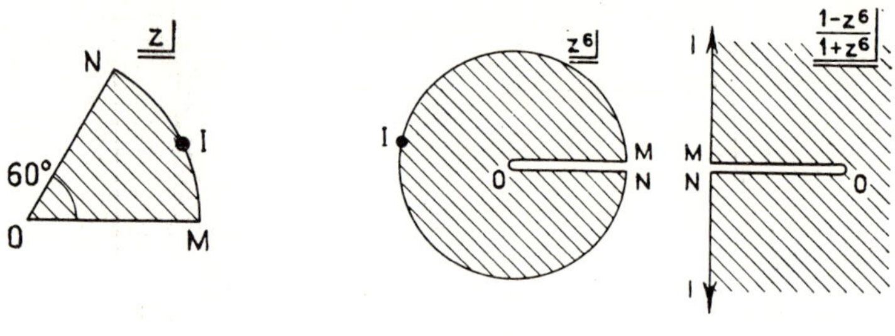

FIGURE 13

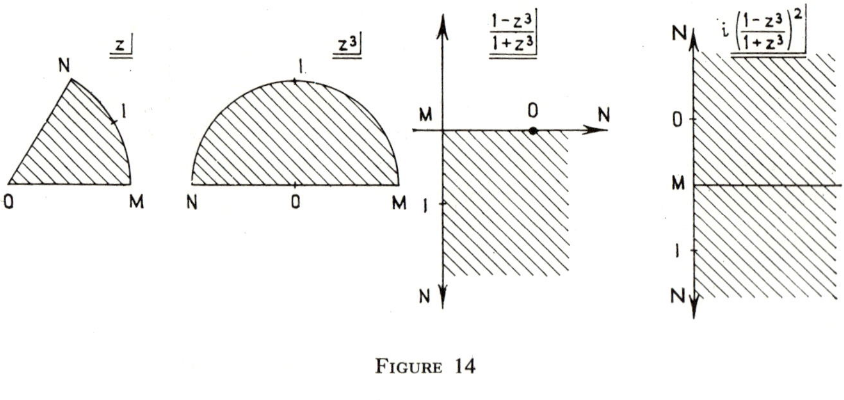

FIGURE 14

7. The final value of a multiple-valued function

Problem. Find the final value of $u = \arctan(1 - z)^{1/2}$ as z describes the line segment from the point $z = 0$ to the point $z = 1 + i$, the initial value of u being $\pi/4$.

Solutions. Let us set

$$v = (1 - z)^{1/2}.$$

For $z = 0$, we choose $v = +1$. As z describes the segment $[0, 1 + i]$, the argument of $1 - z$ varies from 0 (an assigned value, up to $2k\pi$) to $-\pi/2$, and the argument of v varies from 0 to $-\pi/4$. The point v describes an arc of the hyperbola

$$x^2 - 2xy - y^2 = 1.$$

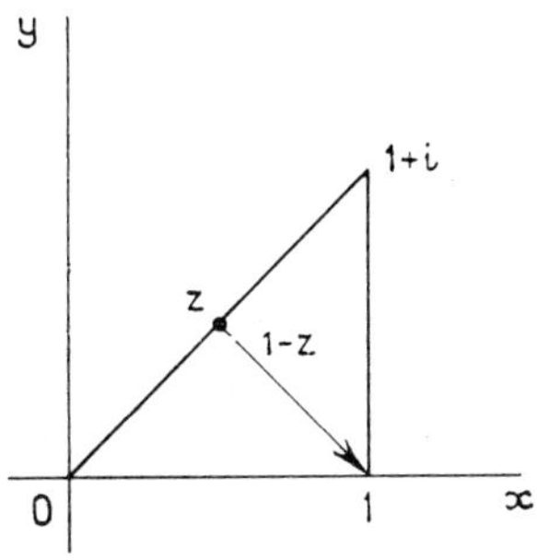

FIGURE 15

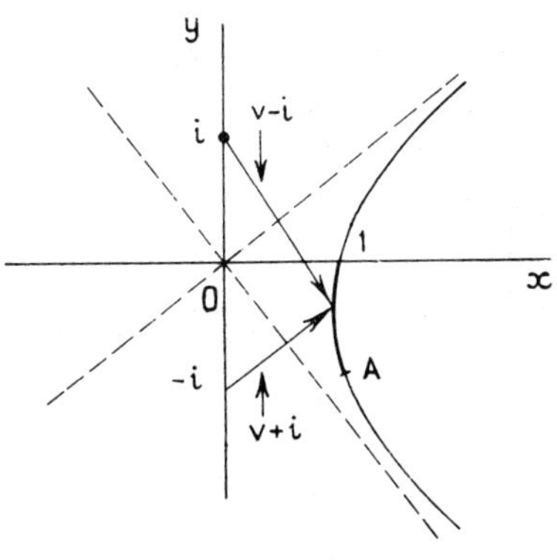

FIGURE 16

This arc extends from the point $v = 1$ to the point $A = \sqrt{-i} = e^{-i\pi/4}$. We then have

$$u = \arctan v = \frac{1}{2i} \log \frac{i - v}{i + v}.$$

We set

$$\frac{v - i}{v + i} = re^{i\theta}, \qquad u = \frac{\theta + \pi}{2} - \frac{i}{2} \log r.$$

By hypothesis, $u = \pi/4$ for $z = 0$, that is, for $v = 1$.

The values of the absolute values and arguments of $v - i$ and $v + i$ that are compatible with all these hypotheses are as follows:

	Absolute value of		Argument of			
	$v - i$	$v + i$	$v - i$	$v + i$	r	θ
Initial	$\sqrt{2}$	$\sqrt{2}$	$-\frac{\pi}{4}$	$\frac{\pi}{4}$	1	$-\frac{\pi}{2}$
Final	$\left[\frac{1}{2} + \left(1 + \frac{1}{\sqrt{2}}\right)^2\right]^{1/2}$ $= \sqrt{2 + \sqrt{2}}$	$\left[\frac{1}{2} + \left(1 - \frac{1}{\sqrt{2}}\right)^2\right]^{1/2}$ $= \sqrt{2 - \sqrt{2}}$	$-\frac{\pi}{2} + \alpha$	α	$\left(\frac{2+\sqrt{2}}{2-\sqrt{2}}\right)^{1/2}$ $= 1+\sqrt{2}$	$-\frac{\pi}{2}$

where α is equal to $\pi/8$.

Thus at the point $z = 1 + i$,

$$u = \frac{\pi}{4} - \frac{i}{2} \log (1 + \sqrt{2}).$$

This is the value assumed by $u = \arctan (1 - z)^{1/2}$ when z leaves the point $z = 0$ and arrives at the point $1 + i$ after describing the line segment $[0, 1 + i)$ when the initial value of u is fixed at $\pi/4$.

The value of u is determined and the expression found gives the real part and the imaginary part.

Remark. The function arctan has the two critical points

$$v = i \qquad \text{and} \qquad v = -i.$$

The function $v = (1 - z)^{1/2}$ has the critical point $z = 1$. The function

$$u = \arctan (1 - z)^{1/2}$$

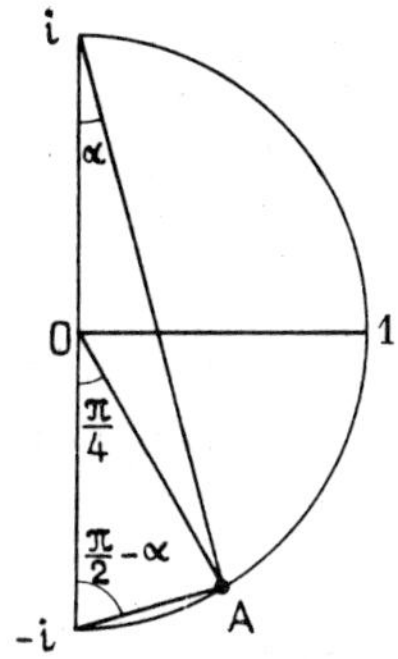

FIGURE 17

has the two critical points $z = 1$ and $z = 2$. The value of u at the point $1 + i$ depends on the initial value and also on the path taken from the origin to the point $1 + i$.

8. The expansion of an implicit function

Suppose that a relationship of the form

$$z = a + x\varphi(z)$$

exists between two complex variables x and z, where a is a given constant and $\varphi(z)$ is holomorphic in a neighborhood of $z = a$.

This relation defines a holomorphic function z of x in a neighborhood of $x = 0$ that assumes the value $z = a$ for $x = 0$.

Suppose that $f(z)$ is a holomorphic function in a neighborhood of $z = a$. Then $f(z(x))$ is a holomorphic function of x in a neighborhood of $x = 0$. We shall prove the general formula

$$f(z) = \sum_{n=0}^{\infty} \frac{x^n}{n!} \frac{d^{n-1}}{da^{n-1}} [f'(a)\varphi^n(a)]$$

for the Taylor-series expansion of $f(z)$ in powers of x in a neighborhood of $x = 0$.

In this exercise we shall examine the proof of a particular case of this formula in detail.

Problems. A. Show that the equation

$$xz^2 - 2z + 2a - x = 0, \tag{1}$$

where a is a real constant such that $0 \leqslant a < 1$, defines an analytic function x of the complex variable z. If z describes a circle of sufficiently small radius with center at a, x describes a closed curve without multiple points that encircles the origin. Find an expression for this curve (in terms of the curve described by the point $1/x$).

B. Show that (1) enables us to define z as a holomorphic function of x in a neighborhood of the origin. What is the value of this function for $x = 0$? What is the radius of convergence of the expansion of z in powers of x?

C. We set $\varphi(z) = (z^2 - 1)/2$, so that the function z of the complex variable x is defined by the equation

$$z = a + x\varphi(z).$$

Let $f(z)$ be a holomorphic function of z in a neighborhood of $z = a$. In a sufficiently small circle with center at the origin, we may write

$$f(z) = A_0 + A_1 x + \cdots + A_n x^n + \cdots .$$

Show that

$$A_n = \frac{1}{n!}\frac{d^{n-1}}{da^{n-1}}\{f'(a)[\varphi(a)]^n\}.$$

D. *An application.* Expand $1/(1 - 2ax + x^2)^{1/2}$ in powers of x.

Solutions. **A.** By hypothesis, a is a real number such that $0 \leqslant a < 1$. The complex variables x and z obey the relation

$$\psi(x, z) = xz^2 - 2z + 2a - x = 0. \tag{1}$$

Thus we have

$$x = 2\frac{z - a}{z^2 - 1}. \tag{2}$$

From this we see that x is a meromorphic function of z with simple poles at $z = \pm 1$.

Suppose that z describes a circle of radius R with center at a. Then we may set

$$z = a + Re^{i\theta}.$$

The point x describes a closed curve γ, which can be obtained from the curve described by $1/x$ by an inversion followed by a reflection of the x-axis. Now, we see immediately that this last curve is an ellipse E, defined parametrically by the equations

$$\left.\begin{aligned} X - a &= \frac{R^2 - (1 - a^2)}{2R} \cos\theta \\ Y &= \frac{R^2 + 1 - a^2}{2R} \sin\theta \end{aligned}\right\} \quad \text{with} \quad \frac{1}{x} = X + iY.$$

The center of this ellipse is the point $(a, 0)$ and one of its axes is the real axis. It encircles the point a. The inverse curve also encircles the origin if the origin is in the interior of E. This will certainly be the case if R is sufficiently small because $1 - a^2$ is positive. In general, the abscissa of the left vertex of the ellipse must be negative; that is,

$$\frac{|R^2 - (1 - a^2)|}{2R} > a.$$

In discussing this inequality, we see that it is verified for

$$R < 1 - a.$$

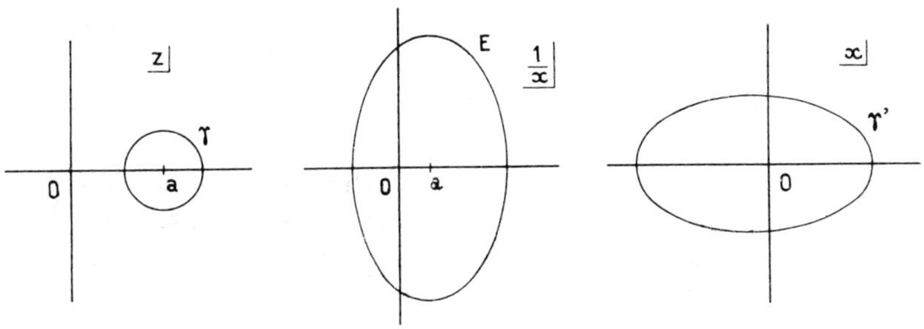

FIGURE 18

Thus if z describes a circle γ of radius less than $1 - a$ with center at a, then x describes a simple closed curve, the inverse of an ellipse, which encircles the origin.

B. Let us see if we can turn (2) around and represent z as an analytic function of x. Formally, we have

$$z = \frac{1 \pm (1 - 2ax + x^2)^{1/2}}{x}.$$

Thus z is a non-single-valued function possessing two critical points, which are the roots of the equation $1 - 2ax + x^2 = 0$, namely, $x = a \pm i(1 - a^2)^{1/2}$. These are conjugate points of absolute value 1.

When $x = 0$, (1) is satisfied only when $z = a$. The implicit-function theorem is applicable to (1) if $\partial\psi/\partial z \neq 0$ when $x = 0$ and $z = a$. Then

$$\frac{\partial\psi}{\partial z} = 2xz - 2.$$

At the point $x = 0$, $z = a$, we have $\partial\psi/\partial z = -2$. Thus there exists a unique function $z(x)$ that is continuous and assumes the value a at $x = 0$ such that $\psi(x, z) = 0$. This function is differentiable and hence holomorphic. Let us denote by $(1 - 2ax + x^2)^{1/2}$ that branch of the radical that is equal to $+1$ at $x = 0$. We have

$$z(x) = \frac{1 - (1 - 2ax + x^2)^{1/2}}{x}. \tag{3}$$

We can verify easily that $z(x)$ does tend to a as $x \to 0$.

The radius of convergence of the expansion of z in powers of x is equal to 1 since the critical points $a \pm i(1 - a^2)^{1/2}$ of the preceding function lie on the circle of radius 1 with center at the origin.

C. We set $\varphi(z) = (z^2 - 1)/2$. Then the function z satisfies the equation

$$z = a + x\varphi(z).$$

Let $f(z)$ be a holomorphic function of the complex variable z in a neighborhood of $z = a$. In a circle of sufficiently small radius with center at the origin, z is a holomorphic function of x. Therefore, $f(z)$ is also a holomorphic function of x. Let us write

$$f(z) = A_0 + A_1x + \cdots + A_nx^n + \cdots.$$

If we set $f(z) = F(x)$, we have

$$A_n = \frac{1}{n!}F^{(n)}(0).$$

Let Γ denote a closed curve bounding a simply connected region containing the origin and contained in the circle of the preceding problem.

According to Cauchy's integral formula as applied to the $(n - 1)$st derivative of $F'(x)$,

$$F^{(n)}(0) = \frac{(n-1)!}{2\pi i}\int_\Gamma \frac{F'(x)}{x^n}\,dx.$$

Since $F'(x) = f'(z)\, dz/dx$, we have

$$A_n = \frac{1}{2ni\pi} \int_\Gamma \frac{f'(z)}{x^n} \frac{dz}{dx}\, dx.$$

Let us choose for Γ the image under the mapping $x = 2(z-a)/(z^2-1)$ of a circle C of sufficiently small radius with center at a. We know from Problem A that this is possible. Then we have

$$\frac{1}{x^n} = \frac{\varphi^n(z)}{(z-a)^n}$$

and

$$A_n = \frac{1}{2ni\pi} \int_C \frac{f'(z)}{(z-a)^n} \varphi^n(z)\, dz.$$

Therefore,

$$A_n = \frac{1}{n!} \frac{d^{n-1}}{da^{n-1}} [f'(a)\varphi^n(a)].$$

Example. Let us choose $f(z) = z$. Then

$$z = \frac{1-(1-2ax+x^2)^{1/2}}{x}.$$

Since $f'(z) = 1$, the expression for A_n reduces to

$$A_n = \frac{1}{n!} \frac{d^{n-1}}{da^{n-1}} [\varphi^n(a)] = \frac{1}{2^n n!} \frac{d^{n-1}}{da^{n-1}} [(a^2-1)^n].$$

Thus we obtain a series expansion of z. It is somewhat more interesting to derive this expansion with respect to a (which can be justified).
We obtain the expansion

$$\frac{1}{(1-2ax+x^2)^{1/2}} = \sum_{n=0}^{\infty} \frac{x^n}{2^n n!} \frac{d^n}{da^n} [(a^2-1)^n].$$

Considered as a function of a, A_n is the *Legendre polynomial* P_n of degree n. The function

$$\frac{1}{(1-2ax+x^2)^{1/2}}$$

is the *generating function* of the Legendre polynomials.

9. Evaluation of an integral by different methods (the case of a single-valued function)

Problems. Evaluate the integral

$$I = \int_C \frac{dz}{2 - \cos z},$$

where C is the line segment joining the points $i \log 3$ and $i \log 3 + 2\pi$.

A. Use the primitive.

B. Use the residue theorem.

Solutions. **A.** If we set $t = \tan(z/2)$, the primitive of the function $(2 - \cos z)^{-1}$ is the function

$$f(z) = \frac{2}{\sqrt{3}} \arctan t\sqrt{3}.$$

Thus we need to evaluate the change in $(2/\sqrt{3}) \arctan t\sqrt{3}$ as z describes the segment given. Now,

$$t = \frac{1}{i}\frac{e^{iz} - 1}{e^{iz} + 1}, \qquad \arctan t\sqrt{3} = \frac{1}{2i} \log \frac{i - t\sqrt{3}}{i + t\sqrt{3}}.$$

Thus, by eliminating t,

$$f(z) = \frac{1}{i\sqrt{3}} \log \frac{e^{iz} - \alpha}{1 - \alpha e^{iz}} \qquad \text{where} \qquad \alpha = 2 - \sqrt{3}.$$

Let us set $u = e^{iz}$. Then

$$f(z) = \frac{1}{i\sqrt{3}} \log \frac{u - \alpha}{1 - \alpha u}.$$

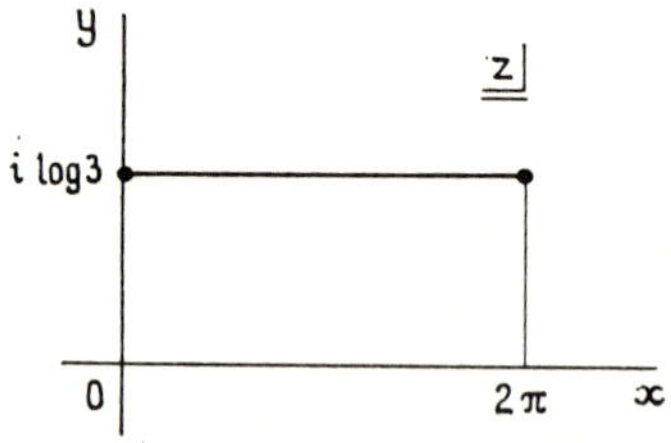

FIGURE 19

Now, as z varies on the given segment, u describes the circle γ of radius $\frac{1}{3}$ with center at 0.

The complex number

$$v = \frac{u - \alpha}{1 - u\alpha}$$

therefore describes the image of that circle (see Figure 20). Since v is a homographic function of u, this curve is a circle γ'. Since the real axes

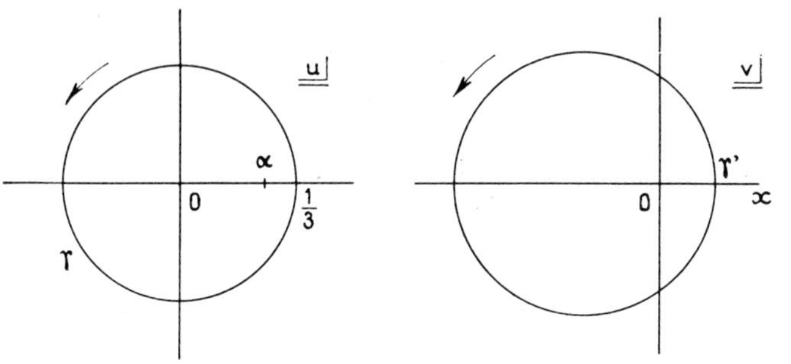

FIGURE 20

correspond to each other, γ' is orthogonal to the real axis. If $u = \frac{1}{3}$, v is real and positive; if $u = -\frac{1}{3}$, then v is real and negative. Thus the circle γ' encircles the origin $v = 0$. The change in $(1/i) \log v$ as v describes γ' is equal to change in the argument of v, namely, 2π. Finally,

$$I = \frac{2\pi}{\sqrt{3}}.$$

B. Let us set $u = e^{iz}$ directly. Since

$$\cos z = \frac{1}{2}\left(u + \frac{1}{u}\right), \qquad dz = \frac{1}{i}\frac{du}{u},$$

we have

$$I = \frac{1}{i}\int_\gamma \frac{du}{2u - \frac{1}{2}(u^2 + 1)} = -\frac{2}{i}\int_\gamma \frac{du}{u^2 - 4u + 1}.$$

The function $u^2 - 4u + 1$ has two zeros: $u_1 = 2 - \sqrt{3}$ and $u_2 = 2 + \sqrt{3}$. The first of these lies inside the circle γ of radius $\frac{1}{3}$ with center at 0, the second outside it. Thus the function $(u^2 - 4u + 1)^{-1}$ has a simple pole u_1 in the interior of γ, and its residue is $(2u_1 - 4)^{-1} = -1/(2\sqrt{3})$. From the *residue theorem*,

$$I = -\frac{2}{i}2i\pi\left(-\frac{1}{2\sqrt{3}}\right) = \frac{2\pi}{\sqrt{3}}.$$

We can avoid use of the residue theorem proper by writing

$$I = -\frac{2}{i}\int_\gamma \frac{du}{(u-u_1)(u-u_2)} = \frac{1}{2i\pi}\int_\gamma \frac{\varphi(u)}{u-u_1}\,du,$$

with

$$\varphi(u) = -\frac{4\pi}{u-u_2}.$$

The function $\varphi(u)$ is holomorphic in the interior of γ. Therefore, we may employ *Cauchy's formula* to I:

$$I = \varphi(u_1) = -\frac{4\pi}{u_1-u_2} = \frac{2\pi}{\sqrt{3}}.$$

10. Fourier series and meromorphic functions

The general theory of Fourier series applies to real functions that are not necessarily analytic or even continuous. However, it happens that the function that is expanded in a Fourier series can be extended holomorphically into a certain domain containing the real axis. In this case, the evaluation of the Fourier coefficients can be made with the aid of integration procedures in the complex plane. However, we can also interpret a Fourier series in x as a Laurent series in e^{ix}. Then the expansion is made directly without individual calculation of the Fourier coefficients.

The expansion of a function $f(z)$ in a Laurent series in a neighborhood of the origin is valid in the open domain between two circles entered at the origin. The function $f(z)$ must be holomorphic (and hence single-valued) between these two circles. It may have poles or other essential singularities in the interior of the smaller circle, but the origin is not necessarily a singular point.

Problems. A. Evaluate the integral

$$I_n = \int_{-\pi}^{\pi} \frac{\cos nx}{2+\cos x}\,dx$$

by making the change of variable $e^{ix} = z$. Use the result to find the Fourier-series expansion of the function $(2 + \cos x)^{-1}$.

B. Expand in a Laurent series the function $\cos nx/(2 + \cos x)$, treated as a function of $z = e^{ix}$, in a circular annulus with center at 0, of radius to be determined. Find by this procedure the Fourier series for $(2 + \cos x)^{-1}$ obtained in Problem A.

C. Reevaluate the integral I_n without going into the complex plane by showing a recursion relation between I_{n-2}, I_{n-1}, and I_n.

(*N. B.*: Compare this exercise with Exercise 11 of Chapter 2.)

Solutions. **A.** Consider the integral

$$I_n = \int_{-\pi}^{\pi} \frac{\cos nx}{2 + \cos x}\, dx \qquad (n \text{ a nonnegative integer}).$$

If we set $z = e^{ix}$, the point z describes in the positive direction the unit circle C. Since

$$\int_{-\pi}^{\pi} \frac{\sin nx}{2 + \cos x}\, dx = 0$$

(note that the integrand is an odd function), we have

$$\begin{aligned} I_n &= \int_{-\pi}^{\pi} \frac{e^{inx}}{2 + \cos x}\, dx \\ &= \frac{1}{i}\int_C \frac{z^n}{2 + \frac{1}{2}\,[z + (1/z)]}\,\frac{dz}{z} \\ &= \frac{2}{i}\int_C \frac{z^n}{z^2 + 4z + 1}\, dz. \end{aligned}$$

We now have an integral of a rational fraction. The poles are the zeros of the denominator:

$$z_1 = -2 + \sqrt{3}, \qquad z_2 = -2 - \sqrt{3}.$$

These are real, negative, and reciprocals of each other. The pole z_1 is in the interior of C and the corresponding residue is

$$\frac{2}{i}\,\frac{z_1{}^n}{2z_1 + 4} = \frac{z_1{}^n}{i\sqrt{3}}.$$

Therefore, we have

$$I_n = \frac{2\pi}{\sqrt{3}} z_1^{\,n}$$

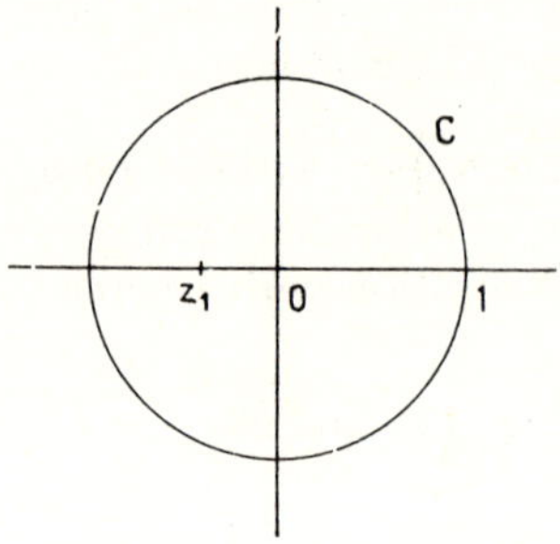

FIGURE 21

The value I_n/π is the nth coefficient of the Fourier-series expansion of $(2 + \cos x)^{-1}$, which is a continuous infinitely differentiable function with period 2π:

$$\frac{\pi}{2 + \cos x} = \frac{I_0}{2} + I_1 \cos x + \cdots + I_n \cos nx + \cdots .$$

Thus we have

$$\frac{1}{2 + \cos x} = \frac{2}{\sqrt{3}} \left[\frac{1}{2} + z_1 \cos x + \cdots + z_1^{\,n} \cos nx + \cdots\right],$$

with

$$z_1 = -(2 - \sqrt{3}).$$

Check. The series in the brackets is the real part of the geometric series

$$\frac{1}{2} + z_1 e^{ix} + \cdots + z_1^{\,n} e^{inx} + \cdots = -\frac{1}{2} + \frac{1}{1 - z_1 e^{ix}} = \frac{1}{2}\frac{1 + z_1 e^{ix}}{1 - z_1 e^{ix}}$$

$$= \frac{1}{2}\frac{(1 + z_1 e^{ix})(1 - z_1 e^{-ix})}{1 - 2z_1 \cos x + z_1^{\,2}} .$$

Therefore, for every real z_1 such that $|z_1| < 1$,

$$\frac{1}{2} + z_1 \cos x + \cdots + z_1^{\,n} \cos nx + \cdots = \frac{1}{2}\frac{1 - z_1^{\,2}}{1 + z_1^{\,2} - 2z_1 \cos x} .$$

If $z_1 = -(2 - \sqrt{3})$, we find in the right-hand member

$$\frac{-3 + 2\sqrt{3}}{8 - 4\sqrt{3} - 2z_1 \cos x} = \frac{\sqrt{3} z_1}{4z_1 + 2z_1 \cos x} = \frac{\sqrt{3}}{2} \frac{1}{2 + \cos x}.$$

B. *Use of the Laurent series.* To expand

$$f(x) = \frac{1}{2 + \cos x}$$

in a Fourier series, we set $e^{ix} = z$, from which we get

$$f = \frac{2z}{z^2 + 4z + 1}.$$

This function can be expanded in a Laurent series between the circles of radius $| z_1 |$ and $| z_2 |$ with centers at the origin.

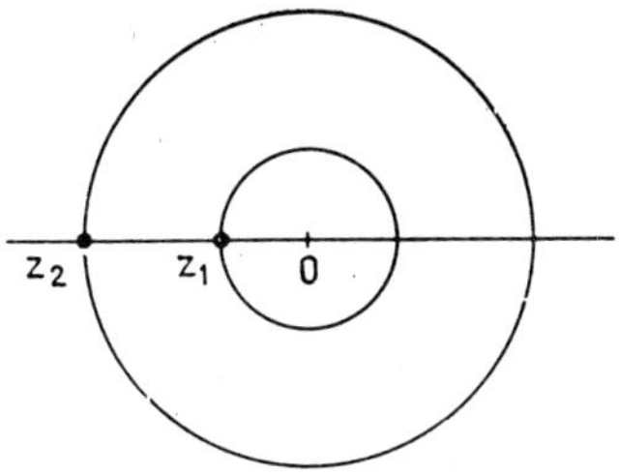

FIGURE 22

Decomposing the rational function f into partial fractions, we have

$$f = \frac{2}{z_1 - z_2}\left(\frac{z_1}{z - z_1} - \frac{z_2}{z - z_2}\right) = \frac{1}{\sqrt{3}}\left(\frac{z_1}{z - z_1} - \frac{z_2}{z - z_2}\right).$$

Since $| z_1/z | < 1$, we have

$$\frac{z_1}{z - z_1} = \frac{z_1}{z} \frac{1}{1 - (z_1/z)} = \frac{z_1}{z} + \cdots + \frac{z_1{}^n}{z^n} + \cdots.$$

Since $| z/z_2 | < 1$, we have

$$-\frac{z_2}{z - z_2} = \frac{1}{1 - (z/z_2)} = 1 + \cdots + \frac{z^n}{z_2{}^n} + \cdots.$$

Therefore,

$$f = \frac{1}{\sqrt{3}} \left[1 + \frac{z}{z_2} + \frac{z_1}{z} + \cdots + \frac{z^n}{z_2{}^n} + \frac{z_1{}^n}{z^n} + \cdots \right].$$

Since $z_2 = 1/z_1$, the general term is

$$\left(z^n + \frac{1}{z^n} \right) z_1{}^n = 2z_1{}^n \cos nx,$$

so that

$$f = \frac{2}{\sqrt{3}} \left[\frac{1}{2} + z_1 \cos nx + \cdots + z_1{}^n \cos nx + \cdots \right].$$

Remark. This expansion is not valid just for real values of x. It converges in the region of validity of the Laurent series, namely,

$$| z_1 | < | z | < | z_2 |.$$

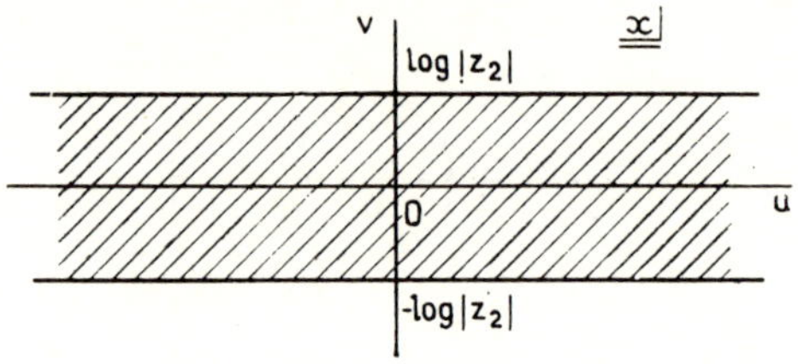

FIGURE 23

If we set $x = u + iv$, we have

$$z = e^{i(u+iv)}, \qquad | z | = e^{-v}.$$

Therefore,

$$\frac{1}{| z_2 |} < e^{-v} < | z_2 | \qquad \text{and} \qquad -\log | z_2 | < v < \log | z_2 |.$$

In the complex x-plane, the region of convergence is a strip symmetric about the x-axis.

C. The integral

$$I_n = \int_{-\pi}^{\pi} \frac{\cos nx}{2 + \cos x} \, dx,$$

can in theory be reduced to an integral of a rational function by the change of variable $\tan(x/2) = t$. However, a more practical way of evaluating it is as follows: First, we have

$$2 \cos (n-1)x(2+\cos x) = 4 \cos (n-1)x + \cos nx + \cos (n-2)x,$$

from which we get

$$2 \cos (n-1)x = 4 \frac{\cos (n-1)x}{2+\cos x} + \frac{\cos nx}{2+\cos x} + \frac{\cos (n-2)x}{2+\cos x},$$

this formula being valid for $n \geqslant 2$. Integrating from $-\pi$ to π, we get

$$I_n + 4I_{n-1} + I_{n-2} = 2 \int_{-\pi}^{\pi} \cos (n-1)\, x\, dx = 0.$$

For $n = 1$,

$$1 = \frac{2}{2+\cos x} + \frac{\cos x}{2+\cos x},$$

from which we get

$$2I_0 + I_1 = \int_{-\pi}^{\pi} dx = 2\pi.$$

Finally,

$$I_0 = \int_{-\pi}^{\pi} \frac{dx}{2+\cos x} = \frac{2}{\sqrt{3}} \left(\arctan \frac{t}{\sqrt{3}}\right)_{-\infty}^{\infty} = \frac{2\pi}{\sqrt{3}} \qquad \left(t = \tan \frac{x}{2}\right)$$

and

$$I_1 = \frac{2\pi}{\sqrt{3}} (\sqrt{3} - 2) = \frac{2\pi}{\sqrt{3}} z_1 .$$

We evaluate I_n with the aid of the linear homogeneous recursion formula

$$I_n + 4I_{n-1} + I_{n-2} = 0. \tag{1}$$

Thus I_n is completely determined once we know I_0 and I_1. Now, one possible form of solution is

$$I_n = r^n,$$

where r is an unknown real or complex number. If we substitute this expression in (1), we obtain

$$r^2 + 4r + 1 = 0.$$

We obtain two possible values for r, which are the numbers z_1 and z_2 that we have already encountered. The general solution of (1) is

$$I_n = \lambda z_1{}^n + \mu z_2{}^n, \tag{2}$$

where λ and μ are two constants such that

$$I_0 = \lambda + \mu, \qquad I_1 = \lambda z_1 + \mu z_2\,.$$

We know that I_n, being the Fourier coefficient of a differentiable function, tends to 0 as $n \to \infty$. Since $|z_2| > 1$, it follows that μ is necessarily 0. Therefore,

$$\lambda = I_0 = \frac{2\pi}{\sqrt{3}} \qquad \text{and} \qquad I_n = \frac{2\pi}{\sqrt{3}} z_1{}^n.$$

We can verify that $\lambda z_1 = 2\pi z_1/\sqrt{3}$ is indeed the value of I_1.

11. Integration of a function with an essential singularity

Problems. Evaluate the integral

$$I = \frac{1}{2i\pi} \int_\gamma \frac{e^{1/z}}{z-1}\, dz$$

over the circle γ of radius R with center at the origin by the following methods:

A. Use the residue theorem directly.

B. First make the change of variable $1/z = t$.

Solutions. **A.** The function $e^{1/z}(z-1)^{-1}$ has a pole at $z = 1$ and an isolated essential singularity at the point $z = 0$. Its integral can be evaluated by use of the residue theorem. At the point $z = 0$, the function $e^{1/z}$ can be expanded in a Laurent series

$$e^{1/z} = 1 + \frac{1}{z} + \frac{1}{2z^2} + \cdots + \frac{1}{n!z^n} + \cdots$$

and the function $(z-1)^{-1}$ can be expanded in a Taylor series

$$\frac{1}{z-1} = -1 - z - z^2 - \cdots - z^k - \cdots.$$

The product of these two can be expanded in a Laurent series. This series is obtained by term-by-term multiplication. The value of the residue c, that is, the coefficient of the term $1/z$, is

$$c = -1 - \frac{1}{2} - \cdots - \frac{1}{n!} - \cdots = 1 - e.$$

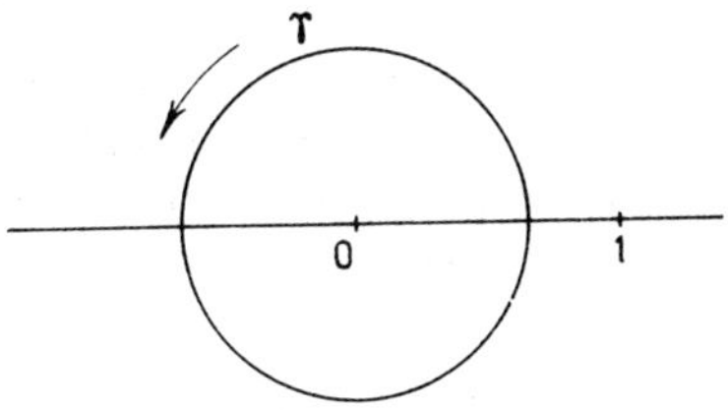

FIGURE 24

Thus if the radius R of the circle γ is less than 1, we have

$$I = 1 - e.$$

If $R > 1$, the pole $z = 1$ is in the interior of γ. The residue corresponding to this pole is equal to e. The integral is equal to $1 - e + e = 1$.

B. Let us first set $z = 1/t$. Then t describes a circle γ' of radius $1/R$ in the negative sense. If we change the sign in front of the integral, γ' is described in the positive sense and

$$I = \frac{1}{2i\pi} \int_{\gamma'} \frac{e^t}{t(1-t)}\, dt.$$

Now, the integrand is a meromorphic function. It has two simple poles, namely, $t = 0$ and $t = 1$, with residue 1 and $-e$, respectively. If $R < 1$, so that $1/R > 1$, both poles are in the interior of γ' and

$$I = 1 - e.$$

If $R > 1$, only the pole $t = 0$ is of significance, and then

$$I = 1.$$

12. Evaluation of a definite integral by different methods (the case of a multiple-valued function)

Problems. Evaluate the integral

$$I = \int_0^\infty \frac{dx}{1 + x^\alpha},$$

where α is a real number exceeding 1, by the following methods:

A. Integrate the function $1/(1 + z^\alpha)$, which in general is multiple-valued, on a closed contour consisting of a segment of the positive real axis, a segment of the ray of polar angle $2\pi/\alpha$, and two circular arcs—one of very small, the other of very large radius.

B. Consider the case in which α is an even integer. Calculate $2I$ by the residue method applied to rational functions.

C. Make the change of variable $x^\alpha = t$ and use the formula for complements.

Solutions. **A.** The general method. Consider the function

$$f(z) = \frac{1}{1 + z^\alpha},$$

where α is a real number exceeding 1. In general, the function z^α, when α is not an integer, is a critical point at the origin. We make it a single-valued function by cutting the complex plane along the positive half of the real axis. We choose the real determination in the upper neighborhood of this semiaxis. Then we integrate $1/(1 + z^\alpha)$ along the curve shown in Figure 25 constituted by the two segments d_1 and d_2 and the two circular arcs γ_1 and γ_2, where γ_1 has a very small radius r and γ_2 has a very large radius R.

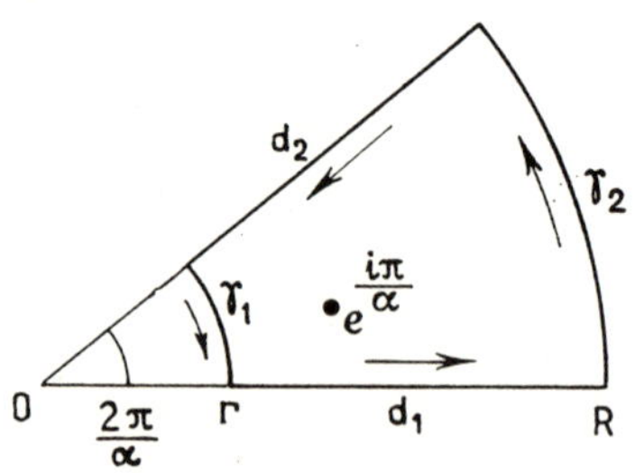

FIGURE 25

Here d_1 is a segment of the real axis, and d_2 is a segment of the ray whose polar angle is $2\pi/\alpha$. In the interior of this contour, $(1 + z^\alpha)^{-1}$ has a simple pole: $z = e^{i\pi/\alpha}$. The residue at this point is the value of $1/\alpha z^{\alpha-1} = z/\alpha z^\alpha$ at $z = e^{i\pi/\alpha}$, namely, $e^{i\pi/\alpha}/(-\alpha)$. The integral around the contour is equal to

$$-\frac{2i\pi}{\alpha} e^{i\pi/\alpha}.$$

On d_1 it tends to $I = \int_0^\infty dx/(1 + x^\alpha)$. On d_2 we set $z = te^{2i\pi/\alpha}$. Then the integral tends to

$$-\int_0^\infty e^{2i\pi/\alpha} \frac{dt}{1 + t^\alpha} = -e^{2i\pi/\alpha} I.$$

On γ_1 and γ_2, Jordan's lemma is applicable. Specifically,

$$| zf(z) | = \frac{| z |}{| 1 + z^\alpha |}$$

tends to 0 as $| z | \to 0$ or as $| z | \to \infty$ (since $\alpha > 1$). Finally,

$$I - e^{2i\pi/\alpha} I = -\frac{2i\pi}{\alpha} e^{i\pi/\alpha} \qquad \text{and} \qquad I = \frac{\pi}{\alpha \sin(\pi/\alpha)}.$$

B. If α is an integer, $f(z)$ is single-valued. The method of Problem A remains valid in this case, in particular if α is odd. If α is an even integer $2p$, we can also write

$$2I = \int_{-\infty}^{\infty} \frac{dx}{1 + x^{2p}}$$

and integrate around the contour shown in Figure 26.

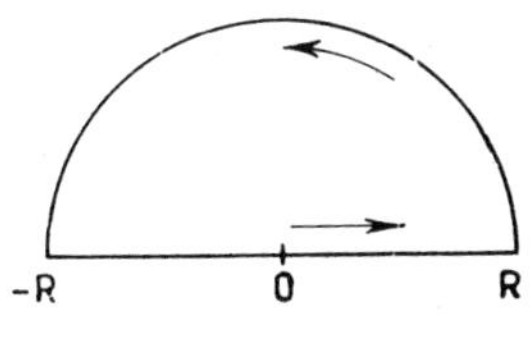

FIGURE 26

On the semicircle, the integral tends to 0 as the radius increases without bound. It remains to calculate the sum of the residues of the poles of the function $1/(1 + z^{2p})$ in the semicircle. The locations of these poles are

$$z_k = \exp\left(i\pi \frac{(2k+1)}{2p}\right) \qquad \text{for} \qquad k = 0, 1, 2, ..., p - 1,$$

and their residues are

$$\frac{1}{2pz_k^{2p-1}} = -\frac{z_k}{2p}.$$

Therefore,

$$2I = -\frac{2i\pi}{2p}(z_0 + z_1 + \cdots + z_{p-1}).$$

Now,

$$\begin{aligned} z_0 + z_1 + \cdots + z_{p-1} &= e^{i\pi/2p} + e^{3i\pi/2p} + \cdots + e^{i\pi(2p-1)/2p} \\ &= e^{i\pi/2p}(1 + e^{i\pi/p} + \cdots + e^{i\pi(p-1)/p}) \\ &= e^{i\pi/2p}\frac{1 - e^{i\pi}}{1 - e^{i\pi/p}} = -\frac{1}{i\sin(\pi/2p)}. \end{aligned}$$

Finally,

$$2I = \frac{\pi}{p}\frac{1}{\sin(\pi/2p)} \qquad \text{or} \qquad I = \frac{\pi}{\alpha\sin(\pi/\alpha)}.$$

Remark. This method is easily extended to the case in which α is any rational number. We need only consider the case in which α is of the form $2p/(2q + 1)$, where p and q are integers. We next show that I is a continuous function of α. Then we derive the value of I in the general case by letting the fraction $2p/(2q + 1)$ approach α, when α is any real number, rational or irrational.

C. If we set $x^\alpha = t$, we have

$$I = \frac{1}{\alpha}\int_0^\infty \frac{t^{(1/\alpha)-1}}{1 + t}\,dt.$$

In this form, I is reduced directly to the complement formula on gamma functions. Assuming this formula known, we have

$$I = \frac{1}{\alpha}\Gamma\left(\frac{1}{\alpha}\right)\Gamma\left(1 - \frac{1}{\alpha}\right) = \frac{1}{\alpha}\frac{\pi}{\sin(\pi/\alpha)}.$$

13. Application of the residue theorem to Bernoulli numbers

In this exercise we shall apply the theory of integration of analytic functions to the evaluation of the Fourier transform of the function $1/(e^{2\pi x} - 1)$ and we shall derive from this an integral representation of the *Bernoulli numbers*.

We recall that the Bernoulli numbers B_n are the products obtained by multiplying the constant terms in the *Bernoulli polynomials* $P_n(x)$ by $n!$. These are defined by

$$P_1(x) = x - \tfrac{1}{2}, \qquad P_n'(x) = P_{n-1}(x), \qquad P_n(1) = P_n(0) \qquad (n \geqslant 2)$$

$$B_n = n! P_n(0).$$

It is shown that

$$\frac{1}{2}\frac{e^a + 1}{e^a - 1} - \frac{1}{a} = \sum_{k=1}^{\infty} \frac{B_{2k}}{(2k)!} a^{2k-1}$$

and that

$$B_1 = -\tfrac{1}{2}, \qquad B_{2k+1} = 0 \qquad (k \geqslant 1).$$

Problems. A. What are the singularities of the function

$$f(z) = \frac{e^{iaz}}{e^{2\pi z} - 1},$$

where a is a real positive number?

B. By integrating the function $f(z)$ over the contour shown in Figure 27, where the points $z = 0$ and $z = i$ are bypassed by quarter-circles of infinitesimal radius, show that

$$\int_0^{\infty} \frac{\sin ax}{e^{2\pi x} - 1}\, dx = \frac{1}{4}\frac{e^a + 1}{e^a - 1} - \frac{1}{2a}.$$

C. Use this result to evaluate the integral

$$\int_0^{\infty} \frac{x^{2n-1}}{e^{2\pi x} - 1}\, dx$$

and to obtain an expression for the Bernoulli numbers.

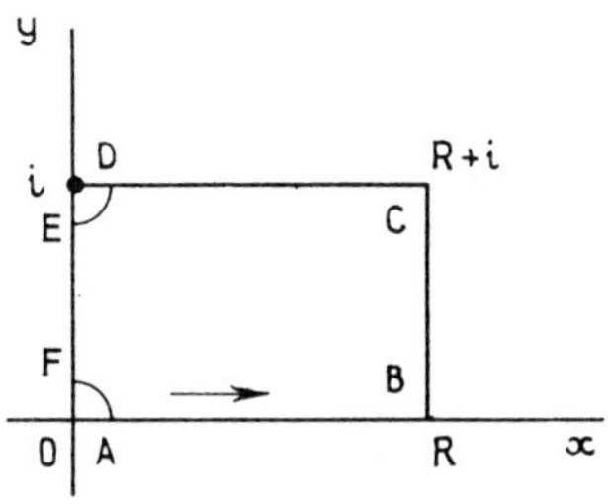

FIGURE 27

Solutions. **A.** Consider the function

$$f(z) = \frac{e^{iaz}}{e^{2\pi z} - 1}.$$

This function is meromorphic. It has poles at points z such that $e^{2\pi z} = 1$, that is, at points at which $2\pi z = 2ik\pi$ or $z = ik$, where k is an integer. These are simple poles. At the pole ik, the residue is equal to

$$\left(\frac{e^{iaz}}{2\pi e^{2\pi z}}\right)_{z=ik} = \frac{e^{-ak}}{2\pi}.$$

B. We integrate $f(z)$ over the contour shown in Figure 27. In and on this contour, $f(z)$ is holomorphic. Therefore, its integral is zero. It consists of six terms added together.

If ϵ is the radius of the little quarter-circles centered at 0 and i, we have

$$\int_{AB} = \int_{\epsilon}^{R} \frac{e^{iax}}{e^{2\pi x} - 1}\, dx.$$

On BC, we may set $z = R + it$, so that

$$\int_{BC} = i\int_0^1 \frac{e^{iaR}e^{-at}}{e^{2\pi R}e^{2\pi it} - 1}\, dt.$$

As $R \to \infty$, this integral tends to 0. To see this, note that it is less in absolute value than

$$K\int_0^1 \frac{dt}{|\, e^{2\pi R}e^{2i\pi t} - 1\,|},$$

where K denotes the maximum value of e^{-at}, which is equal either to 1 or to e^{-a}, depending on the sign of a. Now,

$$|\, e^{2\pi R}e^{2i\pi t} - 1\,| = (1 - 2e^{2\pi R}\cos 2\pi t + e^{4\pi R})^{1/2}$$

increases without bound as $R \to \infty$. It follows easily from this that the integral tends to 0.

On CD, we set $z = x + i$:

$$\int_{CD} = -\int_{\epsilon}^{R} \frac{e^{-\alpha}e^{i\alpha x}}{e^{2\pi x} - 1}\, dx.$$

On the arc DE, we are in a neighborhood of the pole $z = i$. We go part of the way around this pole in the negative sense. As $\epsilon \to 0$, the integral tends to a limit, which is equal to the product of the corresponding residue c multiplied by $-2i\pi/4$, that is, equal to $e^{-a}/2\pi$:

$$\int_{DE} = -\frac{ie^{-a}}{4}.$$

Specifically, if we set $z = i + u$, we need to evaluate

$$\int f(i + u)\, du,$$

where $f(i + u)$ has an expansion of the form

$$f(i + u) = \frac{c}{u} + A_0 + A_1 u + \cdots + A_n u^n + \cdots.$$

If we set $u = \epsilon e^{i\theta}$, we have

$$\int_{DE} \frac{c}{u}\, du = ic \int_0^{-\pi/2} d\theta = -i\frac{\pi}{2}c.$$

Then we have

$$\int_{DE} u^n\, du = i\epsilon^{n+1} \int_0^{-\pi/2} e^{i(n+1)\theta}\, d\theta.$$

This integral does not vanish but it tends to 0 as $\epsilon \to 0$. Since a power series is a continuous function that can be integrated term by term, the integral

$$\int_{DE} (A_0 + A_1 u + \cdots + A_n u^n + \cdots)\, du$$

tends to 0 as $\epsilon \to 0$. The result then follows.

On EF, we set $z = iy$. Then

$$\int_{EF} = -i \int_{\epsilon}^{1-\epsilon} \frac{e^{-ay}}{e^{2\pi iy} - 1}\, dy.$$

Finally, the integral along the arc FA is evaluated in the same way as the integral along DE. The residue is equal to $1/2\pi$, and we find

$$\int_{FA} = \frac{-i}{4}.$$

We see easily that passage to the limit as $R \to \infty$ is possible with each of these integrals. Therefore,

$$\lim_{\epsilon\to 0}\left[\int_\epsilon^\infty \frac{e^{iax}}{e^{2\pi x}-1}\,dx - e^{-a}\int_\epsilon^\infty \frac{e^{iax}}{e^{2\pi x}-1}\,dx - i\int_\epsilon^{1-\epsilon}\frac{e^{-ay}}{e^{2i\pi y}-1}\,dy\right]$$
$$-\frac{ie^{-a}}{4}-\frac{i}{4}=0. \qquad (1)$$

But the three integrals in the bracketed expression do not individually approach limits as $\epsilon \to 0$.

Let us write the imaginary part of (1). It contains the term

$$(1-e^{-\alpha})\int_\epsilon^\infty \frac{\sin ax}{e^{2\pi x}-1}\,dx,$$

which exists in the limit because, unlike $e^{iax}/(e^{2\pi x}-1)$, the function $\sin ax/(e^{2\pi x}-1)$ is continuous at the origin.

On the other hand,

$$-i\frac{1}{e^{2i\pi y}-1} = -i\frac{e^{-2i\pi y}-1}{2(1-\cos 2\pi y)} = -i\frac{\cos 2\pi y - 1 - i\sin 2\pi y}{2(1-\cos 2\pi y)}$$
$$= -\frac{\sin 2\pi y}{2(1-\cos 2\pi y)}+\frac{i}{2}.$$

We need to take the imaginary part, multiply by e^{-ay}, and integrate (in the limit) from 0 to 1. Finally, we have

$$(1-e^{-a})\int_0^\infty \frac{\sin ax}{e^{2\pi x}-1}\,dx + \frac{1}{2}\int_0^1 e^{-ay}\,dy = \frac{1+e^{-a}}{4}.$$

Since

$$\int_0^1 e^{-ay}\,dy = \frac{1-e^{-a}}{a},$$

we finally have

$$\int_0^\infty \frac{\sin ax}{e^{2\pi x}-1}\,dx = \frac{1}{4}\frac{1+e^{-a}}{1-e^{-a}}-\frac{1}{2a}. \qquad (2)$$

This formula yields the *Fourier sine transform* of $1/(e^{2\pi x}-1)$. We can verify that the right-hand member is a continuous odd function.

C. *Application to the Bernoulli numbers.* The right-hand member of (2) is a meromorphic function of a with poles at the points $a = 2k\pi i$, where k is a nonzero integer. Therefore, it can be expanded in powers of a in the circle $|a| < 2\pi$. We define B_{2k} by

$$\frac{1}{2}\frac{e^a + 1}{e^a - 1} - \frac{1}{a} = \sum_{k=1}^{\infty} \frac{B_{2k}}{(2k)!} a^{2k-1}. \tag{3}$$

On the other hand,

$$\sin ax = \sum_{k=1}^{\infty} (-1)^{k-1} \frac{a^{2k-1}x^{2k-1}}{(2k-1)!}.$$

Let us multiply this expression by $1/(e^{2\pi x} - 1)$ and integrate from 0 to $+\infty$. If this integration is legitimate, we get

$$(-1)^{k-1}B_{2k} = 4k \int_0^{\infty} \frac{x^{2k-1}}{e^{2\pi x} - 1}\, dx.$$

We know, for example, that $B_2 = \frac{1}{6}$. From this we derive that

$$\int_0^{\infty} \frac{x}{e^{2\pi x} - 1}\, dx = \frac{1}{24}.$$

We still need to justify the above calculation, that is, to show that the series

$$\sum_{k=1}^{\infty} (-1)^{k-1} \frac{a^{2k-1}}{(2k-1)!} \frac{x^{2k-1}}{e^{2\pi x} - 1}$$

can be integrated termwise from 0 to $+\infty$. Since we are integrating over an unbounded interval, this does not follow directly from the integration properties of uniformly convergent series.

First, let us note that there exists a number M such that

$$\frac{x}{e^{2\pi x} - 1} < Me^{-\pi x},$$

since the function $xe^{\pi x}/(e^{2\pi x} - 1) = x/2 \sinh \pi x$ is continuous and tends to 0 as $x \to \infty$. Now, we have

$$\frac{\sin ax}{e^{2\pi x} - 1} = \sum_{k=0}^{\infty} (-1)^k \frac{a^{2k+1}x^{2k}}{(2k+1)!} \frac{x}{e^{2\pi x} - 1}.$$

The remainder R_k of this alternating series is less in absolute value then

$$\frac{a^{2k+1}x^{2k}}{(2k+1)!} \frac{x}{e^{2\pi x} - 1} < M \frac{a^{2k+1}}{(2k+1)!} x^{2k}e^{-\pi x}.$$

Therefore,

$$\left| \int_0^\infty R_k(x)\, dx \right| < M \frac{a^{2k+1}}{(2k+1)!} \int_0^\infty x^{2k} e^{-\pi x}\, dx$$

$$= M \frac{a^{2k+1}}{(2k+1)!} \cdot \frac{(2k)!}{\pi^{2k+1}} = \frac{M}{2k+1} \left(\frac{a}{\pi}\right)^{2k+1}$$

If $0 \leqslant a < \pi$, the integral of the remainder tends to 0 and the series can be integrated term by term. Therefore, we can equate the expansions of the two members of (2) for every a belonging to the interval $0 \leqslant a < \pi$. It follows that these expansions are identical and have the same coefficients. Note that, in this proof, we do not use the entire interval of convergence $(0, 2\pi)$ of the series expansion of the right-hand member.

14. The Fourier transform of 1/cosh t

To prove the formula on complements,

$$\Gamma(p)\Gamma(1-p) = \int_0^\infty \frac{t^{p-1}}{1+t}\, dt = \frac{\pi}{\sin p\pi} \qquad (0 < p < 1),$$

we integrate $z^{p-1}/(1+z)$ on a suitable contour in the complex plane.

If we replace the real number p by a complex number $a + ib$, we obtain a more general formula which, by means of a simple change of variable, yields the Fourier transform of $1/\cosh t$. Except for a normalizing factor, $1/\cosh t$ is a *probability density*. Thus we obtain from it the *characteristic function*, which is of the same form.

This result was found by a different method in Exercise 14 of Chapter 2 and it will be established again in Exercise 15 of this chapter.

Problem. Evaluate the integral

$$\int_\Gamma \frac{z^{a+ib-1}}{1+z}\, dz,$$

where a and b are two real numbers with $0 < a < 1$ and where Γ is the contour shown in Figure 28. By taking $a = \frac{1}{2}$ and making a change of variable, evaluate the Fourier integral

$$\int_0^\infty \frac{\cos \omega t}{\cosh t}\, dt \qquad (\omega \text{ real}).$$

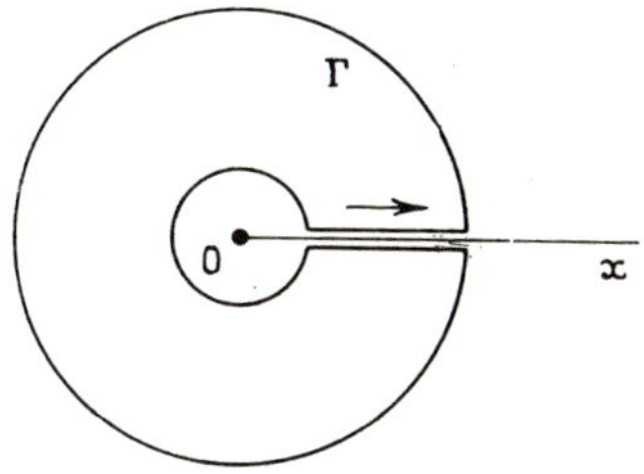

FIGURE 28

Solution. Consider the integral

$$\int_{\Gamma} \frac{z^{a+ib-1}}{1+z}\, dz,$$

where a and b are two real numbers with $0 < a < 1$ and Γ is the contour shown in Figure 28.

The two circles γ_1 and γ_2 are, respectively, very small and very large. Since z^{a+ib-1} possesses a critical point at the origin, we need to make a cut in the complex plane. We make it along the positive semiaxis. In the cut plane, the determination of z^{a+ib-1} is well defined if we choose, for example,

$$z^{a+ib-1} = e^{(a+ib-1)\log x},$$

at a point whose abscissa x is positive above the cut. Here $\log x$ is the real value of the logarithm.

The function $z^{a+ib-1}/(1+z)$ thus defined has a simple pole at $z = -1$. The residue is the value of $z^{a+ib-1} = e^{(a+ib-1)\log z}$ at that point.

We can assign to z the argument 0 at the point $z = x$ (above the cut). Then we necessarily have $-1 = e^{i\pi}$, so that $\log z = i\pi$ at the point $z = -1$. Therefore, we have the residue

$$e^{(a+ib-1)i\pi} = +\, e^{\pi[-b+i(a-1)]} = -e^{\pi(ai-b)}.$$

The integral in question is equal to the product of this residue by $2i\pi$. Now, it is the sum of four integrals (over the four portions of our closed contour). On the circles γ_1 and γ_2, the limiting value of the integral is 0 according to Jordan's lemma. Specifically, since $0 < a < 1$, the expression

$$\left| \frac{z \cdot z^{a+ib-1}}{1+z} \right| = \frac{|z|^a}{|1+z|}$$

tends to 0 both as $|z| \to 0$ and as $|z| \to \infty$. We still have the integrals on either side of the cut. Above we have

$$\int_0^\infty \frac{x^{a+ib-1}}{1+x}\,dx,$$

where x^{a+ib-1} has been defined by $e^{(a+ib-1)\log x}$. Below we need to choose $z = xe^{2i\pi}$. Then

$$z^{a+ib-1} = e^{(2i\pi+\log x)(a+ib-1)} = e^{2\pi(ai-b)}x^{a+ib-1}.$$

Finally, we have

$$[1 - e^{2\pi(ai-b)}]\int_0^\infty \frac{x^{a+ib-1}}{1+x}\,dx = -2i\pi e^{\pi(ai-b)}.$$

For the rest, let us study only the case in which $a = \frac{1}{2}$. We have

$$(1 + e^{-2\pi b})\int_0^\infty \frac{x^{ib-(1/2)}}{1+x}\,dx = 2\pi e^{-\pi b}$$

or

$$\int_0^\infty \frac{x^{ib-(1/2)}}{1+x}\,dx = \frac{\pi}{\cosh b\pi}.$$

Finally, we set $x = e^{2t}$ and $2b = \omega$. We obtain

$$\int_{-\infty}^\infty \frac{e^{i\omega t}}{\cosh t}\,dt = \frac{\pi}{\cosh(\omega\pi/2)}$$

so that

$$\int_0^\infty \frac{\cos\omega t}{\cosh t}\,dt = \frac{\pi}{2\cosh(\omega\pi/2)},$$

where ω is real.

15. The Fourier transform of 1/cosh πx and 1/sinh πx

This exercise supplements the preceding one. It provides the same results for $1/\cosh\pi x$ by a different procedure. Instead of integrating a multiple-valued function, we integrate a meromorphic function.

Problem. Consider the function

$$f(z) = \frac{e^{(\pi+i\omega)z}}{e^{2\pi z}+1},$$

where ω is a real parameter. Find the singularities of $f(z)$. By integrating $f(z)$ over the contour shown in Figure 29, where ϵ is infinitesimal and R is infinitely great, establish a relation between the integrals

$$A = \int_0^\infty \frac{\cos \omega x}{\cosh \pi x}\, dx, \qquad B = \int_0^\infty \frac{\sin \omega x}{\sinh \pi x}\, dx.$$

Evaluate A and B by replacing ω with $-\omega$ in that relation.

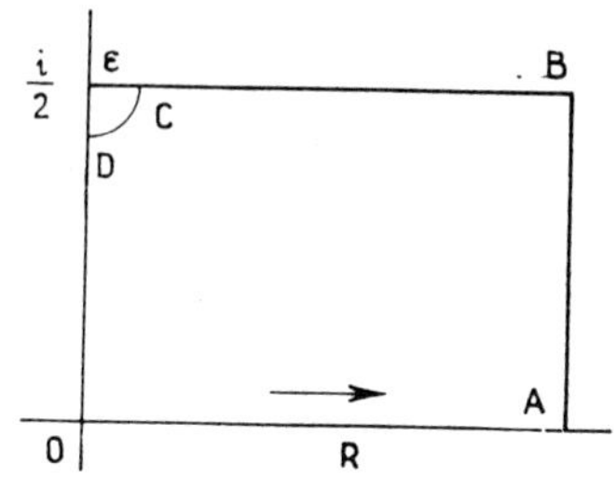

FIGURE 29

Solution. Consider the function

$$f(z) = \frac{e^{(\pi + i\omega)z}}{e^{2\pi z} + 1},$$

where ω is a real parameter. We integrate over the contour γ, avoiding the point $i/2$ by replacing the adjacent portion of our contour with a quarter-circle of radius ϵ.

The poles of the function $f(z)$ are the points z at which

$$2\pi z = i(\pi + 2k\pi),$$

that is, the points at which

$$z = \frac{i}{2} + ik.$$

These are points on the imaginary axis. In the interior of γ, $f(z)$ is holomorphic and

$$\int_\gamma f(z)\, dz = 0.$$

Now, this integral is the sum of five integrals (corresponding to the five portions of the contour).

On OA we have $z = x$ (real), so that

$$\int_{OA} = \int_0^R \frac{e^{(\pi + i\omega)x}}{e^{2\pi x} + 1}\, dx.$$

As $R \to \infty$, the integral $\int_{OA}$ tends to a limit, which is

$$\frac{1}{2}\int_0^\infty \frac{e^{i\omega x}}{\cosh \pi x}\,dx.$$

On AB we set $z = R + iy$. Then

$$\int_{AB} = i\int_0^{1/2} \frac{e^{\pi R - \omega y}\, e^{i(\omega R + \pi y)}}{e^{2\pi R}\, e^{2i\pi y} + 1}\,dy.$$

We have

$$\left|\int_{AB}\right| < \int_0^{1/2} \frac{e^{\pi R - \omega y}}{|\, e^{2\pi R}\, e^{2i\pi y} + 1 \,|}\,dy.$$

The value of y is bounded. We can verify, as in Exercise 13, that this expression tends to 0 as $R \to \infty$.

On BC we set $z = x + i/2$. We obtain

$$\int_{BC} = -ie^{-\omega/2}\int_\epsilon^R \frac{e^{\pi x + i\omega y}}{1 - e^{2\pi x}}\,dx = +i\,\frac{e^{-\omega/2}}{2}\int_\epsilon^R \frac{e^{i\omega y}}{\sinh \pi x}\,dx.$$

On CD we detour around the simple pole $i/2$ by going along a quarter-circle in the negative sense. We know (see Exercise 13) that the integral is equal to the product obtained by multiplying the residue at this pole by $-2i\pi/4$. This residue is equal to

$$\left[\frac{e^{(\pi + i\omega)z}}{2\pi e^{2\pi z}}\right]_{z=i/2} = -\frac{i}{2\pi}\,e^{-\omega/2}.$$

Therefore,

$$\int_{CD} = -\tfrac{1}{4}\,e^{-\omega/2}.$$

On DO we set $z = iy$. We have

$$\int_{DO} = -\,i\int_0^{(1/2)-\epsilon} \frac{e^{(\pi + i\omega)iy}}{e^{2i\pi y} + 1}\,dy = -\frac{i}{2}\int_0^{(1/2)-\epsilon} \frac{e^{-\omega y}}{\cos \pi y}\,dy.$$

Note that passage to the limit as $R \to \infty$ is possible with each of these integrals. On the other hand, the integrals containing ϵ diverge if $\epsilon = 0$. Passage to the limit as $\epsilon \to 0$ can be made only globally:

$$\frac{1}{2}\int_0^\infty \frac{e^{i\omega x}}{\cosh \pi x}\,dx - \frac{1}{4}\,e^{-\omega/2} + \lim_{\epsilon \to 0}\left[\frac{i}{2}\,e^{-\omega/2}\int_\epsilon^\infty \frac{e^{i\omega x}}{\sinh \pi x}\,dx - \frac{i}{2}\int_0^{(1/2)-\epsilon} \frac{e^{-\omega y}}{\cos \pi y}\,dy\right] = 0.$$

We take the real part of this equation and set

$$A = \int_0^\infty \frac{\cos \omega x}{\cosh \pi x}\, dx, \qquad B = \int_0^\infty \frac{\sin \omega x}{\sinh \pi x}\, dx.$$

We get

$$\tfrac{1}{2} A - \tfrac{1}{4} e^{-\omega/2} - \lim_{\epsilon \to 0} \tfrac{1}{2} e^{-\omega/2} \int_\epsilon^\infty \frac{\sin \omega x}{\sinh \pi x}\, dx = 0.$$

It follows from this formula that $\int_\epsilon^\infty (\sin \omega x/\sinh \pi x)\, dx$ tends to a limit, equal to B, as $\epsilon \to 0$. Convergence of the integral is easy to verify. Specifically, we have

$$A = (B + \tfrac{1}{2})\, e^{-\omega/2}.$$

Let us replace ω with $-\omega$. The integral A remains unchanged; B is transformed into $-B$:

$$A = (-B + \tfrac{1}{2})\, e^{\omega/2}.$$

Solution of this linear system in A and B yields

$$A = \frac{1}{2 \cosh (\omega/2)}, \qquad B = \frac{1}{2} \frac{\sinh (\omega/2)}{\cosh (\omega/2)}.$$

16. Evaluation of a definite integral on different closed curves

Problems. Evaluate the integral

$$\int_C \frac{dz}{z(1 - z^2)^{1/2}}$$

when C is

A. The circle of radius $\frac{1}{2}$ with center at 0.

B. The circle of radius 2 with center at 0.

C. The contour shown in Figure 30.
Choose with precision a determination of $(1 - z^2)^{1/2}$ in the cut plane.

Solutions. **A.** Let us denote by Ω the complex plane minus the two closed rays

$$(-\infty, -1] \qquad \text{and} \qquad [1, +\infty).$$

In Ω there exist two holomorphic determinations of the symbol

$$(1 - z^2)^{1/2}.$$

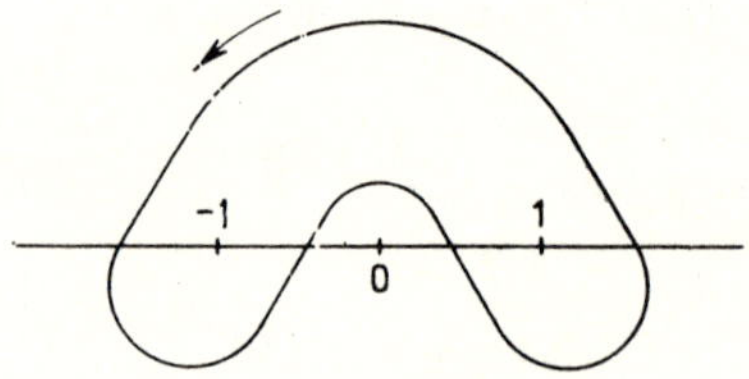

FIGURE 30

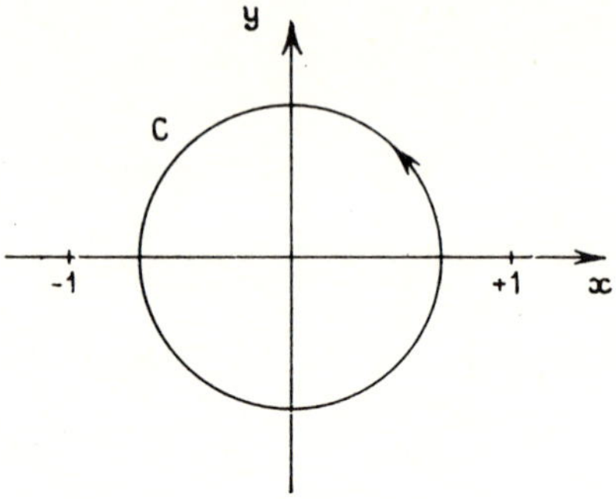

FIGURE 31

For $x = 0$, one of these is equal to 1, the other to -1. Let C denote a circle of radius $r < 1$ with center at 0.

If $f(z)$ is holomorphic in Ω and if C is described in the positive sense, then

$$\int_C \frac{f(z)}{z}\,dz = 2i\pi f(0)$$

by virtue of Cauchy's formula. Thus according to the determination chosen for $(1 - z^2)^{1/2}$,

$$\int_C \frac{dz}{z(1 - z^2)^{1/2}} = \pm 2i\pi.$$

B. Let us denote by Ω the complement with respect to the complex plane of the closed circle of radius 1 with center at 0.

In Ω there exist two holomorphic determinations of $(1 - z^2)^{1/2}$. For $z = 2$, for example, one of these is equal to $i\sqrt{3}$, the other to $-i\sqrt{3}$. Let C be a circle of radius exceeding 1 with center at 0 (or, more generally, any simple closed curve whose interior contains the unit circle).

If $f(z)$ is holomorphic in Ω, the value of the integral

$$\int_C \frac{f(z)}{z}\, dz$$

is independent of C, as we know from Cauchy's theorem applied to a doubly connected region. Now, if C_r is a circle with center at 0 and we let its radius r become infinite, then the integral $\int_C [f(z)/z]\, dz$ tends to 0 by virtue of Jordan's lemma provided $|f(z)|$ tends to 0 as $|z|$ tends to infinity. This is the case with $|1/(1 - z^2)^{1/2}|$. Thus the value of this integral is 0.

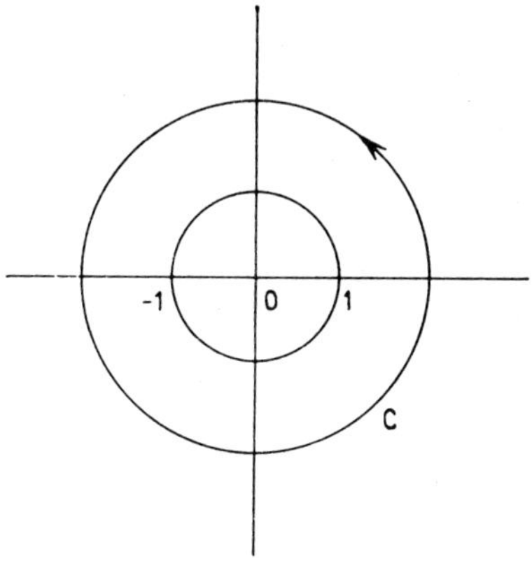

FIGURE 32

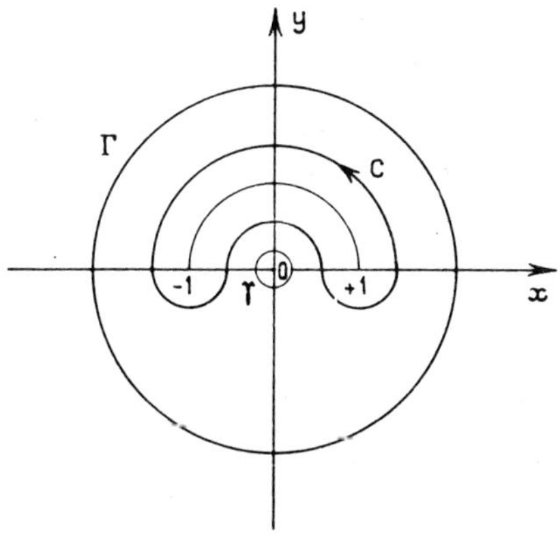

FIGURE 33

C. Let Ω denote the complement with respect to the complex plane of the closed unit semicircle lying in the upper half of that plane (i.e., the half corresponding to complex numbers with nonnegative imaginary parts). In Ω there exist two holomorphic determinations of

$$(1 - z^2)^{1/2}$$

(just as in Problem A). Their values are ± 1 for $z = 0$.

Let C be the contour shown in Figure 33 or, more generally, any simple closed curve whose interior contains the unit semicircle but not the origin.

Let $f(z)$ be a function that is holomorphic in Ω. Consider two circles γ and Γ centered at the origin with radii r and R sufficiently small and sufficiently large respectively that they both lie outside C with the interior of Γ containing both C and γ. Then, according to Cauchy's theorem applied to a multiply connected region, we have

$$\int_\Gamma \frac{f(z)}{z}\,dz - \int_C \frac{f(z)}{z}\,dz - \int_\gamma \frac{f(z)}{z}\,dz = 0,$$

where the integrations around C, γ, and Γ are in the positive sense. Now,

$$\int_\gamma \frac{f(z)}{z}\,dz = 2\pi i f(0).$$

We saw in Problem B that, for $f(z) = 1/(1 - z^2)^{1/2}$,

$$\int_\Gamma \frac{f(z)}{z}\,dz = 0.$$

Thus

$$\int_C \frac{f(z)}{z}dz = -2\pi i f(0) = \mp 2\pi i.$$

Therefore, for a single determination of $(1 - z^2)$, the value of the integral is the negative of the integral evaluated in Problem A.

An alternative solution. In the second case, let us set $z = 1/t$. Then t describes a circle C' of radius less than 1 with center at 0 in the negative sense. By making a change of sign, we need to calculate

$$\int_{C'} \frac{dt}{(t^2 - 1)^{1/2}},$$

where the integration around C' is in the positive sense. Now, we can choose a holomorphic determination of $(t^2 - 1)^{1/2}$ in the interior of C'. Thus the integral vanishes by virtue of Cauchy's theorem.

In the third case, we can deform the contour in such a way as to make it homologous to two closed curves C_1 and C_2 encircling the point 0, themselves homologous to two circles taken one in the positive sense, the other in the negative.

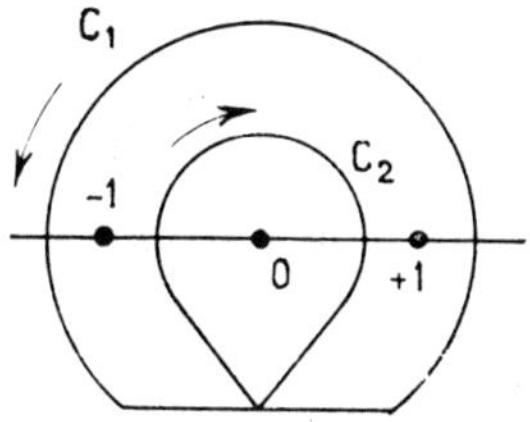

FIGURE 34

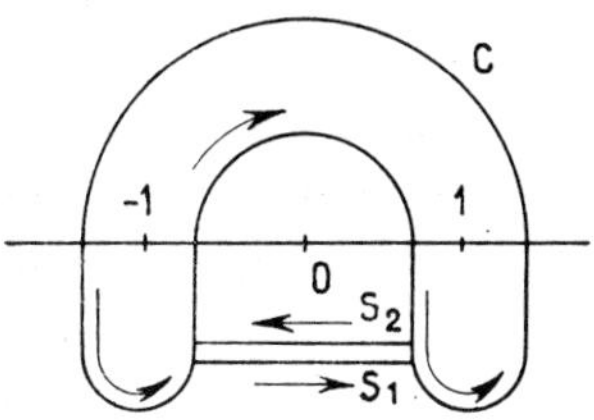

FIGURE 35

We can also (and this amounts to the same thing) complete C with two small segments S_1 and S_2 taken in opposite senses to each other. Over these segments the integrals cancel each other. If we combine S_1, S_2, and C, we can then decompose this union into two closed curves C_1 and C_2 much like those used in the preceding method. Now, we have seen that the integral over C_1 vanishes. The integral over C_2 is identical to that of Problem A except for the sense taken around the curve.

17. Integration of a multiple-valued function

Problem. Let α denote a real number between 0 and 1. Find the singularities of the function $f(z) = \log(1 - z)/z^{1+\alpha}$.

Integrate this function over the contour shown in Figure 36 and evaluate the integrals

$$\int_0^\infty \frac{\log(x+1)}{x^{1+\alpha}}\,dx, \qquad \int_0^\infty \frac{\log|1-x|}{x^{1+\alpha}}\,dx.$$

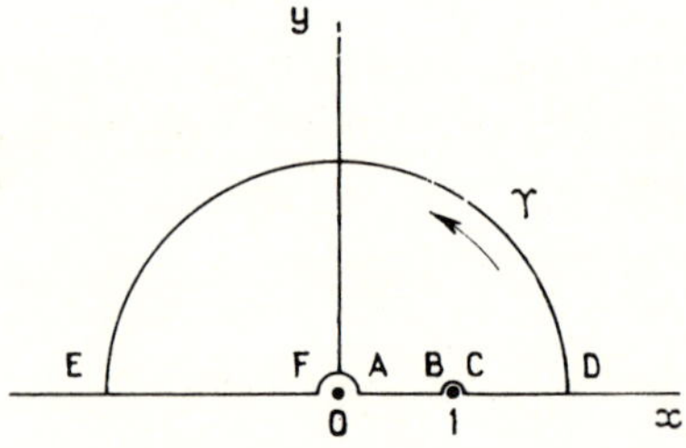

FIGURE 36

Solution. The function

$$f(z) = \frac{\log(1 - z)}{z^{1+\alpha}} \qquad (0 < \alpha < 1)$$

has the critical points $z = 0$ and $z = 1$.

In the interior of the contour of integration, we can choose a single-valued (and even holomorphic) determination of $f(z)$. We need only assign to $f(z)$ the real value $x^{-1-\alpha} \log(1 - x)$ at the point x on the real axis such that $0 < x < 1$.

Let us study the value of $\log(1 - z)$ on the real axis.

If we set $1 - z = re^{i\theta}$, we have

on AB:

$$r = 1 - x, \qquad \theta = 0, \qquad \log(1 - z) = \log r = \log(1 - x) \text{(by hypothesis)}$$

on CD:

$$r = x - 1, \qquad \theta = -\pi, \qquad \log(1 - z) = \log(x - 1) - i\pi$$

on EF:

$$r = |x| + 1, \qquad \theta = 0, \qquad \log(1 - z) = \log(|x| + 1).$$

Similarly, the values of $z^{1+\alpha}$ are

on AB:

$$x^{1+\alpha}$$

on CD:

$$x^{1+\alpha}$$

on EF:

$$(\rho e^{i\pi})^{1+\alpha} = -\rho^{1+\alpha} e^{i\pi\alpha} \qquad \text{or} \qquad \rho = |x| = -x.$$

Let us now integrate $f(z)$. According to Cauchy's theorem, the integral vanishes. On the semicircle of infinitely large radius, the integral tends to 0 by virtue of Jordan's lemma. Specifically, we have

$$|zf(z)| = \frac{|\log(1-z)|}{|z|^\alpha}.$$

Let us now set

$$1 - z = re^{i\theta}, \qquad \log(1-z) = \log r + i\theta,$$

where θ varies from 0 to π. As $r \to \infty$, the value of $|\log(1-z)|$ becomes infinitely great and is equivalent to $\log r$. Its quotient by $|z|^\alpha$, which is equivalent to r^α, tends to 0.

The same conclusion holds on the little circles with centers at 0 and 1. Thus we have in the limit

$$\int_{EF} f(z)\,dz + \int_{AB} f(z)\,dz + \int_{CD} f(z)\,dz = 0$$

or

$$-e^{-i\pi\alpha}\int_0^\infty \frac{\log(\rho+1)}{\rho^{1+\alpha}}\,d\rho + \int_0^1 \frac{\log(1-x)}{x^{1+\alpha}}\,dx + \int_1^\infty \frac{\log(x-1) - i\pi}{x^{1+\alpha}}\,dx = 0.$$

Separating the real and imaginary parts, we have

$$-\cos\pi\alpha\int_0^\infty \frac{\log(\rho+1)}{\rho^{1+\alpha}}\,d\rho + \int_0^1 \frac{\log(1-x)}{x^{1+\alpha}}\,dx + \int_1^\infty \frac{\log(x-1)}{x^{1+\alpha}}\,dx = 0,$$

$$\sin\pi\alpha\int_0^\infty \frac{\log(\rho+1)}{\rho^{1+\alpha}}\,d\rho - \pi\int_1^\infty \frac{dx}{x^{1+\alpha}} = 0.$$

Since

$$\int_1^\infty \frac{dx}{x^{1+\alpha}} = \frac{1}{\alpha},$$

we have

$$\int_0^\infty \frac{\log(\rho+1)}{\rho^{1+\alpha}}\,d\rho = \frac{\pi}{\alpha\sin\pi\alpha}.$$

Then

$$\int_0^\infty \frac{\log|1-x|}{x^{1+\alpha}}\,dx = \frac{\pi\cos\pi\alpha}{\alpha\sin\pi\alpha}.$$

18. A property of elliptic integrals

If $p(z)$ is a fourth-degree polynomial, the function $[p(z)]^{1/2}$ has four critical points [provided the zeros of $p(z)$ are distinct]. We make this function single-valued by making two cuts, each connecting a pair of these zeros. If $p(z)$ is a third-degree polynomial, we can still think of it as having four critical points, the fourth being at infinity. Here we shall use the example of a third-degree polynomial to prove a property of the elliptic integral

$$\int \frac{dz}{\sqrt{(pz)}}$$

taken over arcs connecting pairs of critical points as described above.

Note that this property was proved in Exercise 5 of Chapter 2 by means of a real homographic transformation.

Problem. Let a, b, and c be three real numbers such that $a < b < c$. By integrating the function $f(z) = 1/[(z-a)(z-b)(z-c)]^{1/2}$ over a convenient contour in the complex plane, prove the identity

$$\int_a^b \frac{dx}{[(x-a)(b-x)(c-x)]^{1/2}} = \int_c^\infty \frac{dx}{[(x-a)(b-x)(c-x)]^{1/2}}.$$

Solution. The function

$$f(z) = \frac{1}{[(z-a)(z-b)(z-c)]^{1/2}}$$

has the three critical points a, b, and c. By cutting the complex plane along the segments $[a, b]$ and $[c, \infty)$, we make a single-valued determination.

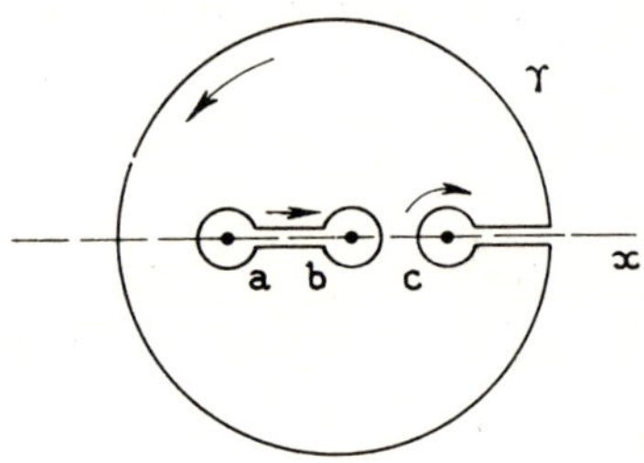

FIGURE 37

If, for example, a, b, and c are real numbers such that $a < b < c$, we cut along the segment $[a, b]$ of the real axis and along the ray extending along the real axis to the right of c. We encircle these cuts with loops and integrate along the contour γ shown in Figure 37.

We choose the determination of $f(z)$ that is real and positive at real points z lying above the cut $[a, b]$. Since $f(z)$ is holomorphic in the interior of the domain bounded by γ, the integral over γ is 0.

From Jordan's lemma, the integral over the large circle tends to 0 because the function

$$|zf(z)| = \frac{|z|}{[|(z-a)(z-b)(z-c)|]^{1/2}}$$

tends to 0 as $|z| \to \infty$. Similarly, on the little circle centered at a, the function

$$|(z-a)f(z)| = \frac{(|z-a|)^{1/2}}{(|z-b||z-c|)^{1/2}}$$

tends to 0 as $|z-a| \to 0$. The integrals over the three little circles tend to 0 as the radii of these circles tend to 0.

At a point z lying below the cut $[a, b]$, the value of the function $f(z)$ is opposite to the value that it assumes at the corresponding point above the cut. The integral over the loop encircling $[a, b]$ is, therefore,

$$\int_a^b f(x)\,dx - \int_a^b [-f(x)]\,dx = 2\int_a^b f(x)\,dx$$

for $f(x)$ real and positive.

Let us now find the value of $f(z)$ at a point z lying above the cut $[c, \infty]$. At a point z_0 lying between a and b above the cut, $f(z)$ is real and positive. As we go from z_0 to z, we add $-\pi$ to the arguments of

$$z - b \quad \text{and} \quad z - c.$$

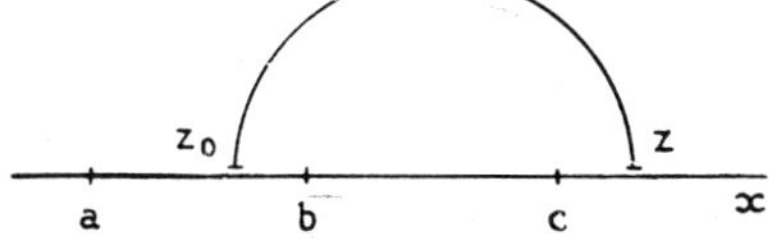

FIGURE 38

Thus we multiply

$$z - b \quad \text{and} \quad z - c$$

by $e^{-i\pi}$, the product by $e^{-2i\pi}$, and the square root $f(z)$ by $e^{-i\pi} = -1$. Thus, at z, the function $f(z)$ is real and negative.

On the loop around the cut $[c, \infty]$, we have

$$\int f(z)\,dz = -2\int_c^\infty \frac{dx}{[(x-a)(b-x)(c-x)]^{1/2}}.$$

In sum, we have

$$\int_a^b \frac{dx}{[(x-a)(b-x)(c-x)]^{1/2}} - \int_c^\infty \frac{dx}{[(x-a)(b-x)(c-x)]^{1/2}} = 0.$$

This is the result required.

*19. Properties of the elliptic function θ(z)

In this exercise we shall consider a holomorphic function $\theta(z)$ that depends on a nonreal parameter t that satisfies the two functional equations

$$\theta(z+\pi) = \theta(z), \qquad \theta(z+\pi t) = \text{const} \times e^{-2iz}\theta(z).$$

The logarithmic derivative of θ is such that

$$\frac{\theta'(z+\pi t)}{\theta(z+\pi t)} = -2i + \frac{\theta'(z)}{\theta(z)}.$$

It follows that $\varphi(z) = (d/dz)[\theta'(z)/\theta(z)]$ is a periodic function with period πt. Now, $\theta(z)$ is also periodic with period π. The same is true of $\varphi(z)$. Thus $\varphi(z)$ is a *doubly periodic* meromorphic function, and the ratio of the two periods is the complex number t. It is an *elliptic function.* The purpose of this exercise is to prove certain properties of $\theta(z)$ when this function is defined by a Fourier-series expansion.

We note also that θ is a solution of the partial differential equation

$$\frac{1}{i}\frac{\partial\theta}{\partial t} = -\frac{\pi}{4}\frac{\partial^2\theta}{\partial z^2}.$$

When t is a purely imaginary number $t = it'$, this equation becomes the *heat-flow equation*

$$\frac{\partial\theta}{\partial t'} = \frac{\pi}{4}\frac{\partial^2\theta}{\partial z^2}.$$

In this problem we shall use the following result: If $\theta(z)$ is holomorphic in a domain $\mathscr{D}$ and does not vanish on the boundary Γ of $\mathscr{D}$, the integral

$$\frac{1}{2\,i\pi}\int_\Gamma \frac{\theta'(z)}{\theta(z)}\,dz,$$

which is the variation on Γ of the argument of $\theta(z)$, is equal to the number of zeros of $\theta(z)$ in the interior of $\mathscr{D}$.

Problems. A. Denote by q a complex number whose absolute value is less than 1. Consider the function

$$\theta(z) = 1 - 2q \cos 2z + \cdots + (-1)^n 2q^{n^2} \cos 2nz + \cdots$$

of the complex variable z. Show that $\theta(z)$ is a periodic function with period π that is holomorphic in every bounded domain.

B. Set $q = e^{i\pi t}$, where t is a complex number with positive imaginary part. Prove the formula

$$\theta(z + \pi t) = -\frac{1}{q} e^{-2iz}\theta(z).$$

C. Show that $\pi t/2$ is a zero of $\theta(z)$. Show that $\theta(z)$ has, in general, exactly one zero in the interior of the parallelogram in the complex plane whose vertices are the points

$$z_0, \qquad z_0 + \pi, \qquad z_0 + \pi t, \qquad z_0 + \pi + \pi t,$$

where z_0 is a given complex number. Find the zeros of $\theta(z)$.

Write $\theta(z) = \sum_{n=-\infty}^{\infty} (-1)^n e^{i\pi t n^2} e^{2inz}$.

D. Suppose that Γ is a parallelogram with vertices at the points

$$-\frac{\pi}{2}, \quad \frac{\pi}{2}, \quad -\frac{\pi}{2} + \pi t, \quad \frac{\pi}{2} + \pi t.$$

Show that

$$\int_\Gamma \frac{\theta'(z)}{\theta(z)} e^{-2inz}\, dz = 2i\pi e^{-ni\pi t},$$

where n is an arbitrary integer. Show that the integral on the left reduces to

$$\int_{-\pi/2}^{\pi/2} \frac{\theta'(z)}{\theta(z)} e^{-2i\pi z}\, dz.$$

From this, find the Fourier series expansion of the function $\theta'(z)/\theta(z)$.

Solutions. **A.** Let q be a complex number such that $|q| < 1$. Consider the function $\theta(z)$ defined by the series

$$\theta(z) = 1 - 2q \cos 2z + \cdots + (-1)^n 2q^{n^2} \cos 2nz + \cdots, \tag{1}$$

where $z = x + iy$ is complex.

We have

$$2 \cos 2nz = e^{2inz} + e^{-2inz} = e^{-2ny}e^{2inx} + e^{2ny}e^{-2inx},$$
$$2 \mid \cos 2nz \mid < e^{2ny} + e^{-2ny} < 2e^{2n|y|}.$$

The general term in this series is less in absolute value than

$$2\,|\,q\,|^{n^2}e^{2n|y|} < 2[|\,q\,|^n e^{2|y|}]^n.$$

As $|\,q\,|^n$ tends to 0, we can, for each fixed value of y, find a value of n sufficiently large that $|\,q\,|^n e^{2|y|}$ will be less than a given positive number r that is less than 1. Then the general term of the series is majorized by $2r^n$ and the series converges.

If we differentiate the series term by term, an analogous reasoning shows that the series obtained by differentiating converges uniformly in every bounded domain. Therefore, $\theta(z)$ is differentiable at every point z. In other words, it is a holomorphic function.

Finally, $\theta(z + \pi)$ is obviously equal to $\theta(z)$.

Thus $\theta(z)$ is an *even holomorphic function that is periodic with period* π.

B. We set $q = e^{i\pi t}$. Since $|\,q\,| < 1$, the imaginary part of t is positive. We note first that it is convenient to write

$$\theta(z) = \sum_{n=-\infty}^{\infty} (-1)^n q^{n^2} e^{2inz} = \sum_{n=-\infty}^{\infty} (-1)^n e^{i\pi t n^2} e^{2inz}.$$

Then

$$\begin{aligned}\theta(z + \pi t) &= \sum_{n=-\infty}^{\infty} (-1)^n e^{i\pi t n^2} e^{2in(z+\pi t)} \\ &= \sum_{n=-\infty}^{\infty} (-1)^n e^{i\pi t(n^2+2n)} e^{2inz} \\ &= -\sum_{n=-\infty}^{\infty} (-1)^{n+1} e^{i\pi t(n+1)^2} e^{2i(n+1)z} e^{-i\pi t} e^{-2iz},\end{aligned}$$

or, finally,

$$\theta(z + \pi t) = -\frac{1}{q}\, e^{-2iz}\theta(z). \tag{2}$$

We recall that

$$\theta(z + \pi) = \theta(z). \tag{3}$$

C. For $z = -\pi t/2$, we have from the preceding formula

$$\theta\left(\frac{\pi t}{2}\right) = -\frac{1}{q}\, e^{i\pi t}\theta\left(-\frac{\pi t}{2}\right) = -\theta\left(-\frac{\pi t}{2}\right).$$

But $\theta(z)$ is an even function. Therefore, $\theta(\pi t/2) = 0$, so that $\pi t/2$ is a zero of $\theta(z)$.

Consider the parallelogram Γ in the complex plane with vertices at the points

$$A(z_0), \qquad B(z_0 + \pi),$$
$$D(z_0 + \pi t), \qquad C(z_0 + \pi + \pi t).$$

It is always possible to choose z_0 in such a way that no zero of $\theta(z)$ will lie on Γ since the zeros of a holomorphic function are isolated.

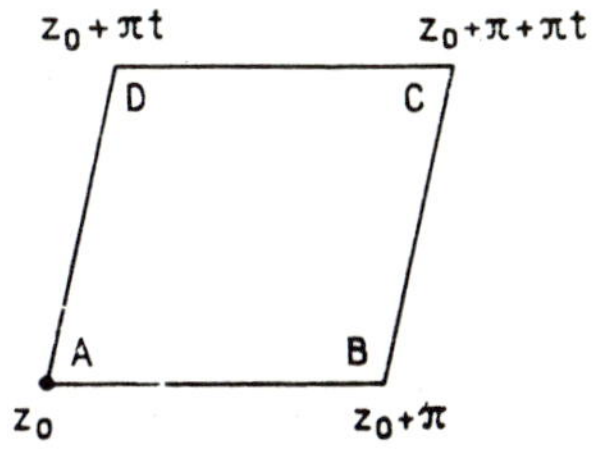

FIGURE 39

We know that the number of zeros of a holomorphic function θ that lie in the interior of a closed curve Γ is given by the integral

$$N = \frac{1}{2i\pi}\int_\Gamma \frac{\theta'(z)}{\theta(z)}\,dz,$$

where the integration over Γ is in the positive sense. The integral

$$\int_\Gamma \frac{\theta'(z)}{\theta(z)}\,dz$$

is the sum of the four integrals over the individual sides of the parallelogram. Let us first consider the integrals over the horizontal sides:

(a) $$\int_{AB} \frac{\theta'(z)}{\theta(z)}dz = \int_{z0}^{z_0+\pi} \frac{\theta'(z)}{\theta(z)}\,dz,$$

(b) $$\int_{CD} \frac{\theta'(z)}{\theta(z)}\,dz.$$

If we set $z = it + u$, this integral becomes

$$-\int_{z0}^{z_0+\pi} \frac{\theta'(u+\pi t)}{\theta(u+\pi t)}\,du.$$

But

$$\theta(u+\pi t) = -\frac{1}{q}\,e^{-2iu}\theta(u),$$

from which we get

$$\frac{\theta'(u+\pi t)}{\theta(u+\pi t)} = -2i + \frac{\theta'(u)}{\theta(u)}.$$

Therefore,

$$\int_{CD} \frac{\theta'(z)}{\theta(z)}\, dz = -\int_{AB} \frac{\theta'(u)}{\theta(u)}\, du + 2i\pi.$$

For the union of the two sides AB and CD, the integral

$$\frac{1}{2i\pi} \int \frac{\theta'(z)}{\theta(z)}\, dz$$

has the value 1.

Since $\theta(z)$ is of period π, the integrals over the sides AD and BC cancel each other out. Finally, $N = 1$. Thus there is exactly one zero of $\theta(z)$ in the entire parallelogram Γ.

Since $\pi t/2$ is a zero of $\theta(z)$, every complex number of the form

$$\frac{\pi t}{2} + n_1\pi,$$

where n_1 is an integer, is also a zero by virtue of the periodicity of $\theta(z)$.

But if z is a zero, the formula

$$\theta(z+\pi t) = -\frac{1}{q}\, e^{-2iz}\theta(z)$$

shows that $z + \pi t$ is also a zero. Thus, more generally, $z + n_2\pi t$, where n_2 is an integer, is a zero. All the complex numbers

$$\frac{\pi t}{2} + n_1\pi + n_2\pi t \tag{4}$$

are zeros of $\theta(z)$. Now, every parallelogram Γ contains one such number. Thus we have found *all* the zeros of $\theta(z)$.

D. We choose $z_0 = -\pi/2$. Thus we obtain a particular parallelogram Γ.

We shall evaluate the integral

$$\int_\Gamma \frac{\theta'(z)}{\theta(z)}\, e^{-2niz}\, dz.$$

In the interior of Γ, the function $\theta(z)$ vanishes at a point z_0 which is of the form $\pi t/2 + k\pi$, where k is a particular integer. The residue at this point is

$$\exp(-2niz_0) = \exp\left[-2ni\left(\frac{\pi t}{2} + k\pi\right)\right] = e^{-ni\pi t}.$$

Therefore, we have

$$\int_\Gamma \frac{\theta'(z)}{\theta(z)} e^{-2niz}\, dz = 2i\pi e^{-ni\pi t}.$$

Since $\theta'(z)e^{-2niz}/\theta(z)$ is a periodic function with period π, the integrals over the sides BC and AD cancel each other out. On CD we set $z = \pi t + u$. Then

$$\int_{CD} = -\int_{-\pi/2}^{\pi/2} \frac{\theta'(\pi t + u)}{\theta(\pi t + u)} e^{-2ni(\pi t + u)}\, du.$$

Now,

$$\frac{\theta'(u + \pi t)}{\theta(u + \pi t)} = -2i + \frac{\theta'(u)}{\theta(u)}.$$

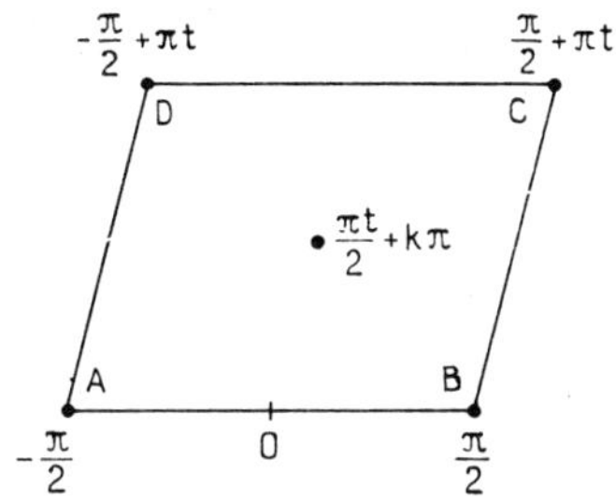

FIGURE 40

Thus we have

$$\int_{CD} = -e^{-2ni\pi t}\int_{AB} \frac{\theta'(u)}{\theta(u)} e^{-2niu}\, du + 2ie^{-2ni\pi t}\int_{-\pi/2}^{\pi/2} e^{-2niu}\, du.$$

If $n \neq 0$, the last integral vanishes. Finally,

$$\int_\Gamma \frac{\theta'(z)}{\theta(z)} e^{-2niz}\, dz = (1 - e^{-2ni\pi t})\int_{-\pi/2}^{\pi/2} \frac{\theta'(z)}{\theta(z)} e^{-2niz}\, dz = 2i\pi e^{-ni\pi t},$$

or

$$\int_{-\pi/2}^{\pi/2} \frac{\theta'(z)}{\theta(z)} e^{-2niz}\, dz = \frac{\pi}{\sin n\pi t} = 2i\pi \frac{q^n}{q^{2n} - 1}. \tag{5}$$

Let us confine ourselves to the real values of z. The function $\theta'(z)/\theta(z)$ is continuous, differentiable, and periodic with period π, and it has a Fourier-series expansion of the form

$$\frac{\theta'(z)}{\theta(z)} = \sum_{n=-\infty}^{\infty} c_n e^{2niz}, \tag{6}$$

where c_n is given by the formula

$$c_n = \frac{1}{\pi} \int_{-\pi/2}^{\pi/2} \frac{\theta'(z)}{\theta(z)} e^{-2niz}\, dz.$$

Thus for $n \neq 0$,

$$c_n = \frac{2iq^n}{q^{2n} - 1}.$$

For $n = 0$,

$$\int_{-\pi/2}^{\pi/2} \frac{\theta'(z)}{\theta(z)}\, dz = \log \left[\frac{\theta(\pi/2)}{\theta(-\pi/2)}\right].$$

Since θ is periodic with period π, we have $\theta(\pi/2)/\theta(-\pi/2) = 1$ and $c_0 = 0$.

Let us give this expansion a real form (this will be a real expansion if q is real). Since $\theta'(z)/\theta(z)$ is an odd function, the terms in $\cos 2nz$ disappear, and we find

$$\frac{\theta'(z)}{\theta(z)} = 4 \sum_{n=1}^{\infty} \frac{q^n}{1 - q^{2n}} \sin 2nz = 2i \sum_{n=1}^{\infty} \frac{\sin 2nz}{\sin \pi tn}.$$

We thus obtain the Fourier-series expansion of $\theta'(z)/\theta(z)$.

CHAPTER 5

Ordinary Differential Equations

All the exercises in this chapter except the first and the last two are devoted to linear differential equations and systems, some with constant coefficients, others with variable coefficients. Some involve only integration techniques, such as the method of variation of parameters or lowering the order of an equation when we know particular solutions of it. Others are oriented to boundary-value problems: The equation contains a parameter λ and we seek to find whether, for certain values of λ, there exist solutions belonging to a particular vector space E of functions, for example, the space of twice differentiable functions that vanish at two given points. The corresponding values of λ are the eigenvalues of a certain linear operator in E and the corresponding eigenfunctions satisfy orthogonality conditions with respect to some specific measure (or rather, some particular weight).

Thus we obtain families of orthogonal functions. A study of these orthogonal functions could have been the object of exercises in Chapter 2. However, it seemed preferable to collect them in the chapter on differential equations.

Among the families of orthogonal functions, the most classical ones are the family of orthogonal polynomials. Three exercises are devoted to Hermite's polynomials, defined in the first case by their differential equation, in the second by their orthogonality, and in the third by their generating function. It has not seemed necessary to increase the number of exercises of this type and it would have been beyond the scope of the book to give an "exercise" on the general theory of orthogonal polynomials. Several other examples of orthogonal systems are offered, one of them resulting from the study of an integral equation.

This chapter contains several exercises on Bessel functions. Some of these are also exercises on the Laplace transformation. What is usually called "symbolic" or "operational calculus" can be regarded from the primitive point of view of Heaviside and justified by the methods of modern functional analysis. However, from the standpoint of applications, it is convenient to interpret it with the aid of the Laplace transformation. This is what is done here. Certain exercises are devoted to the real direct Laplace transformation $\mathscr{L}x(t) = \int_0^\infty e^{-pt}x(t)\,dt$, where p is real and positive. We also give several examples on the solution of Laplace's differential equations (linear equations with first-degree coefficients) by means of the general transformation $\int_C e^{-pt}x(t)\,dt$ where $x(t)$ is an analytic function of the complex variable t and C is a curve, closed or otherwise, in the complex plane.

It did not seem worthwhile to devote a special chapter to the calculus of variations. The last two exercises on differential equations are very simple exercises on the calculus of variations with fixed limits. They lead to differential equations or systems with boundary conditions. One of these is preceded with a purely algebraic exercise on the search for minima.

This chapter, which precedes and prepares the way for the one on partial differential equations, is already a synthesizing chapter. In it we use constantly the fundamental concepts that constituted the subject matter of the first four chapters: linear algebra (vector spaces of functions), uniform convergence (integration and differentiation under the integral sign), Fourier series, improper multiple integrals, integration in the complex plane, Taylor's series. On several occasions, we even use the concept of absolute convergence of a double series, which is not essentially different from the concept of absolute convergence of an improper double integral.

Finally, we point out that examples of the solution of differential equations and a study of first integrals will be given in Chapter 6 in connection with first-order partial differential equations.

1. Study of a singular differential equation

Suppose that $f(t, x)$ is a function defined in a neighborhood of the origin. Consider the differential equation

$$\frac{dx}{dt} = f(t, x).$$

If $f(t, x)$ is continuous and, in addition, satisfies a Lipschitz condition

$$|f(t, x) - f(t, x')| \leqslant A \mid x - x' \mid$$

in a neighborhood of the origin, the above differential equation has exactly one solution passing through the origin. This is true, in particular, when f has a continuous partial derivative $\partial f/\partial x$ in a neighborhood of the origin. If f is not continuous, existence and uniqueness are not guaranteed. However, continuity of f does not in itself ensure validity of the existence theorem. In this exercise, we shall construct a continuous function that does not satisfy a Lipschitz condition in a neighborhood of the origin, and the differential equation has infinitely many solutions passing through the origin.

Problems. Define a function $f(t, x)$ by

$$\begin{aligned} f(t, x) &= 2t && \text{if } \ x \geqslant t^2 \\ &= \frac{2x}{t} && \text{if } \ |x| \leqslant t^2 \\ &= -2t && \text{if } \ x \leqslant -t^2. \end{aligned}$$

A. Show that $f(t, x)$ is continuous everywhere in the tx-plane, including the origin.

B. Solve the differential equation

$$\frac{dx}{dt} = f(t, x).$$

Show that an infinitely number of curves representing solutions of this equation pass through the origin.

C. Show that $f(t, x)$ does not satisfy a Lipschitz condition in any neighborhood of the origin, that is, that there does not exist a positive constant A such that for sufficiently small values of $|x|$, $|x'|$, and $|t|$,

$$|f(t, x) - f(t, x')| \leqslant A \mid x - x' \mid.$$

Solutions. **A.** Study of the function $f(t, x)$. The curves $x = \pm t^2$ partition the tx-plane into four regions, which we shall call I, I′, II, and II′ as shown in Figure 1.
In each of these regions, *excluding the boundaries*, the function $f(t, x)$ is a continuous function of the variables t and x:

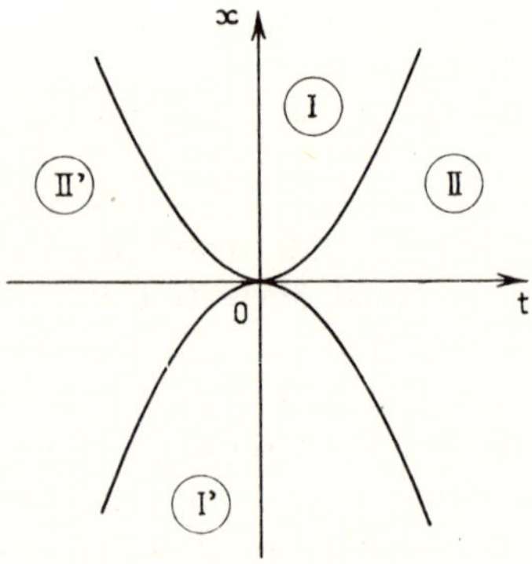

FIGURE 1

In I and I′ the surface representing f as a function of t and x is a plane.

In II and II′ the function f is represented by a conic surface generated by straight lines resting on the t-axis.

Now, it will be sufficient to show that this function is continuous on the boundaries between these regions, that is, on the parabolas $x = \pm t^2$.

Let (t, x) approach (t_0, x_0), where $x_0 = \pm t_0^2 \neq 0$. Then:

In region I $f(t, x)$ tends to $2t_0$ (here x_0 is positive).

In region I′ $f(t, x)$ tends to $-2t_0$ (here x_0 is negative).

In II and II′ $f(t, x)$ tends to $2x_0/t_0$, that is, to $2t_0$ if $x_0 > 0$ and to $-2t_0$ if $x_0 < 0$.

The function $f(t, x)$ tends to the same limit regardless of the path by which (t, x) approaches (t_0, x_0).

To see what happens at the origin, we note that

$$|f(t, x)| \leqslant \frac{2t^2}{|t|} = 2|t|$$

everywhere in regions II and II′. Therefore, everywhere in the plane,

$$|f(t, x)| \leqslant 2 \max\{|t|, |x|\}.$$

From this we conclude immediately that $|f(t, x)|$ tends to zero as (t, x) tends to the origin. Therefore, the function $f(t, x)$ is continuous everywhere in the plane.

B. *Solution of the differential equation.* (a) *In region* I. Here

$$\frac{dx}{dt} = 2t,$$

so that

$$x = t^2 + C,$$

where C is a constant. Therefore, the integral curves are parabolas that can be obtained from the boundary $x = t^2$ by a translation parallel to the x-axis.

As C varies from 0 to $+\infty$, these parabolas sweep the entire region I.

(b) *In region* I′. Here, we have the family

$$\frac{dx}{dt} = -2t,$$

which is symmetric to the preceding family, so that

$$x = -t^2 + C \qquad (C \leqslant 0).$$

(c) *In regions* II *and* II′,

$$\frac{dx}{dt} = \frac{2x}{t},$$

so that

$$\frac{dx}{x} = \frac{2dt}{t},$$

and hence

$$x = Ct^2.$$

These integral curves are parabolas obtained from the boundary $x = t^2$ by multiplying the ordinate by a constant C.

As C varies from -1 to $+1$, these parabolas completely sweep regions II and II′. All of them pass through the origin.

Finally, through every point in the plane other than the origin, there passes one and only one integral curve. But an infinite number of integral curves pass through the origin.

C. *Lipschitz conditions.* We shall verify for this example the theorem on the uniqueness of a solution of a differential equation by showing that the function $f(t, x)$ defined above verifies a Lipschitz condition at every point M_0 of the tx-plane different from the origin.

(a) *In regions I and I′* (excluding the boundaries). Let M_0 be an interior point of I or I'. Then there exists a rectangle in the tx-plane centered at M_0 that is entirely contained in the region. We have

$$f(t, x_1) = f(t, x_2) = 2t,$$

so that

$$|f(t, x_1) - f(t, x_2)| = 0$$

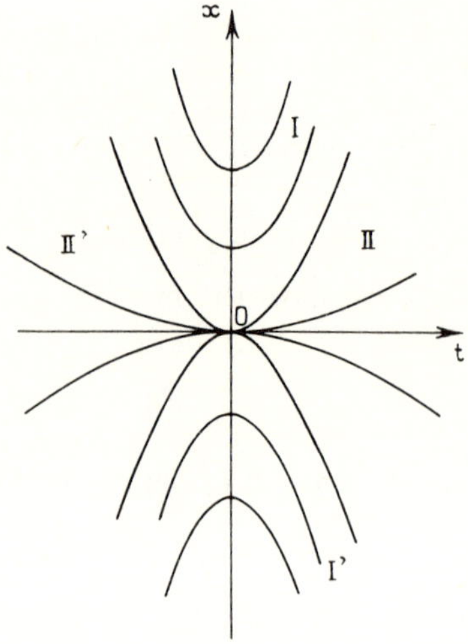

FIGURE 2

for all points $M_1(t, x_1)$ and $M_2(t, x_2)$ in that rectangle. Thus a Lipschitz condition is satisfied at M_0 with Lipschitz constant A equal to zero.

(b) *In regions II and II'* (excluding the boundaries). Again, there exists a rectangle centered at M_0 that lies entirely in the region in question. It is defined by

$$t_0 - \tau \leqslant t \leqslant t_0 + \tau, \qquad x_0 - \chi \leqslant x \leqslant x_0 + \chi,$$

and the interval $[t_0 - \tau, t_0 + \tau]$ does not contain 0. We have

$$|f(t, x_1) - f(t, x_2)| = \frac{2}{|t|}(x_1 - x_2),$$

$$|f(t, x_1) - f(t, x_2)| \leqslant \frac{2}{|t_0 - \tau|}|x_1 - x_2| \qquad \text{in II},$$

$$\leqslant \frac{2}{|t_0 + \tau|}|x_1 - x_2| \qquad \text{in II}'.$$

Thus a Lipschitz condition is verified in the rectangle in question with Lipschitz constant

$$A = \frac{2}{t_0 - \tau} \qquad \text{or} \qquad \frac{2}{|t_0 + \tau|}.$$

(c) *On the boundaries* (exclusive of the origin). Every rectangle centered at M_0 will have a nonempty intersection with II or II'. Consider a rectangle not containing the origin and let T be the absolute value of the abscissa of whichever side is closer to the origin. Just as with case (b),

$$|f(t, x_1) - f(t, x_2)| \leqslant \frac{2}{T} \, |x_1 - x_2| \, .$$

Thus a Lipschitz condition is verified in this rectangle.

At every point of the tx-plane other than the origin, the conditions under which the differential equation will have a unique solution are verified. Therefore, there passes exactly one integral curve through such a point and it was found in Problem B by direct integration.

(d) *At the origin.* Finally, let us show that a Lipschitz condition is not satisfied at the origin. In every rectangular neighborhood of the origin

$$-\tau \leqslant t \leqslant \tau, \qquad -\chi \leqslant x \leqslant \chi,$$

there exist points of region II arbitrarily close to the origin.

One can always find points M_1 and M_2 such that

$$|f(t, x_1) - f(t, x_2)| = \frac{2}{t} \, | x_1 - x_2 | \, ,$$

where t is any number between 0 and τ.

For any constant A, we can choose t sufficiently small that $2/t > A$. Therefore, no constant A exists such that Lipschitz' inequality will be satisfied in a neighborhood of the origin.

2. Study of the differential equation $y'' - y - 1/x$. Application to the evaluation of a definite integral

Problems. A. Solve the differential equation

$$y'' + y = \frac{1}{x} \tag{E}$$

by the method of variation of parameters. Use the result to show that the function

$$y_1(x) = \int_0^\infty \frac{\sin t}{t + x} \, dt$$

is a particular solution of it for $x > 0$.

B. Use the second mean-value theorem to show that the integral $y_1(x)$ converges uniformly for $0 \leqslant x \leqslant x_1$ (where x_1 is any fixed positive number), that the function $y_1(x)$ is continuous from the right at $x = 0$, and that $y_1(x)$ tends to zero as $x \to \infty$.

C. Show that the integral

$$y_2(x) = \int_0^\infty \frac{e^{-tx}}{1+t^2}\, dt$$

is twice differentiable under the integral sign and that it verifies the differential equation (E) for $x > 0$, that it is continuous for $x \geqslant 0$, and that it tends to 0 as $x \to \infty$.

D. Show that (E) possesses exactly one solution that tends to 0 as $x \to \infty$. Show from this that $y_1(x) = y_2(x)$ if $x > 0$.

Show finally that this conclusion remains valid if $x = 0$ and then evaluate the integral $\int^\infty (\sin t/t)\, dt$.

Solutions. **A.** Solution of the differential equation (E). Two linearly independent solutions of the corresponding homogeneous equation are

$$f_1 = \sin x, \qquad f_2 = \cos x.$$

Therefore, the general solution of the homogeneous equation is

$$y(x) = C_1 \sin x + C_2 \cos x.$$

To get the general solution of the nonhomogeneous equation, we apply the method of variation of parameters. We set

$$y = u_1 f_1 + u_2 f_2$$

and we impose on these unknown functions u_1 and u_2 the requirement that they satisfy the system of linear equations

$$u_1' f_1 + u_2' f_2 = 0, \qquad u_1' f_1' + u_2' f_2' = \frac{1}{x}.$$

When we solve this system and substitute the values of f_1 and f_2, we obtain

$$u_1' = \frac{\cos x}{x}, \qquad u_2' = -\frac{\sin x}{x}.$$

From this, we have the *particular* solution of (E):

$$y = \sin x \int_{x_0}^x \frac{\cos v}{v}\, dv - \cos x \int_{x_0}^x \frac{\sin v}{v}\, dv = \int_{x_0}^x \frac{\sin (x - v)}{v}\, dv,$$

which is valid if x and x_0 have the same sign (without this restriction, the first integral diverges).

If x is positive, we may take $x_0 = \infty$, since the integral

$$\int_{\infty}^{x} \frac{\sin (x - v)}{v}\, dv = \int_{0}^{\infty} \frac{\sin t}{t + x}\, dt = y_1(x)$$

converges. Therefore, the general solution of (E) is

$$\boxed{y(x) = y_1(x) + C_1 \sin x + E_2 \cos x \qquad (\text{for } x > 0).} \tag{1}$$

B. *Study of the particular solution* $y_1(x)$. It is easily shown by use of the second mean-value theorem that the integral defining $y_1(x)$ is uniformly convergent for $x \geqslant x_0$, where x_0 is a positive number. But this does not imply that $y_1(x)$ is continuous at $x = 0$ because the function $(t + x)^{-1} \sin t$ is discontinuous at the point $t = 0$, $x = 0$. To show this we write

$$y_1(x) = \cos x \int_{x}^{\infty} \frac{\sin v}{v}\, dv - \sin x \int_{x}^{\infty} \frac{\cos v}{v}\, dv.$$

The first integral is continuous from the right at $x = 0$ and hence so is the product obtained by multiplying it by $\cos x$. The second integral does not converge when $x = 0$. But we may write

$$\int_{x}^{\infty} = \int_{x}^{x_0} + \int_{x_0}^{\infty},$$

where x_0 is a fixed positive number. It will be sufficient to show that

$$\sin x \int_{x}^{x_0} \frac{\cos v}{v}\, dv$$

tends to a limit as $x \to 0$. Now, the absolute value of the integral is less than $\int_{x}^{x_0} (dv/v) = \log x_0 - \log x$, and hence its product with the function $\sin x$ tends to 0 as $x \to 0$. Therefore, $y_1(x)$ is a *continuous function* of x for all $x \geqslant 0$.

To find the limit of $y_1(x)$ as $x \to \infty$, we integrate by parts:

$$y_1(x) = -\left(\frac{\cos t}{t + x}\right)_0^{\infty} + \int_0^{\infty} \cos t . d\frac{1}{t + x} = \frac{1}{x} + \int_0^{\infty} \cos t d\left(\frac{1}{t + x}\right).$$

From this, we get

$$| y_1(x) | \leqslant \frac{1}{x} + \int_0^{\infty} d\left(-\frac{1}{t + x}\right) = \frac{2}{x}.$$

Thus $y_1(x)$ tends to zero as $x \to \infty$.

C. *Study of the function* $y_2(x)$. (a) *Continuity*. The function

$$K(x, t) = \frac{e^{-tx}}{1 + t^2}$$

is a continuous function of t and x in the region $t \geqslant 0$, $x \geqslant 0$. It is majorized by the integrable function $1/(1 + t^2)$, which is independent of x. The integral defining $y_2(x)$ is therefore uniformly convergent and we may assert that $y_2(x)$ is continuous for $x \geqslant 0$. Integration by parts enables us to majorize $y_2(x)$:

$$y_2(x) = -\left[\frac{e^{-tx}}{x(1 + t^2)}\right]_0^\infty + \frac{1}{x}\int_0^\infty e^{-tx}\, d\left(\frac{1}{1 + t^2}\right) = \frac{1}{x} + \frac{1}{x}\int_0^\infty e^{-tx}\, d\left(\frac{1}{1 + t^2}\right),$$

$$|\, y_2(x)| \leqslant \frac{1}{x} + \frac{1}{x}\int_0^\infty d\left(-\frac{1}{1 + t^2}\right) = \frac{2}{x}.$$

Thus $y_2(x)$ tends to zero as $x \to \infty$.

(b) *Differentiability*. The functions

$$\frac{\partial K(x, t)}{\partial x} = -\frac{te^{-tx}}{1 + t^2}, \qquad \frac{\partial^2 K(x, t)}{\partial x^2} = \frac{t^2 e^{-tx}}{1 + t^2}$$

are continuous functions of x and t in the region $t \geqslant 0$, $x \geqslant 0$. Their absolute values are majorized, respectively, by the integrable functions

$$-\frac{t \exp(tx_0)}{1 + t^2} \qquad \text{and} \qquad \frac{t^2 \exp(-tx_0)}{1 + t^2},$$

in the interval $x \geqslant x_0 > 0$. The integrals

$$\int_0^\infty -\frac{te^{-tx}}{1 + t^2}\, dt \qquad \text{and} \qquad \int_0^\infty \frac{t^2 e^{-tx}}{1 + t^2}\, dt$$

exist and are uniformly convergent for $x \geqslant x_0$. Therefore, $y_2(x)$ is twice differentiable and

$$y_2'(x) = \int_0^\infty -\frac{te^{-tx}}{1 + t^2}\, dt,$$

$$y_2''(x) = \int_0^\infty \frac{t^2 e^{-tx}}{1 + t^2}\, dt \qquad \text{for} \quad x \geqslant x_0 > 0.$$

Since

$$y_2''(x) + y_2(x) = \int_0^\infty \frac{1 + t^2}{1 + t^2} e^{-tx}\, dt = \frac{1}{x} \qquad \text{for} \quad x > 0,$$

$y_2(x)$ is a solution of (E) for $x > 0$.

D. *Identification of* $y_1(x)$ *and* $y_2(x)$. Let us go back to expression (1) of the general solution of (E). As x increases without bound, $y_1(x)$ tends to zero and the term $C_1 \sin x + C_2 \cos x$ has a limit only when $C_1 = C_2 = 0$. Thus there is only one solution of (E) that tends to zero as $x \to \infty$ and that one is $y_1(x)$. Since $y_2(x)$ is also a solution of (E) and it too tends to zero, it follows that

$$y_1(x) = y_2(x) \qquad \text{for} \quad x > 0.$$

Now it follows from the result of Problems B and C that the two functions are continuous for $x \geqslant 0$. Since they coincide for $x < 0$, they also coincide for $x = 0$ since, otherwise, at least one of them would be discontinuous at $x = 0$. Therefore,

$$y_1(0) = \int_0^\infty \frac{\sin t}{t}\, dt = y_2(0) = \int_0^\infty \frac{1}{1+t^2}\, dt = \frac{\pi}{2}$$

or

$$\int_0^\infty \frac{\sin t}{t}\, dt = \frac{\pi}{2}.$$

Remark 1. This integral, which is useful in connection with Fourier series and integrals, can be evaluated in many other ways, for example, by integration of e^{iz}/z in the complex plane, by use of the properties of the Dirichlet integral, and by various methods of differentiating under the integral sign.

Remark 2. The identity

$$\int_0^\infty \frac{\sin t}{x+t}\, dt = \int_0^\infty \frac{e^{-xu}}{1+u^2}\, du \qquad (x > 0)$$

can be proved without use of the differential equation. It would be sufficient to integrate $e^{ixz}/(1+z)$ around a closed contour formed by the positive half of the real axis, the positive half of the imaginary axis, and a quarter of a circle of very large radius in the first quadrant.

3. Solution of a linear differential equation. An application

Problems. A. Show that the integral

$$I(x) = \int_0^\infty \frac{\sin tx}{t(1+t^4)}\, dt, \tag{1}$$

where x is a positive parameter, can be differentiated four times under the integral sign.

B. Then show that $I(x)$ satisfies the differential equation

$$\frac{d^4y}{dx^4} + y = \frac{\pi}{2}. \tag{2}$$

Now solve this equation and assign values to the constants of integration that will equate the solution to $I(x)$.

Solutions. **A.** *Differentiability of* $I(x)$. We set

$$K(x, t) = \frac{\sin tx}{t(1 + t^4)}.$$

Then

$$\frac{\partial K(x, t)}{\partial x} = \frac{\cos tx}{1 + t^4}, \qquad \frac{\partial^2 K}{\partial x^2} = \frac{-t \sin tx}{1 + t^4},$$

$$\frac{\partial^3 K}{\partial x^3} = \frac{-t^2 \cos tx}{1 + t^4}, \qquad \frac{\partial^4 K}{\partial x^4} = \frac{t^3 \sin tx}{1 + t^4}.$$

These five functions are continuous functions of x and t in the region $t \geqslant 0$, $x \geqslant 0$.

In the interval $0 \leqslant x \leqslant A$, the first four of these functions are majorized independently of x by the integrable functions

$$|K| < \frac{A}{1 + t^4}, \qquad \left|\frac{\partial K}{\partial x}\right| < \frac{1}{1 + t^4},$$

$$\left|\frac{\partial^2 K}{dx^2}\right| < \frac{t}{1 + t^4}, \qquad \left|\frac{\partial^3 K}{\partial x^3}\right| < \frac{t^2}{1 + t^4}.$$

For the first of these, we use the inequality

$$\left|\frac{\sin tx}{t}\right| = x \left|\frac{\sin tx}{tx}\right| \leqslant x \leqslant A.$$

The corresponding integrals are uniformly convergent.

The uniform convergence of the last integral in an interval of the form $\epsilon \leqslant x \leqslant A$ (where $\epsilon > 0$) can be shown by using the second mean-value theorem:

$$\int_p^q \frac{t^3 \sin tx}{1 + t^4}\, dt = \frac{p^3}{1 + p^4} \int_p^\xi \sin xt\, dt$$

$$= \frac{p^3}{1 + p^4} \frac{\cos xp - \cos x\xi}{x}, \qquad \text{where} \quad p \leqslant \xi \leqslant q.$$

$$\left|\int_p^q \frac{t^3 \sin tx}{1 + t^4}\, dt\right| \leqslant \frac{p^3}{p^4} \frac{2}{x} \leqslant \frac{2}{p\epsilon}, \qquad \text{in the interval} \quad \epsilon \leqslant x \leqslant A,$$

$$\left|\int_p^\infty \frac{t^3 \sin tx}{1 + t^4}\, dt\right| \leqslant \frac{2}{p\epsilon}.$$

Now $2(p\epsilon)^{-1}$ tends to zero independently of x (for $\epsilon \leqslant x \leqslant A$) as p increases without bound. This is the definition of uniform convergence of the integral.

The hypotheses of the theorem on differentiating under the integral sign are satisfied for the first four derivatives of $K(x, t)$. Therefore, $I(x)$ is four times differentiable and, in particular,

$$\frac{d^4I}{dx^4} = \int_0^\infty \frac{t^3 \sin xt}{1 + t^4}\, dt.$$

B. *An expression for $I(x)$.* We have

$$\frac{d^4I}{dx^4} + I = \int_0^\infty \frac{1 + t^4}{1 + t^4} \frac{\sin xt}{t}\, dt = \int_0^\infty \frac{\sin xt}{t}\, dt = \frac{\pi}{2} \qquad \text{(since } x > 0\text{)}.$$

Thus $I(x)$ satisfies the differential equation

$$\frac{d^4y}{dx^4} + y = \frac{\pi}{2}. \tag{2}$$

A particular solution of this equation is obviously $y = \pi/2$. The general solution of the corresponding homogeneous equation is

$$y_0(x) = C_1 \exp\left[\left(\frac{1+i}{\sqrt{2}}\right) x\right] + C_2 \exp\left[\left(\frac{1-i}{\sqrt{2}}\right) x\right] + C_3 \exp\left[-\left(\frac{1+i}{\sqrt{2}}\right) x\right]$$
$$+ C_4 \exp\left[-\left(\frac{1-i}{\sqrt{2}}\right) x\right]$$

or, in terms of real quantities,

$$y_0(x) = A \cosh\frac{x}{\sqrt{2}} \cos\frac{x}{\sqrt{2}} + B \cosh\frac{x}{\sqrt{2}} \sin\frac{x}{\sqrt{2}} + C \sinh\frac{x}{\sqrt{2}} \cos\frac{x}{\sqrt{2}}$$
$$+ D \sinh\frac{x}{\sqrt{2}} \sin\frac{x}{\sqrt{2}}.$$

Thus the general solution of (2) is $y(x) = y_0(x) + \pi/2$.

To determine the constants of integration corresponding to $I(x)$, we note first that if C_1 and C_2 are nonzero, then $y_0(x)$ and $y_0'(x)$ are not bounded as $x \to \infty$.

Since

$$|I'(x)| = \left|\int_0^\infty \frac{\cos tx}{1 + t^4}\, dt\right| < \int_0^\infty \frac{dt}{1 + t^4} < \infty,$$

we must have

$$C = C_2 = 0.$$

On the other hand, because of the uniform convergence, $I(x)$ and $I'(x)$ are continuous at $x = 0$ and

$$I(0) = 0, \qquad I'(0) = \int_0^\infty \frac{dt}{1 + t^4} = \frac{\pi\sqrt{2}}{4}.$$

Thus we have the relations

$$C_3 + C_4 + \frac{\pi}{2} = 0, \qquad -\frac{(1+i)}{\sqrt{2}}\, C_3 - \frac{(1-i)}{\sqrt{2}}\, C_4 = \frac{\pi\sqrt{2}}{4},$$

which yields

$$C_3 = C_4 = -\frac{\pi}{4},$$

from which we get

$$I(x) = \frac{\pi}{2}\left\{1 - \frac{1}{2}\exp\left[-\left(\frac{1+i}{\sqrt{2}}\right)x\right] - \frac{1}{2}\exp\left[-\left(\frac{1-i}{\sqrt{2}}\right)x\right]\right\}$$

or, better,

$$\boxed{I(x) = \frac{\pi}{2}\left[1 - \exp\left(-\frac{x}{\sqrt{2}}\right)\cos\frac{x}{\sqrt{2}}\right] \qquad (x > 0).}$$

4. Reduction of the order of a linear differential system

Problem. Consider a homogeneous linear differential system of the form

$$\frac{dX}{dt} = \mathbf{A}X, \tag{1}$$

where $\mathbf{A}$ is a matrix with n rows and n columns whose elements are functions of t. The unknown X is a vector in n-dimensional vector space.

Suppose that we know p linearly independent solutions $X_1, X_2, \ldots, X_p$ of this system (where $p < n$).

Let us supplement the p vectors $X_1, \ldots, X_p$ with $n - p$ arbitrary vectors $Y_{p+1}, \ldots, Y_n$ in such a way as to form a basis for the n-dimensional vector space and let us do this for every t in a certain interval. We denote by $\mathbf{M}$ the matrix whose columns are constituted by the components of the p vectors X_j and the $n - p$ vectors Y_j. Finally, let us make the transformation

$$X = \mathbf{M}Z.$$

This Z verifies a linear differential system. Show that this system is solved by the first p vectors of the basis and that its solution leads to a system of lower order than the original system.

Solutions. We denote by $(X, Y, \ldots)$ the matrix (not necessarily square) whose columns are constituted by the components of the vectors $X, Y, \ldots$.

Thus we write

$$\mathbf{M} = (X_1, X_2, \ldots, X_p, Y_{p+1}, \ldots, Y_n).$$

Let us now recall the following result concerning the definition of square matrices: Let $\mathbf{M}$ denote a square matrix:

$$\mathbf{M} = \begin{pmatrix} m_{11} & \cdots & m_{1j} & \cdots & m_{1n} \\ \cdot & \cdot & \cdot & \cdot & \cdot \\ m_{n1} & \cdots & m_{nj} & \cdots & m_{nn} \end{pmatrix}.$$

The components of the vector transformed by $\mathbf{M}$ of the basis vector

$$x_1 = 0, \ldots, x_{j-1} = 0, x_j = 1, x_{j+1} = 0, \ldots, x_n = 0$$

are the elements $m_{ij} \cdots m_{nj}$ of the jth column of $\mathbf{M}$.

If we choose for $\mathbf{M}$ the matrix defined above, we see that $X_1, X_2, \ldots, X_p$ are the transforms by $\mathbf{M}$ of the first p vectors of the basis $E_1, \ldots, E_p$.

Now let us seek to solve the differential equation

$$X' = \mathbf{A}X \tag{1}$$

by making the change of variable

$$X = \mathbf{M}Z. \tag{2}$$

We obtain a new linear system in Z of the form

$$\mathbf{M}Z' + \mathbf{M}'Z = \mathbf{AM}Z \qquad \text{or} \qquad \mathbf{M}Z' = (\mathbf{AM} - \mathbf{M}')Z.$$

By hypothesis, the matrix $\mathbf{M}$ has an inverse. Therefore,

$$Z' = \mathbf{M}^{-1}(\mathbf{AM} - \mathbf{M}')Z,$$

which we may write

$$Z' = \mathbf{B}Z, \tag{3}$$

where $\mathbf{B}$ is a new matrix.

If Z is a solution of (3), X is a solution of (1). Now, to every solution X_j (for $j = 1, ..., p$) of (1) there corresponds a basis vector E_j. The vectors $E_1, ..., E_p$ are therefore solutions of (3). If this is the case, we have

$$\mathbf{B}E_j = 0, \qquad (j = 1, 2, ..., p).$$

Thus the first j columns of $\mathbf{B}$ are made up of zeros. The system $Z' = \mathbf{B}Z$ can be put in the form

$$\begin{aligned}
x_1' &= b_{1p+1}x_{p+1} + \cdots + b_{1n}x_n, \\
x_2' &= b_{2p+1}x_{p+1} + \cdots + b_{2n}x_n, \\
&\cdots\cdots\cdots\cdots\cdots\cdots \\
x_{p+1}' &= b_{p+1p+1}x_{p+1} + \cdots + b_{p+1n}x_n, \\
&\cdots\cdots\cdots\cdots\cdots\cdots \\
x_n' &= b_{np+1}x_{p+1} + \cdots + b_{nn}x_n.
\end{aligned} \tag{4}$$

The last $n - p$ equations contain only the last $n - p$ components. Thus solution of (4) leads to a linear differential system in an $(n - p)$-dimensional space. The order of the system is reduced from n to $n - p$. Having calculated $x_{p+1}, ..., x_n$, we then calculate $x_1, ..., x_p$ by quadratures.

Example. Consider the matrix

$$\mathbf{A} = \begin{pmatrix} -6 & 5 & 3 \\ -8 & 7 & 4 \\ -2 & 1 & 1 \end{pmatrix}.$$

It has a simple eigenvalue $\lambda = 0$ with the eigenvector

$$1, \quad 0, \quad 2$$

and the double eigenvalue $\lambda = 1$ with the eigenvector

$$1, \quad 2, \quad -1.$$

Thus the linear system $X' = \mathbf{A}X$ has the two vector solutions

$$\begin{aligned}
X_1 &: \quad 1, \quad 0, \quad 2, \\
X_2 &: \quad e^t, \quad 2e^t, \quad -e^t.
\end{aligned}$$

Let us supplement X_1 and X_2 with the vector

$$Y_3: \quad 1, \quad 0, \quad 0,$$

which is independent of X_1 and X_2, and let us set

$$\mathbf{M} = \begin{pmatrix} 1 & e^t & 1 \\ 0 & 2e^t & 0 \\ 2 & -e^t & 0 \end{pmatrix}.$$

Let us make the change of variable

$$x = u + e^t v + w, \qquad y = 2e^t v, \qquad z = 2u - e^t v. \tag{5}$$

The initial system becomes

$$\begin{aligned} u' + e^t(v + v') + w' &= e^t v - 6w, \\ 2e^t(v + v') &= 2e^t v - 8w, \\ 2u' - e^t(v + v') &= -e^t v - 2w, \end{aligned}$$

or

$$u' = -3w, \qquad v' = -4we^{-t}, \qquad w' = w.$$

This system can be broken into two systems: the "system"

$$w' = w,$$

which contains only one equation with a scalar unknown, and the system

$$u' = -3w, \qquad v' = -4we^{-t},$$

which, once w is known, yields u and v by integration.

Here we have

$$w = C_1 e^t,$$

so that

$$\begin{aligned} u' &= -3C_1 e^t, & u &= -3C_1 e^t + C_2, \\ v' &= -4C_1, & v &= -4C_1 t + C_2. \end{aligned}$$

We now return to x, y, and z by formulas (5).

The presence of a nonexponential term $C_1 t$ reflects the fact that the matrix $\mathbf{A}$, which has a double eigenvalue, is not diagonalizable.

Other methods of solving the system $X' = \mathbf{A}X$. Since the matrix $\mathbf{A}$ is constant (independent of t) in this particular case, we may proceed by a change of basis and reduce the matrix to the Jordan form or, more simply, to a triangular form. We can also use the Laplace transformation.

5. A second-order linear differential equation with given boundary conditions

Problems. A. Show that there exist real numbers μ such that the differential equation

$$x^2y'' + xy' + \mu^2 y = 0 \tag{E}$$

has real solutions defined for $x > 0$ that are not identically zero but that do vanish at $x = 1$ and $x = l$ (where l is a given positive number distinct from 1). Find the general solution.

B. Suppose $y_1(x)$ and $y_2(x)$ are two solutions of the type referred to in the preceding problem corresponding to two distinct values μ_1 and μ_2 of μ. Find a nonnegative function $c(x)$ such that

$$\int_1^l y_1(x)y_2(x)c(x)\,dx = 0.$$

Solutions. **A.** *Solutions of* (E). Equation (E) is an Euler equation. We seek particular solutions of the form

$$y = x^r,$$

where r verifies the equation

$$r(r-1) + r + \mu^2 = 0,$$

that is,

$$r^2 + \mu^2 = 0.$$

(a) If $\mu = 0$, the general solution of (E) is

$$y(x) = C_1 \log x + C_2 .$$

This is a monotonic function of x for $x > 0$. The only solution such that $y(1) = y(l) = 0$.

(b) If $\mu \neq 0$, the general solution of (E) is

$$y(x) = C_1 x^{i\mu} + C_2 x^{-i\mu}, \qquad (x > 0)$$

or, by setting $A = i(C_1 - C_2)$ and $B = C_1 + C_2$,

$$y(x) = A \sin(\mu \log x) + B \cos(\mu \log x), \qquad (x > 0).$$

For $y(1)$ to be equal to 0, it is necessary that $B = 0$. For $y(l)$ to be equal to 0, it is necessary that

$$A \sin(\mu \log l) = 0.$$

(a) If $\mu \neq k\pi/\log l$, where k is an integer, A must be 0. In this case, the only solution such that $y(1) = y(l) = 0$ is the identically zero solution.

(b) If $\mu = k\pi/\log l$, where k is a nonzero integer, there are nontrivial solutions such that $y(1) = y(l) = 0$. These are obviously defined up to a constant factor A:

$$\boxed{y(x) = A \sin\left(\frac{k\pi}{\log l} \log x\right) \qquad (x > 0).}$$

B. *The orthogonality of eigenfunctions.* Let us first put (E) in the reduced (also called *self-adjoint*) form. If we divide by x, we have

$$xy'' + y' + \frac{\mu^2}{x} y = 0 \qquad \text{or} \qquad \frac{d}{dx}(xy') + \frac{\mu^2}{x} y = 0.$$

The coefficient μ^2/x of y is continuous in the interval $[1, l]$.

Let us now consider two solutions y_1 and y_2 of the form (1) corresponding to two distinct values μ_1 and μ_2 of μ. We have

$$\frac{d}{dx}(xy_1') + \frac{\mu_1^2}{x} y_1 = 0, \qquad \frac{d}{dx}(xy_2') + \frac{\mu_2^2}{x} y_2 = 0.$$

Let us multiply the first of these equations by y_2 and the second by $-y_1$ and then add them:

$$y_2 \frac{d}{dx}(xy_1') - y_1 \frac{d}{dx}(xy_2') + \frac{y_1 y_2}{x}(\mu_1^2 - \mu_2^2) = 0.$$

Calculation shows that the first two terms in this expression represent the derivative of $x(y_2 y_1' - y_1 y_2')$. If we integrate from 1 to l, we obtain

$$[x(y_2 y_1' - y_1 y_2']_1^l + (\mu_1^2 - \mu_2^2) \int_1^l \frac{y_1 y_2}{x}\, dx = 0.$$

The functions y_1 and y_2 vanish for $x = 1$ and $x = l$. Since their derivatives are continuous, the first term vanishes. Since $\mu_1 \neq \mu_2$, we have

$$\int_1^l \frac{y_1 y_2}{x}\, dx = 0.$$

The real functions y_1 and y_2 are orthogonal on the interval $[0, l]$ with weight $1/x$.

Of course, we can verify by direct calculation that

$$\int_1^l \sin\left(\frac{k_1\pi}{\log l}\log x\right)\sin\left(\frac{k_2\pi}{\log l}\log x\right)\frac{dx}{x} = 0$$

if $k_1 \neq k_2$. We need only make the change of variable $\log x = t \log l$.

The functions $y_k(x)$ are the eigenfunctions of the given differential equation for the given boundary-value problem, that is, in the vector space of real twice continuously differentiable functions that vanish at $x = l$ and $x = 1$. The numbers $k\pi/\log l$ are the eigenvalues. The eigenfunctions corresponding to two distinct eigenvalues are orthogonal relative to a "measure," which depends on the differential operator in question and which in the present case has weight $1/x$.

6. Study of a linear integral equation

In this exercise we shall study an equation of the form

$$\lambda f(x) = \int_0^1 K(x, y) f(y)\, dy,$$

in which $K(x, y)$ is a continuous symmetric kernel that can be expanded in a Fourier series. This equation is, moreover, equivalent to a linear second-order differential equation whose solutions, thanks to the boundary conditions, belong to a well-specified vector space of functions.

In connection with this exercise, we shall give some applications of Parseval's formula and we shall interpret them in terms of norms and scalar products in the Hilbert space L^2.

Problems. Suppose that $K(x, y)$ is the function, defined in the square $0 \leqslant x \leqslant 1$, $0 \leqslant y \leqslant 1$, that is equal to

$$\begin{aligned} &x(1-y) \quad \text{if} \quad 0 \leqslant x \leqslant y \leqslant 1,\\ &y(1-x) \quad \text{if} \quad 0 \leqslant y \leqslant x \leqslant 1. \end{aligned}$$

A. Show that

$$K(x, y) = \frac{2}{\pi^2}\sum_{n=1}^{\infty}\frac{\sin n\pi x \sin n\pi y}{n^2}.$$

B. Show that $f(x)$ is a real function of a real variable x and that f is continuous on the interval $0 \leqslant x \leqslant 1$. We define

$$g(x) = \int_0^1 K(x, y) f(y)\, dy.$$

Show that $g(x)$ vanishes for $x = 0$ and $x = 1$ and that it has a second derivative equal to $-f(x)$.

C. We set

$$c_n = \sqrt{2} \int_0^1 f(x) \sin n\pi x\, dx.$$

Use Parseval's equality to show that

$$\int_0^1 [g(x)]^2\, dx = \frac{1}{\pi^4} \sum_{n=1}^{\infty} \frac{c_n^{\,2}}{n^4}.$$

From this, derive the inequalities

$$\int_0^1 [g(x)]^2\, dx \leqslant \frac{1}{\pi^4} \int_0^1 [f(x)]^2\, dx,$$

$$\int\int K(x, y) f(x) f(y)\, dx\, dy \leqslant \frac{1}{\pi^2} \int_0^1 [f(x)]^2\, dx,$$

where the double integral is over the square $0 \leqslant x \leqslant 1$, $0 \leqslant y \leqslant 1$. For what functions $f(x)$ does equality hold? Interpret these inequalities by introducing a norm and a scalar product in the space L^2 of functions that are square-integrable over the interval [0, 1].

D. Consider the integral equation

$$\lambda f(x) = \int_0^1 K(x, y) f(y)\, dy,$$

where λ is a real nonzero parameter. Show that there exist values of λ for which this equation has nonzero solutions. Find these values and the corresponding solutions.

Solutions. **A.** *Expression of $K(x, y)$ in the form of a series.* The function $K(x, y)$ is symmetric in x and y:

$$K(x, y) = K(y, x).$$

If we fix y and then consider K as a function of x, it is continuous and differentiable on [0, 1] except at $x = 0$, $x = 1$, and $x = y$. Therefore, we can expand it in a Fourier sine series. According to Jordan's theorem as applied to sine series, this series converges in [0, 1] to $K(x, y)$.

The Fourier coefficients are

$$k_n(y) = 2\int_0^1 K(t, y) \sin n\pi t \, dt$$
$$= 2\int_0^y t(1-y) \sin n\pi t \, dt + 2\int_y^1 y(1-t) \sin n\pi t \, dt.$$

Integrating by parts, we obtain

$$k_n(y) = -2\frac{y(1-y)}{n\pi}\cos n\pi y + \frac{2(1-y)}{n^2\pi^2}\sin n\pi y + 2\frac{y(1-y)}{n\pi}\cos n\pi y + \frac{2y}{n^2\pi^2}\sin n\pi y$$
$$= \frac{2}{n^2\pi^2}\sin n\pi y.$$

Therefore, for every value of y, we have

$$K(x, y) = \frac{2}{\pi^2}\sum_{n=1}^{\infty}\frac{\sin n\pi y \sin n\pi x}{n^2}. \tag{1}$$

We note that this expression is indeed symmetric in x and y.

B. *The relation between the functions $f(x)$ and $g(x)$.* From the definition of $K(x, y)$, we have

$$g(x) = (1-x)\int_0^x yf(y)\,dy + x\int_x^1 (1-y)f(y)\,dy. \tag{2}$$

The function $g(x)$ is differentiable and its derivative is given by

$$\frac{dg}{dx} = -\int_0^x yf(y)\,dy + (1-x)\,xf(x) + \int_x^1 (1-y)f(y)\,dy - x(1-x)f(x)$$
$$= -\int_0^x yf(y)\,dy + \int_x^1 f(y)\,dy - \int_x^1 yf(y)\,dy \tag{3}$$
$$= \int_x^1 f(y)\,dy - \int_0^1 yf(y)\,dy.$$

The last term on the right is a constant. If we again differentiate with respect to x, we obtain

$$\frac{d^2g}{dx^2} = -f(x). \tag{4}$$

In expression (2), we see immediately that

$$g(0) = g(1) = 0. \tag{5}$$

Therefore, the function $g(x)$ is the solution of the differential equation

$$g''(x) = -f(x) \tag{6}$$

that vanishes at $x = 0$ and $x = 1$.

C. *Inequalities involving $f(x)$ and $g(x)$.* The function $f(x)$ is continuous on $[0, 1]$. Therefore, its absolute value is bounded by some number M. The series

$$\sum_{n=1}^{\infty} \frac{2 \sin n\pi x \sin n\pi y}{\pi^2 n^2} f(y)$$

converges uniformly with respect to y on $[0, 1]$. Therefore, we may integrate this series term by term:

$$\int_0^1 K(x, y) f(y)\, dy = \sum_{n=1}^{\infty} \int_0^1 \frac{2 \sin n\pi x \sin n\pi y}{\pi^2 n^2} f(y)\, dy.$$

From this, we get

$$g(x) = \sum_{n=1}^{\infty} \frac{\sqrt{2}}{\pi^2} \frac{c_n}{n^2} \sin n\pi x. \tag{7}$$

Since

$$| c_n | \leqslant \sqrt{2} \int_0^1 |f(y)|\, dy \leqslant \sqrt{2}\, M,$$

the sine series (7) converges uniformly to an odd periodic function of period 2 that coincides with $g(x)$ in the interval $[0, 1]$. We may apply Parseval's formula to it:

$$2 \int_0^1 | g(x)|^2\, dx = \frac{2}{\pi^4} \sum_{n=1}^{\infty} \frac{c_n{}^2}{n^4}. \tag{8}$$

On the other hand, we can expand the function $f(x)$ in a sine series. Its Fourier coefficients are

$$a_n = 2 \int_0^1 f(t) \sin n\pi t\, dt = \sqrt{2}\, c_n .$$

The series $\sum_{n=1}^{\infty} a_n \sin n\pi x$ converges in mean square to $f(x)$, and Parseval's equality yields

$$2\int_0^1 |f(x)|^2\, dx = \sum_{n=1}^{\infty} a_n^{\,2} = 2\sum_{n=1}^{\infty} c_n^{\,2}. \tag{9}$$

Since

$$\frac{c_n^{\,2}}{n^4} \leqslant c_n^{\,2}, \tag{10}$$

we derive from (8) and (9) the first inequality required:

$$\int_0^1 |g(x)|^2\, dx \leqslant \frac{1}{\pi^4}\sum_{n=1}^{\infty} c_n^{\,2} = \frac{1}{\pi^4}\int_0^1 |f(x)|^2\, dx. \tag{11}$$

To prove the second inequality

$$\iint K(x, y) f(x) f(y)\, dx\, dy \leqslant \frac{1}{\pi^2}\int_0^1 |f(x)|^2\, dx, \tag{12}$$

we note first that we can make the function $g(x)$ appear in the double integral by integrating first with respect to y:

$$\iint K(x, y) f(x) f(y)\, dx\, dy = \int_0^1 f(x)\, g(x)\, dx.$$

From the results of Problem B,

$$\begin{aligned}\int_0^1 f(x)\, g(x)\, dx &= -\int_0^1 g''(x)\, g(x)\, dx = -[g'(x)\, g(x)]_0^1 + \int_0^1 [g'(x)]^2\, dx \\ &= \int_0^1 [g'(x)]^2\, dx.\end{aligned}$$

Therefore, the double integral is positive. If we apply Schwarz' inequality to the simple integral $\int_0^1 f(x) g(x)\, dx$, we have

$$\iint K(x, y) f(x) f(y)\, dx\, dy \leqslant \left\{\int_0^1 [f(x)]^2\, dx \int_0^1 [g(x)]^2\, dx\right\}^{1/2}$$

and, from (11), we finally obtain inequality (12):

$$\iint K(x, y) f(x) f(y)\, dx\, dy = \int_0^1 [g'(x)]^2\, dx \leqslant \frac{1}{\pi^2}\int_0^1 [f(x)]^2\, dx$$

For equality to hold in (11) and (12), it is first necessary that equality hold in (10), that is,

$$\frac{c_n^2}{n^4} = c_n^2,$$

which means that $c_n = 0$ for $n \geqslant 2$, c_1 being able to take any value whatever. Since the trigonometric system is a complete system, the only continuous functions (or, more generally, square-integrable functions) for which we can have equality in (11) and (12) are functions of the form

$$f(x) = c_1 \sin \pi x.$$

From Problem B, we immediately have

$$g(x) = \frac{c_1}{\pi^2} \sin \pi x = \frac{1}{\pi^2} f(x),$$

which yields

$$\int_0^1 [g(x)]^2\, dx = \frac{1}{\pi^4} \int_0^1 [f(x)]^2\, dx$$

and

$$\int_0^1 f(x)\, g(x)\, dx = \frac{1}{\pi^2} \int_0^1 [f(x)]^2\, dx.$$

In (11) and (12), we have equality for the functions $f(x) = c_1 \sin \pi x$ and only for such functions.

Interpretation. If f is continuous, it belongs to the space L^2 of real functions such that $\int_0^1 [f(x)]^2\, dx$ exists. This space is a vector space. A *norm* $\| f \|$ is defined on it by

$$\| f \|^2 = \int_0^1 f^2\, dx.$$

For any two elements f and g of L^2, a *scalar product* of f and g is defined by

$$(f, g) = \int_0^1 f(x)\, g(x)\, dx.$$

The formula

$$g(x) = \int_0^1 K(x, y) f(y)\, dy$$

maps f into g by a linear operation $\mathbf{K}$. We may write $g = \mathbf{K}f$.

Formula (11) expresses the fact that

$$\| g \| \leqslant \frac{1}{\pi^2} \| f \| \qquad \text{or} \qquad \| \mathbf{K}f \| \leqslant \frac{1}{\pi^2} \| f \|.$$

Formula (12) expresses the fact that

$$(\mathbf{K}f, f) \leqslant \frac{1}{\pi^2} \|f\|^2.$$

These equalities hold for $f(x) = c \sin \pi x$.

D. If

$$\lambda f(x) = \int_0^1 K(x, y) f(y)\, dy,$$

it follows from Problem B that

$$\lambda f'' = -f.$$

Thus f is a solution of the differential equation

$$f'' + \frac{f}{\lambda} = 0,$$

and it verifies the boundary conditions

$$f(0) = f(1) = 0.$$

We distinguish two cases:

(a) $\lambda < 0$. We set $1/\lambda = -\omega^2$, so that

$$f(x) = ae^{\omega x} + be^{-\omega x},$$

where a and b are two constants such that

$$a + b = 0, \qquad ae^{\omega} + be^{-\omega} = 0.$$

In general, these two linear equations have the unique solution $a = b = 0$. The only possible exception would be for ω to verify the equation

$$e^{\omega} - e^{-\omega} = 0 \qquad \text{or} \qquad e^{2\omega} = 1.$$

But this implies that $\omega = 0$, which is incompatible with the equation $1/\lambda = -\omega^2$. For $\lambda < 0$, the integral equation has only the solution $f = 0$.

(b) $\lambda > 0$. Then

$$f(x) = a \cos \omega x + b \sin \omega x \qquad \text{where} \quad \frac{1}{\lambda} = \omega^2.$$

The constants a and b verify the conditions

$$a = 0 \qquad \text{and} \qquad b \sin \omega = 0.$$

Since a and b cannot vanish simultaneously, ω must be a multiple of π. Therefore, let us write $\omega = n\pi$. Thus

$$\lambda = \frac{1}{n^2\pi^2},$$

the given equation has the nonzero solution

$$f(x) = b \sin n\pi x,$$

which is defined up to a constant factor.

The values $\lambda = 1/n^2\pi^2$ are called the *eigenvalues* of the operator $-d^2/dx^2$ in the vector space of twice differentiable functions on the interval $[0, 1]$ and that vanish at $x = 0$ and $x = 1$.

Remark 1. For $\lambda = 0$, all functions orthogonal to the kernel $K(x, y)$ considered as a function of y are nonzero solutions of the integral equation. These are even functions that can be expanded in cosine series.

Remark 2. If $\mathbf{K}f - \lambda f$, inequality (12) in Problem C expresses the fact that

$$|\lambda| \, \|f\| \leqslant \frac{1}{\pi^2} \|f\| \qquad \text{or} \qquad |\lambda| \leqslant \frac{1}{\pi^2}.$$

This immediately gives an upper bound of the set of eigenvalues of the integral equation.

7. The differential equation of the Hermite polynomials

The study of Hermite polynomials $P_n(x)$ constitutes the subject of three exercises. In the first exercise, these polynomials are defined as solutions of a linear second-order differential equation depending on an integer n. This equation can be solved by the generalized Laplace transformation. From it we can derive the form of the $P_n(x)$ and their orthogonality with weight e^{-x^2} over the interval $(-\infty, +\infty)$.

In the following exercise, the polynomials $P_n(x)$ are defined in terms of the orthogonality condition. We then establish various recursion relations and we again find, in particular, the differential equation referred to above.

In the third exercise, the polynomials $P_n(x)$ are defined as coefficients in the Taylor-series expansion of a generating function. We again derive the orthogonality properties and we evaluate the norm (in the Hilbert-space sense) of a polynomial P_n.

In all these exercises we shall introduce more or less explicitly a real normed vector space E (Hilbert space) defined by the real functions $f(x)$ such that the integral

$$\int_{-\infty}^{\infty} f^2(x)e^{-x^2}\,dx$$

exists.

This integral will be the square of our norm of f and the integral

$$(f, g) = \int_{-\infty}^{\infty} f(x)g(x)e^{-x^2}\,dx$$

will be the scalar product of the two functions f and g. In particular, we shall say that two functions are *orthogonal* if their scalar product vanishes. The Hermite polynomials are mutually orthogonal.

We shall prove that the set of Hermite polynomials constitutes an orthogonal basis in the space E. The representations that will be used are made relative to this basis, that is, as linear combinations of the polynomials P_n.

There exist numerous other normed vector spaces of polynomials. For example, consider the space of functions f defined on $[0, +\infty]$ such that the integral

$$\int_0^{\infty} f^2(x)e^{-x}\,dx$$

exists. Then consider the subspace of this space consisting of polynomials $L_n(x)$ that are orthogonal on the semi-infinite interval $[0, +\infty]$, that is, polynomials such that, for $m \neq n$,

$$\int_0^{\infty} L_n(x)L_m(x)e^{-x}\,dx = 0.$$

These polynomials, known as Laguerre polynomials, satisfy the following relations:

$$xL_n'' + (1 - x)L_n' + nL_n = 0,$$

$$L_{n+1} - (2n + 1 - x)L_n + n^2L_{n-1} = 0,$$

$$L_n(x) = e^x \frac{d^n}{dx^n}(x^n e^{-x}),$$

$$\frac{1}{1-t}\exp\left(-\frac{xt}{1-t}\right) = \sum_{n=0}^{\infty} \frac{t^n}{n!} L_n(x) \qquad (|t| < 1).$$

They can be studied in the same way as the Hermite polynomials.

Problems. A. In the differential equation

$$y'' + (2n + 1 - x^2)\, y = 0, \tag{1}$$

where n is a nonnegative integer, make the substitution

$$y = \exp(-x^2/2)z.$$

This will yield a new differential equation (E).

B. To solve (E), set

$$z(x) = \int_C e^{ux} f(u)\, du,$$

where C is a contour in the complex u-plane and $f(u)$ is an unknown analytic function.

Construct the differential equation satisfied by $f(u)$. Solve this differential equation. Show, by choosing for C a suitable closed contour and applying the residue theorem, that the polynomials

$$P_n(x) = (-1)^n \exp(x^2) \frac{d^n}{dx^n} \exp(-x^2) \tag{2}$$

satisfy (E).

C. Set

$$y_n(x) = \exp(-x^2/2)P_n(x).$$

Then, $y_n(x)$ is a solution of the differential equation (1). Show that

$$\int_{-\infty}^{\infty} y_k(x) y_n(x)\, dx = 0 \qquad \text{if} \quad k \neq n.$$

Solution. **A.** *The transformed equation.* When we set $y = \exp(-x^2/2)z$, differentiation yields

$$y' = \exp(-x^2/2)z' - x \exp(-x^2/2)z,$$

$$y'' = \exp(-x^2/2)z'' - 2x \exp(-x^2/2)z' - \exp(-x^2/2)(1 - x^2)z.$$

When we substitute these expressions in (1), we obtain immediately

$$z'' - 2xz' + 2nz = 0. \tag{E}$$

B. *Solution of* (E). We use the generalized Laplace transformation. We set

$$z(x) = \int_C e^{ux} f(u)\, du.$$

Assuming that we can differentiate under the integral sign, we have

$$z'(x) = \int_C ue^{ux} f(u)\, du, \qquad z''(x) = \int_C u^2 e^{ux} f(u)\, du.$$

If we integrate xz' by parts, we get

$$xz'(x) = [e^{ux} uf(u)]_C - \int_C e^{ux} [uf(u)]'\, du.$$

If we substitute the expressions that we have obtained into (E), we get

$$[-2e^{ux} uf(u)]_C \quad \int_C e^{ux} \{(u^2 + 2n) f(u) - [-2uf(u)]'\}\, du = 0. \tag{3}$$

This equation will be verified if the two terms individually vanish. The integral will vanish if $f(u)$ satisfies the differential equation

$$-[2uf(u)]' = (u^2 + 2n) f(u),$$

which we may write

$$\frac{[uf(u)]'}{uf(u)} = -\frac{u}{2} - \frac{n}{u}.$$

This can be integrated directly to yield

$$\log[uf(u)] = -\frac{u^2}{4} - n \log u + K.$$

Since $f(u)$ is defined up to a constant factor, we may choose $K = 0$. Then

$$f(u) = \frac{\exp(-u^2/4)}{u^{n+1}}.$$

The function $f(u)$ is a holomorphic function in the complex plane except at the origin, where it has a pole of order $n + 1$. Since it is uniform, we may take a closed contour as the contour C. The first term of (3) then vanishes and (3) is verified. For $z(x)$ not to vanish, it is sufficient that the contour encircle the origin. Then, the value of $z(x)$ is given by the residue theorem:

$$z(x) = \int_C \frac{\exp[ux - (u^2/4)]}{u^{n+1}}\, du = e^{x^2} \int_C \frac{\exp\{-[x - (u/2)]^2\}}{u^{n+1}}\, du.$$

If we set $\zeta = x - (u/2)$, we have

$$z(x) = \exp(x^2) \int_{C'} \frac{\exp(-\zeta^2)}{(\zeta - x)^{n+1}} \frac{d\zeta}{(-2)^n}.$$

The contour C' encircles the point $\zeta = x$, and the formula for the nth derivative of a holomorphic function shows that

$$\frac{n!}{2i\pi}\int_{C'}\frac{\exp(-\zeta^2)}{(\zeta - x)^{n+1}}\,d\zeta = \frac{d^n[\exp(-x^2)]}{dx^n}\,.$$

Thus the function

$$z(x) = \frac{2i\pi}{n!}\frac{(-1)^n}{2^n}\exp(x^2)\frac{d^n}{dx^n}[\exp(-x^2)] = \frac{2i\pi}{2^n n!}P_n(x)$$

is indeed a particular solution of (E), and it is easy to see that it is a polynomial of degree n.

If we do not specify what the contour C is, we need to apply the residue theorem to the evaluation of $z(x)$. Up to a factor of $2i\pi$, $z(x)$ is equal to the coefficient of u^n in the Taylor expansion of $\exp(ux - u^2/4)$ in a neighborhood of the origin. Now,

$$\exp(ux) = 1 + ux + \cdots + u^k\frac{x^k}{k!} + \cdots,$$

$$\exp(-u^2/4) = 1 - \frac{u^2}{4} + \cdots + \frac{(-1)^p}{p!}\frac{u^{2p}}{4^p} + \cdots.$$

When we multiply these two series, we see that the coefficient of u^n is

$$\frac{x^n}{n!} - \frac{1}{4}\frac{x^{n-2}}{(n-2)!} + \cdots + \frac{x^{n-2q}}{(n-2q)!}\frac{(-1)^q}{q!\,4^q} + \cdots,$$

the last term depending on whether n is odd or even. Thus $z(x)$ is indeed a polynomial of degree n with the coefficients shown above.

The polynomials found in this way are known up to a constant factor. The Hermite polynomials proper are defined precisely by

$$P_n(x) = (-1)^n\exp(x^2)\frac{d^n}{dx^n}[\exp(-x^2)] = (2x)^n + \cdots.$$

C. *The orthogonality of particular solutions.* If we set $y_n(x) = \exp(-x^2/2)P_n(x)$, then $y_n(x)$ is a solution of

$$y_n'' + (2n + 1 - x^2)y_n = 0,$$

and $y_k(x)$ is a solution of

$$y_k'' + (2k + 1 - x^2)y_k = 0.$$

If we multiply the first of these by y_k and the second by $-y_n$ and then add the results, we obtain

$$y_n''y_k - y_k''y_n + 2(n-k)y_ny_k = 0. \tag{4}$$

The first term is the derivative of $y_n'y_k - y_k'y_n$.

On the other hand, the functions y_n, y_k, and y_ky_n are integrable on $(-\infty, +\infty)$ and they tend to zero as $x \to \pm\infty$. If we integrate (4) from $-\infty$ to $+\infty$, we get

$$(y_n'y_k - y_k'y_n)_{-\infty}^{\infty} + 2(n-k)\int_{-\infty}^{\infty} y_ny_k\,dx = 0.$$

The first term vanishes and, since $n \neq k$,

$$\int_{-\infty}^{\infty} y_n(x)y_k(x)\,dx = 0 \qquad \text{if} \quad k \neq n.$$

In terms of the polynomials $P_n(x)$, this equation becomes

$$\int_{-\infty}^{\infty} \exp(-x^2)\,P_k(x)P_n(x)\,dx = 0 \qquad \text{if} \quad k \neq n.$$

Thus the Hermite polynomials are orthogonal on $(-\infty, +\infty)$ with weight $\exp(-x^2)$.

8. Definition of Hermite polynomials in terms of the orthogonality condition (sequel to Exercise 7)

Problems. The sequence of Hermite polynomials $P_0(x)$, $P_1(x)$, ..., $P_n(x)$ is defined by

(a) $P_n(x)$ is a polynomial of degree n.

(b) The coefficient of x^n is 2^n.

(c) For $m \neq n$.

$$\int_{-\infty}^{\infty} P_m(x)P_n(x)\exp(-x^2)\,dx = 0. \tag{1}$$

A. Evaluate successively P_0, P_1, P_2, and P_3.

B. Suppose that $Q(x)$ is an Hermite polynomial of degree less than some integer n. Show that

$$\int_{-\infty}^{\infty} P_n(x)Q(x)\exp(-x^2)\,dx = 0.$$

C. Show that we can always write P_{n+1} in the form

$$P_{n+1} = 2xP_n + a_nP_n + b_nP_{n-1} + \cdots + l_nP_0, \tag{2}$$

where $a_n, \ldots, l_n$ are constant coefficients all of which vanish except possibly a_n and b_n.

D. Show that, for $m \leqslant n - 2$,

$$\int_{-\infty}^{\infty} \exp(-x^2)\, P_m(x)P_n'(x)\, dx = 0.$$

Show that we can always write P_n' in the form

$$P_n' = 2nP_{n-1} + \alpha_nP_{n-2} + \beta_nP_{n-3} + \cdots + \lambda_nP_0 \tag{3}$$

and that all the coefficients $\alpha_n, \beta_n, \ldots, \lambda_n$ are zero. From this, show that $a_n = 0$ and $b_n = -2n$ in (2).

E. Show that

$$P_n'' - 2xP_n' + 2nP_n = 0.$$

Solutions. **A.** A polynomial of degree n admits $n + 1$ arbitrary coefficients. The relation

$$\int_{-\infty}^{+\infty} P_n(x)P_m(x) \exp(-x^2)\, dx = 0 \qquad (n \neq m)$$

enables us to calculate by induction these coefficients up to a multiplicative constant.

If we require that the coefficient of x^n be equal to 2^n, then $P_n(x)$ is completely determined.

To carry out the actual evaluation, we shall use the formulas

$$\int_{-\infty}^{+\infty} x^{2p+1} \exp(-x^2)\, dx = 0,$$

$$\int_{-\infty}^{+\infty} x^{2p} \exp(-x^2)\, dx = \Gamma\left(p + \frac{1}{2}\right) = \frac{(2p)!}{2^{2p}p!}\sqrt{\pi}.$$

Evaluation of P_0. It follows immediately from the last requirement that $P_0 = 1$.

Evaluation of P_1. $P_1 = 2x + a$ and $\int_{-\infty}^{+\infty} (2x + a) \exp(-x^2)\, dx = 0$, from which we get $a = 0$.

Evaluation of P_2. $P_2 = 4x^2 + bx + c$. From our last two results and the orthogonality conditions, we must have

$$\int_{-\infty}^{+\infty} (4x^2 + bx + c) \exp(-x^2)\, dx = 0,$$

$$\int_{-\infty}^{+\infty} x(4x^2 + bx + c) \exp(-x^2)\, dx = 0.$$

Therefore,

$$\int_{-\infty}^{+\infty} (4x^2 + c) \exp(-x^2)\, dx = 0 \qquad \text{and} \qquad c = \frac{-4\Gamma(\frac{3}{2})}{\Gamma(\frac{1}{2})} = -2.$$

Hence $b = 0$.

Evaluation of P_3. $P_3 = 8x^3 + dx^2 + ex + f$. We must have

(a) $$\int_{-\infty}^{+\infty} (8x^3 + dx^2 + ex + f) \exp(-x^2)\, dx = 0;$$

that is,

$$\int_{-\infty}^{+\infty} (dx^2 + f) \exp(-x^2)\, dx = 0.$$

(b) $$\int_{-\infty}^{+\infty} (8x^3 + dx^2 + ex + f)x \exp(-x^2)\, dx = 0;$$

that is,

$$\int_{-\infty}^{+\infty} (8x^4 + ex^2) \exp(-x^2)\, dx = 0.$$

(c) $$\int_{-\infty}^{+\infty} (8x^3 + dx^2 + ex + f)(x^2 - \tfrac{1}{2}) \exp(-x^2)\, dx = 0;$$

that is,

$$\int_{-\infty}^{+\infty} (dx^2 + f)(x^2 - \tfrac{1}{2}) \exp(-x^2)\, dx = 0.$$

We find that

$$d = f = 0 \qquad \text{and} \qquad e = -12.$$

In sum,

$$P_0 = 1, \qquad P_1 = 2x, \qquad P_2 = 4x^2 - 2; \qquad P_3 = 8x^3 - 12x.$$

B. Suppose that $Q(x)$ is a polynomial of degree $m < n$.

We can always represent $Q(x)$ as a linear combination of the polynomials $P_k(x)$:

$$Q(x) = k_m P_m(x) + k_{m-1} P_{m-1}(x) + \cdots + k_1 P_1(x) + k_0 P_0 .$$

If we calculate

$$\int_{-\infty}^{+\infty} P_n(x)Q(x)\exp(-x^2)\,dx,$$

we get a sum of $m+1$ terms all of which vanish by virtue of the orthogonality of the P_k. Thus

$$\int_{-\infty}^{+\infty} P_n(x)Q(x)\exp(-x^2)\,dx = 0.$$

C. The term of degree $n+1$ is $2^{n+1}x^{n+1}$ both in the polynomial P_{n+1} and in the polynomial $2xP_n$. Thus $P_{n+1} - 2xP_n$ is a polynomial of degree n, which we can represent as a linear combination of the polynomials $P_0, P_1, \ldots, P_n$:

$$P_{n+1} = 2xP_n + a_nP_n + b_nP_{n-1} + \cdots + l_nP_0. \tag{2}$$

Consider an integral of the form

$$(f, g) = \int_{-\infty}^{+\infty} fg\exp(-x^2)\,dx,$$

as the scalar product of two functions f and g belonging to the space L^2 of real functions $f(x)$ such that the integral

$$\int_{-\infty}^{+\infty} f^2(x)\exp(-x^2)\,dx$$

exists. Using this terminology, let us take the scalar product of each member of (2) and P_k, where k is an integer not exceeding $n-2$.

All the scalar products

$$(P_k, P_0), (P_k, P_1), \ldots, (P_k, P_n), (P_k, P_{n+1})$$

vanish except (P_k, P_k). The scalar product (P_k, xP_n) also vanishes, since

$$(P_k, xP_n) = (xP_k, P_n)$$

and $xP_k = Q$ is a polynomial of degree less than n. Therefore, the coefficient of P_k vanishes and (2) reduces to

$$P_{n+1} = 2xP_n + a_nP_n + b_nP_{n-1}. \tag{4}$$

Thus there exists a linear recursion relation between any three Hermite polynomials.

D. Analogously, we may put P_n' in the form

$$P_n' = 2nP_{n-1} + \alpha_n P_{n-2} + \beta_n P_{n-3} + \cdots + \lambda_n P_0 . \tag{3}$$

Now, if $k \leqslant n - 2$, we have $(P_k, P_n') = 0$. To see this, note that

$$\begin{aligned}(P_k, P_n') &= \int_{-\infty}^{\infty} P_k P_n' \exp(-x^2)\, dx \\ &= [P_k P_n \exp(-x^2)]_{-\infty}^{\infty} - \int_{-\infty}^{\infty} P_n \exp(-x^2)(P_k' - 2xP_k)\, dx.\end{aligned}$$

The term outside the integral sign is zero and

$$(P_k, P_n') = (2xP_k, P_n) - (P_k', P_n).$$

If $k \leqslant n - 2$, xP_k is a polynomial of degree less than n. Hence, $(2xP_k, P_n) = 0$. Similarly, $(P_k', P_n) = 0$. Our assertion then follows.

Let us take the scalar product of each side of (3) with P_k, where $k = 0, 1, \ldots, n - 2$. All these scalar products vanish except (P_k, P_k). Therefore, the coefficient of P_k is zero and we are left with

$$P_n' = 2nP_{n-1} . \tag{5}$$

Let us return to formula (4). Differentiation yields

$$P_{n+1}' = (2x + a_n)P_n' + 2P_n + b_n P_{n-1}' .$$

This formula, together with (5), yields

$$2(n + 1)P_n = (2x + a_n)2nP_{n-1} + 2P_n + b_n 2(n - 1)P_{n-2} .$$

This relation can also be written

$$P_n = (2x + a_n)P_{n-1} + \frac{n-1}{n} b_n P_{n-2} .$$

By equating this value of P_n with that given by (4) (with n replaced by $n - 1$), we see that

$$a_n = a_{n-1} , \qquad b_{n-1} = \frac{n-1}{n} b_n .$$

From this we see that a_n and b_n/n are constants. Let us evaluate them with the aid of P_0, P_1, and P_2:

$$P_2 = (2x + a_2)P_1 + b_1 P_0$$

or

$$4x^2 - 2 = 2x \cdot 2x + b_1 , \qquad b_1 = -2, \qquad b_n = -2n, \qquad a_2 = 0.$$

Similarly, we have in general

$$P_{n+1} = 2xP_n - 2nP_{n-1},$$

which is a recursion relation between any three consecutive Hermite polynomials.

E. We have

$$P_n' = 2nP_{n-1}, \qquad P_n'' = 2nP_{n-1}' = 4n(n-1)P_{n-2}.$$

Now

$$P_n - 2xP_{n-1} + 2(n-1)P_{n-2} = 0.$$

Therefore,

$$P_n - 2x\frac{P_n'}{2n} + \frac{2(n-1)P_n''}{4n(n-1)} = 0 \qquad \text{or} \qquad P_n'' - 2xP_n' + 2nP_n = 0.$$

This is the differential equation for Hermite polynomials (see Exercise 7).

9. A generating function for Hermite polynomials (sequel to Exercises 7 and 8)

Problems. Consider the expansion

$$\exp(-t^2 + 2tx) = \sum_{n=0}^{\infty} \frac{t^n}{n!} \tag{1}$$

where x is a real parameter.

A. Show that this expansion converges for every value of the complex number t, that the coefficients $P_n(x)$ are polynomials in x of degree n, and that

$$P_n(x) = (-1)^n \exp(x^2) \frac{d^n}{dx^n} \exp(-x^2).$$

B. Prove the relations

$$P_{n+1} - 2xP_n + 2nP_{n-1} = 0 \qquad \text{and} \qquad P_n' = 2nP_{n-1}$$

by differentiating with respect to t and x, respectively, and show that therefore

$$P_n'' - 2xP_n' + 2nP_n = 0.$$

Assume that the series (1) can be differentiated term by term with respect to x.

C. Set $t = e^{i\theta}$, where θ is a real number. The series (1) then becomes a Fourier series. By applying Parseval's formula to this series, show that

$$\left| \frac{P_n(x)}{n!} \right| < e^{(1/2)+|x|}.$$

By extending this reasoning to the series obtained by differentiating (1), show that the series is uniformly convergent with respect to x for every t such that $|t| < 1$. From this, justify the calculations in Problem B.

D. Take the product of the two expansions

$$\exp(-t^2 + 2tx) = \sum_{n=0}^{\infty} \frac{t^n}{n!} P_n(x), \qquad \exp(-u^2 + 2ux) = \sum_{n=0}^{\infty} \frac{u^n}{n!} P_n(x).$$

By multiplying this product by $\exp(-x^2)$ and integrating with respect to x from $-\infty$ to $+\infty$, prove the formulas

$$\int_{-\infty}^{\infty} P_n(x)P_k(x) \exp(-x^2)\, dx = 0, \quad (n \neq k),$$

$$\int_{-\infty}^{\infty} [P_n(x)]^2 \exp(-x^2)\, dx = 2^n n! \sqrt{\pi}.$$

Solutions. **A.** The function $\exp(-t^2 + 2tx)$ is holomorphic everywhere in the complex t-plane. Therefore, it can be expanded in a series of powers of t with an infinite radius of convergence. To find the coefficients $P_n(x)$, we can multiply the two expansions

$$\exp(-t^2) = 1 - t^2 + \frac{t^4}{2} + \cdots,$$

$$\exp(2tx) = 1 + 2tx + 2t^2x^2 + \cdots$$

term by term. Note that $t^2 - 2tx = (t - x)^2 - x^2$ and that

$$\exp[-(t - x)^2] = \sum_{n=0}^{\infty} \frac{t^n}{n!} P_n(x) \exp(-x^2).$$

Let us set $t = -t'$. We see that $\Sigma\, (-1)^n (t'^n/n!) P_n(x) \exp(-x^2)$ is the expansion of $\exp[-(t' + x)^2]$. From the formula for the coefficients in a Taylor series,

$$(-1)^n P_n(x) \exp(-x^2) = \frac{d^n}{dx^n} \exp(-x^2),$$

or

$$P_n(x) = (-1)^n \exp(x^2) \frac{d^n}{dx^n} \exp(-x^2).$$

Thus $P_n(x)$ is the Hermite polynomial of degree n.

B. The power series (1) can be differentiated term by term with respect to t. We obtain

$$2(x - t) \exp(-t^2 + 2tx) = \sum_{n=0}^{\infty} \frac{t^n}{n!} P_{n+1}(x)$$

or

$$2(x - t) \sum_{n=0}^{\infty} \frac{t^n}{n!} P_n(x) = \sum_{n=0}^{\infty} \frac{t^n}{n!} P_{n+1}(x).$$

Equating coefficients of $t^n/n!$, we obtain

$$2xP_n - 2nP_{n-1} = P_{n+1}$$

or

$$P_{n+1} - 2xP_n = 2nP_{n-1} = 0. \tag{2}$$

Thus we again have the recursion relation between three successive Hermite polynomials.

Let us now differentiate (1) with respect to x. If we accept the validity of this operation, we obtain

$$2t \exp(-t^2 + 2tx) = \sum_{n=0}^{\infty} \frac{t^n}{n!} P_n'(x),$$

or

$$2t \sum_{n=0}^{\infty} \frac{t^n}{n!} P_n(x) = \sum_{n=0}^{\infty} \frac{t^n}{n!} P_n'(x).$$

Equating coefficients of $n^t/n!$, we obtain

$$P_n' = 2nP_{n-1} . \tag{3}$$

If, in (2) we replace P_n with

$$\frac{P'_{n+1}}{2(n + 1)}$$

and P_{n-1} with

$$\frac{P''_{n+1}}{4n(n+1)},$$

we obtain the differential equation

$$P''_{n+1} - 2xP'_{n+1} + 2(n+1)P_{n+1} = 0. \tag{4}$$

C. We shall now justify the term-by-term differentiation of the series in (1).

Let us choose for t a complex number whose absolute value is 1: $t = e^{i\theta}$. We have

$$\exp(-e^{2i\theta} + 2xe^{i\theta}) = \sum_{n=0}^{\infty} \frac{e^{ni\theta}}{n!} P_n(x).$$

The reader will recognize the right-hand member as a Fourier series in θ. Now, the left-hand member, which is continuous function of θ, is square integrable over the interval $0 \leqslant \theta \leqslant 2\pi$. Therefore, we can apply Parseval's formula:

$$\frac{1}{2\pi} \int_0^{2\pi} |f(\theta)|^2 \, d\theta = P_0^2 + \cdots + \frac{P_n^2(x)}{(n!)^2} + \cdots .$$

Now

$$|f(\theta)| = e^{-\cos 2\theta + 2x\cos\theta}$$

and the exponent, $-\cos 2\theta + 2x \cos \theta$, is less than $1 + 2\,|\,x\,|$.

Thus the exponential is itself less than $e^{1+2|x|}$, which is independent of θ, and its mean value on $[0, 2\pi]$ is therefore less than that value. Each term in the series on the right, being less than the sum, has to be less than $e^{1+2|x|}$. Finally,

$$\frac{|\,P_n(x)|}{n!} < e^{(1/2)+|x|}.$$

Note that this is a strict inequality.

Let us now return to the series $\sum_{n=0}^{\infty} (t^n/n!)P_n(x)$. The general term in this series is majorized by $|\,t^n\,|\, e^{(1/2)+|x|}$. If $|\,x\,|$ is always less than some fixed number x_0, it is majorized by an expression of the form $A\,|\,t\,|^n$, where A is a constant. Now, the series whose general term is $A\,|\,t\,|^n$ converges for $|\,t\,| < 1$. Therefore, the given series (1) converges uniformly with respect to x for every fixed t whose absolute value is less than 1.

Now consider the function e^{-t^2+2tx}. Let us differentiate it with respect to x:

$$\frac{d}{dx}\exp(-t^2+2tx) = 2t\exp(-t^2+2tx)$$

$$= 2t\sum_{n=0}^{\infty}\frac{t^n}{n!}P_n(x)$$

$$= 2\sum_{n=1}^{\infty}\frac{t^n}{(n-1)!}P_{n-1}(x).$$

But we have seen that $|P_{n-1}(x)|/(n-1)!$ is less than a fixed number. Therefore, the series that represents $d\exp(-t^2+2tx)/dx$ is, for any fixed t that is less in absolute value than 1, uniformly convergent with respect to x. Since it does represent the series derived from (1) by term-by-term differentiation, we have

$$2\sum_{n=1}^{\infty}\frac{t^n}{(n-1)!}P_{n-1}(x) = \sum_{n=0}^{\infty}\frac{t^n}{n!}P_n'(x),$$

so that

$$P_n'(x) = 2nP_{n-1}(x).$$

D. We can multiply the two power series

$$\exp[-t^2+2tx-(x^2/2)] = \sum_{n=0}^{\infty}\frac{t^n}{n!}P_n(x)\exp(-x^2/2),$$

$$\exp(-u^2+2ux-(x^2/2)] = \sum_{k=0}^{\infty}\frac{u^k}{k!}P_k(x)\exp(-x^2/2).$$

We obtain

$$\exp[-t^2-u^2+2(t+u)x-x^2] = \sum_{n,k}\frac{t^nu^k}{n!k!}P_n(x)P_k(x)\exp(-x^2). \qquad (5)$$

If we integrate this equation from $-\infty$ to $+\infty$ with respect to x, assuming the validity of term-by-term integration, we get

$$\int_{-\infty}^{\infty}\exp[-x^2+2(t+u)x-t^2-u^2]\,dx = \sum_{n,k}\frac{t^nu^k}{n!k!}(P_n, P_k),$$

where

$$(P_n, P_k) = \int_{-\infty}^{\infty} P_n(x)P_k(x) \exp(-x^2)\, dx$$

denotes the scalar product of P_n and P_k in the space of functions that are square-integrable with weight $\exp(-x^2)$. Now,

$$x^2 - 2(t + u)x + t^2 + u^2 = (x - t - u)^2 - 2ut.$$

Since

$$\int_{-\infty}^{\infty} \exp[-(x - t - u)^2]\, dx = \sqrt{\pi},$$

we obtain

$$\sqrt{\pi} e^{2ut} = \sum_{n,k} \frac{t^n u^k}{n!k!} \; P_n, P_k).$$

The left-hand member can be expanded in powers of u and t. However, the expansion contains only terms of the form $(ut)^n$:

$$\sqrt{\pi} e^{2ut} = \sqrt{\pi}\left(1 + 2ut + \cdots + \frac{2^n}{n!} u^n t^n + \cdots\right).$$

By comparing the two expansions, we see that

$$(P_n, P_k) = 0 \qquad \text{if} \quad n \neq k,$$

which means that the polynomials P_k are orthogonal, and

$$\frac{(P_n, P_n)}{(n!)^2} = \sqrt{\pi}\, \frac{2^n}{n!}.$$

Thus

$$\int_{-\infty}^{\infty} P_n^2 \exp(-x^2)\, dx = 2^n n!\, \sqrt{\pi}.$$

The polynomials P_n are not normalized. The sequence of ratios of the form

$$\frac{P_n}{(2^n n!\, \sqrt{\pi})^{1/2}}$$

is an orthonormal sequence.

It remains for us to justify this calculation. We saw in Problem C that

$$\left|\frac{P_n}{n!}\right| < e^{(1/2)+|x|}.$$

Therefore,

$$\left| \frac{P_n}{n!} \frac{P_k}{k!} \right| \exp(-x^2) < \exp(1 + 2 \mid x \mid - x^2).$$

Now, the problem posed is one of integrating a double series term by term with respect to a parameter. This double series is absolutely convergent and it is uniformly convergent with respect to x in every finite interval. Since we are integrating with respect to x from $-\infty$ to $+\infty$, we need to make an adaptation of the classical reasoning.

Consider any partial sum in the series. Since the series converges absolutely, we can choose any particular sum, for example, the sum of terms such that $n < N$ and $k < K$. Let R_{NK} denote the corresponding remainder and let S_{NK} denote the remainder of like rank in the series $\Sigma_{n,k} \mid t^n u^k \mid$. Then

$$\mid R_{NK} \mid < S_{NK} \exp(1 + 2 \mid x \mid - x^2)$$

and

$$\int_{-\infty}^{\infty} \mid R_{NK} \mid dx < S_{NK} \int_{-\infty}^{\infty} \exp(1 + 2 \mid x \mid - x^2)\, dx$$

The integral on the right exists. It is a definite number. If $\mid t \mid < 1$ and $\mid u \mid < 1$, we can choose N and K sufficiently great that S_{NK} is less than any given positive number ϵ. Under these conditions, the integral of the remainder of the series (5) can be made arbitrarily small, which justifies the term-by-term integration.

10. The zeros of Bessel functions

Problems. By setting $y = ux^{-1/2}$, Bessel's differential equation

$$x^2 y'' + xy' + (x^2 - n^2)\, y = 0$$

(where n is a nonnegative integer) becomes

$$u'' + \left[1 + \left(\frac{1}{4} - n^2\right) \frac{1}{x^2}\right] u = 0. \qquad (\mathrm{E}_n)$$

A. Let $u(x)$ denote a solution of the equation (E_0) (that is, the above equation with $n = 0$). And let $v(x)$ denote a solution of the equation

$$v'' + v = 0. \qquad (\mathrm{E}')$$

Show that if α and β are two numbers such that $\beta > \alpha$ and both have the same sign, then

$$(u'v - uv')_\alpha^\beta + \frac{1}{4}\int_\alpha^\beta \frac{uv}{x^2}\,dx = 0.$$

From this, show that, if α and β are two consecutive zeros of $v(x)$, the function $u(x)$ must vanish at least once in the interval (α, β) and that the Bessel function $J_0(x)$ possesses a zero in every interval of the form $(a, a + \pi)$ for every value of a.

B. Use the formula

$$J_{n+1} = -x^n \frac{d}{dx}\frac{J_n}{x^n}$$

to show that, if x_0 and x_1 are two consecutive zeros of the function J_n both of the same sign, then the function J_{n+1} has at least one zero between x_0 and x_1.

Solutions. **A.** *Distribution of the zeros of $J_0(x)$.* Let us multiply (E_0) by $v(x)$ and (E′) by $-u(x)$ and add the results. This yields

$$u''v - v''u + \frac{1}{4x^2}uv = 0.$$

The sum of the first two terms is the derivative of $(u'v - uv')$. If we integrate with respect to x between two real numbers α and β, we obtain

$$(u'v - uv')_\alpha^\beta + \frac{1}{4}\int_\alpha^\beta \frac{uv}{x^2}\,dx = 0.$$

If α and β are zeros of $v(x)$, we have simply

$$u(\beta)v'(\beta) - u(\alpha)v'(\alpha) = \frac{1}{4}\int_\alpha^\beta \frac{uv}{x^2}\,dx. \tag{1}$$

Since $u(x)$ and $v(x)$ are solutions of (E_0) and (E′), they are continuous and differentiable. Therefore, if α and β are two *consecutive* zeros of $v(x)$, the following three statements hold:

(a) $v(x)$ is of constant sign in the open interval $\alpha < x < \beta$.

(b) $v'(\alpha)$ either is zero or it has the sign of $v(x)$ in that interval.

(c) $v'(\beta)$ either is zero or has the opposite sign to $v(x)$ in that interval.

Let us suppose now that $u(x)$ does not vanish identically and does not change sign in the interval $\alpha < x < \beta$. For definiteness, let us suppose that $u(x) \geqslant 0$.

The right-hand member of (1) is nonzero and has the sign of $v(x)$. The left-hand member of (1) either is zero or has the sign opposite to that of $v(x)$. Thus we have a contradiction. Analogous reasoning would show that $u(x)$ cannot be negative. Therefore, $u(x)$ changes sign at least once in the interval. Since the function $u(x)$ is continuous, it has at least one zero between α and β.

Now, the general solution of (E′) is

$$v(x) = A \cos x + B \sin x.$$

For any number a, we have $v(a + \pi) = -v(a)$. If a is a zero of $v(x)$, it is also a zero of $a + \pi$, and it follows from the properties of trigonometric functions that a and $a + \pi$ are consecutive zeros. Also, there exists a solution $v(x)$ of (E′) that has a for a zero. Consequently, the Bessel function $J_0(x)$ has a zero in every interval of the form $(a, a + \pi)$, where a and $a + \pi$ have the same sign.

B. Let us now study the zeros of $J_n(x)$. The function $J_n(x)/x^n$ is continuous and differentiable for all x, including $x = 0$. The zeros of $J_n(x)$ are the same as those of $J_n(x)/x^n$ except that $J_n(0)$ may be 0. Rolle's theorem states that between two zeros of $J_n(x)/x^n$, the function $(d/dx)(J_n(x)/x^n)$ vanishes at least once. Therefore, between any two nonzero zeros of J_n that have the same sign, there is at least one zero of J_{n+1}. In every interval in which $J_n(x)$ possesses p zeros, we can be sure that $J_{n+1}(x)$ has at least $p - 1$ zeros.

11. The asymptotic nature of $J_0(t)$ as $t \to \infty$

Numerical calculation shows that the general behavior of $J_0(t)$ is that of an attenuated sinusoid. We can make this qualitative remark more precise by showing that $J_0(t)$ does indeed approach such a sinusoid for large values of t. The equation of this sinusoid is

$$y = \left(\frac{2}{\pi t}\right)^{1/2} \cos\left(t - \frac{\pi}{4}\right)$$

and we may write

$$J_0(t) = \left(\frac{2}{\pi t}\right)^{1/2} \left[\cos\left(t - \frac{\pi}{4}\right) + \epsilon(t)\right],$$

where the function $\epsilon(t)$ tends to zero as $t \to \infty$. Although it is possible to make precise the behavior of $\epsilon(t)$ in the form of an asymptotic expansion, the present exercise consists in proving the above formula without going into detail involving the concept of asymptotic expansion.

Problems. **A.** Prove that the function

$$\varphi(\omega) = \frac{1}{(1-\omega^2)^{1/2}} - \frac{1}{[2(1-\omega)]^{1/2}} \tag{1}$$

is continuous for $0 \leqslant \omega \leqslant 1$. Show that there exist constants α and β such that

$$|\varphi(\omega)| \leqslant \alpha, \quad |\varphi'(\omega)| \leqslant \frac{\beta}{(1-\omega)^{1/2}} \qquad \text{for} \quad 0 \leqslant \omega < 1.$$

B. Consider the Bessel function

$$J_0(t) = \frac{2}{\pi}\int_0^1 \frac{\cos \omega t}{(1-\omega^2)^{1/2}}\, d\omega. \tag{2}$$

Replace

$$\frac{1}{(1-\omega^2)^{1/2}} \qquad \text{with} \qquad \frac{1}{[2(1-\omega)]^{1/2}} + \varphi(\omega).$$

This will put $J_0(t)$ in the form of the sum of two integrals $A(t)$ and $B(t)$. Show, by means of the change of variable $\omega = 1 - s^2/t$ that

$$A(t) = \frac{2\sqrt{2}}{\pi\sqrt{t}}\int_0^{\sqrt{t}} (\cos t \cos s^2 + \sin t \sin s^2)\, ds.$$

Then show that

$$A(t) = \left(\frac{2}{\pi t}\right)^{1/2}\left[\cos\left(t - \frac{\pi}{4}\right) + \epsilon_1(t)\right], \tag{3}$$

where the function $\epsilon_1(t)$ tends to zero as $t \to \infty$.

C. Show that (3) is valid also for $J_0(t)$; that is, show that, for large values of t, the function $J_0(t)$ can be put in the form

$$J_0(t) = \left(\frac{2}{\pi t}\right)^{1/2}\left[\cos\left(t - \frac{\pi}{4}\right) + \epsilon(t)\right],$$

where $\epsilon(t)$ tends to zero as $t \to \infty$.

Solutions. **A.** *Study of the function* $\varphi(\omega)$. We have

$$\varphi(\omega) = \frac{1}{(1-\omega^2)^{1/2}} - \frac{1}{[2(1-\omega)]^{1/2}}.$$

This function is continuous for $0 \leqslant \omega < 1$. To study the behavior of $\varphi(\omega)$ in a neighborhood of $\omega = 1$, we set $1 - \omega = s$:

$$\varphi(\omega) = \frac{1}{(2s)^{1/2}}\left(\frac{1}{[1-(s/2)]^{1/2}} - 1\right).$$

As $s \to 0$, the expression

$$s \to 0, \frac{1}{[1-(s/2)]^{1/2}} - 1$$

is equivalent to $s/4$. Therefore $\varphi(\omega)$ tends to 0. Since $\varphi(\omega)$ is continuous on the closed interval $0 \leqslant \omega \leqslant 1$, it is bounded. Thus there exists a constant α such that $|\varphi(\omega)| < \alpha$.

Let us now examine $\varphi'(\omega)$. We have

$$\varphi'(\omega) = \frac{\omega}{(1-\omega^2)^{3/2}} - \frac{1}{2\sqrt{2}}\frac{1}{(1-\omega)^{3/2}}$$
$$= \frac{1}{(2s)^{3/2}}\left(\frac{1-s}{[1-(s/2)]^{3/2}} - 1\right).$$

The function

$$\frac{1}{s}\left(\frac{1-s}{[1-(s/2)]^{3/2}} - 1\right)$$

is continuous for $0 < s \leqslant 1$. In a neighborhood of $s = 0$, the expression in braces has the expansion (cut short)

$$-\tfrac{1}{4}s + \cdots.$$

Thus when this expression is multiplied by $1/s$, the product tends to a limit. Since this function is continuous for $0 \leqslant s \leqslant 1$, it is bounded and there exists a constant β such that

$$|\varphi'(\omega)| < \frac{\beta}{s^{1/2}} \quad \text{or} \quad |\varphi'(\omega)| < \frac{\beta}{(1-\omega)^{1/2}}.$$

B. *An asymptotic expression for the function* $A(t)$. We have

$$J_0(t) = \frac{2}{\pi}\int_0^1 \cos\omega t\left(\frac{1}{[2(1-\omega)]^{1/2}} + \varphi(\omega)\right) d\omega = A(t) + B(t),$$

where

$$A(t) = \frac{2}{\pi} \int_0^1 \frac{\cos \omega t}{[2(1-\omega)]^{1/2}} \, d\omega.$$

The substitution $\omega = 1 - s^2/t$ yields

$$d\omega = -\frac{2s\,ds}{t}, \qquad \cos \omega t = \cos(t - s^2),$$
$$[2(1-\omega)]^{1/2} = [2(s^2/t)]^{1/2} = s(2/t)^{1/2}.$$

If $\omega = 0$, $s = \sqrt{t}$ and if $\omega = 1$, $s = 0$.
We have

$$A(t) = \frac{2}{\pi}\left(\frac{2}{t}\right)^{1/2} \int_0^{\sqrt{t}} \cos(t - s^2)\, ds$$
$$= \frac{2}{\pi}\left(\frac{2}{t}\right)^{1/2} \int_0^{\sqrt{t}} (\cos t \cos s^2 + \sin t \sin s^2)\, ds.$$

The integrals $I = \int_0^\infty \cos s^2\, ds$ and $J = \int_0^\infty \sin s^2\, ds$ (Fresnel's integrals) are convergent and

$$I = J = \frac{1}{2}\left(\frac{\pi}{2}\right)^{1/2}$$

Thus the integrals $\int_{\sqrt{t}}^\infty \cos s^2\, ds$ and $\int_{\sqrt{t}}^\infty \sin s^2\, ds$ tend to zero as $t \to \infty$.

Since $\cos t$ and $\sin t$ are bounded functions, the integral

$$\int_{\sqrt{t}}^\infty (\cos t \cos s^2 + \sin t \sin s^2)\, ds = \eta(t)$$

also tends to zero as t increases without bound. Thus we have

$$A(t) = \frac{2}{\pi}\left(\frac{2}{t}\right)^{1/2} [I \cos t + J \sin t - \eta(t)]$$
$$= \frac{2}{\pi}\left(\frac{2}{t}\right)^{1/2} \frac{1}{2}\left(\frac{\pi}{2}\right)^{1/2} [\cos t + \sin t + \epsilon_1(t)],$$

where $\epsilon_1(t)$ tends to 0 as $t \to \infty$: that is,

$$\boxed{A(t) = \left(\frac{2}{\pi t}\right)^{1/2} \left[\cos\left(t - \frac{\pi}{4}\right) + \epsilon_1(t)\right].}$$

C. *An asymptotic expression for $J_0(t)$.* We still need to study the behavior of $B(t)$ as $t \to \infty$, where

$$B(t) = \frac{2}{\pi} \int_0^1 \varphi(\omega) \cos \omega t\, d\omega.$$

Integration by parts yields

$$B(t) = \frac{2}{\pi}\left[\frac{\varphi(\omega)\sin\omega t}{t}\right]_0^1 - \frac{2}{\pi}\int_0^1 \frac{\varphi'(\omega)\sin\omega t}{t}\,d\omega.$$

We can majorize $|B(t)|$ by considering the majorizations of $|\varphi(\omega)|$ and $|\varphi'(\omega)|$:

$$|B(t)| \leqslant \frac{2}{\pi}\frac{2\alpha}{t} + \frac{2}{\pi t}\int_0^1 \frac{\beta\,d\omega}{(1-\omega)^{1/2}} = \frac{4}{\pi t}(\alpha+\beta).$$

If we set

$$\epsilon_2(t) = B(t)\left(\frac{\pi t}{2}\right)^{1/2},$$

we have

$$|\epsilon_2(t)| \leqslant 2(\alpha+\beta)\left(\frac{2}{\pi t}\right)^{1/2}.$$

Thus $\epsilon_2(t)$ tends to zero as t increases without bound.

Returning to $J_0(t) = A(t) + B(t)$, we obtain

$$J_0(t) = \left(\frac{2}{\pi t}\right)^{1/2}\left[\cos\left(t-\frac{\pi}{4}\right) + \epsilon_1(t) + \epsilon_2(t)\right].$$

Finally,

$$\boxed{J_0(t) = \left(\frac{2}{\pi t}\right)^{1/2}\left[\cos\left(t-\frac{\pi}{4}\right) + \epsilon(t)\right],}$$

where $\epsilon(t)$ tends to zero as t increases without bound.

12. The motion of a simple pendulum of variable length

The purpose of this exercise is to give an example, among many others, of the application of Bessel functions to a problem in mechanics. The problem that we shall treat deals with a pendulum whose length varies in proportion to the time. It is easy to set up such a pendulum in practice, and we wish to find the oscillations whose period and amplitude vary as a function of the length of the pendulum, that is, as a function of time.

Problems. A simple pendulum consists of a unit mass attached to the end of a massless string of length ρ. Units are chosen so that the relation between ρ and t is

$$\rho = gt ,$$

where g is the acceleration due to gravity.

A. Show that the differential equation for small oscillations may be written

$$t \frac{d^2\theta}{dt^2} + 2 \frac{d\theta}{dt} + \theta = 0, \tag{1}$$

where θ is the angle that the string makes with the vertical.
Show that a particular solution of this equation is

$$\theta_1 = \frac{J_1(2\sqrt{t})}{\sqrt{t}}, \tag{2}$$

where J_1 is Bessel's function of order 1.

B. Knowing this particular solution θ_1, express the general solution of (1) in quadrature. Show that every solution of (1) that is bounded in a neighborhood of $t = 0$ is the product of θ_1 and a constant.

Solutions. **A.** *The differential equation for small oscillations.* To get the differential equation for the motion of the point M, we need only write the fundamental equation of dynamics for the projection in the perpendicular direction to the string. The projection onto the direction of the string would give the tension in the string.
The projection of the acceleration of the point M is

$$\Gamma^\theta = 2 \frac{d\rho}{dt} \frac{d\theta}{dt} + \rho \frac{d^2\theta}{dt^2} .$$

The projection of the force is $-g \sin \theta$. Thus we have

$$2g \frac{d\theta}{dt} + gt \frac{d^2\theta}{dt^2} = -g \sin \theta.$$

From this we get the differential equation of the motion:

$$t \frac{d^2\theta}{dt^2} + 2 \frac{d\theta}{dt} + \sin \theta = 0.$$

The equation for small oscillations (that is, with small values of θ) is obtained by replacing $\sin\theta$ with θ:

$$t\frac{d^2\theta}{dt^2} + 2\frac{d\theta}{dt} + \theta = 0. \tag{1}$$

Since t is nonnegative, the change of variable $u = 2\sqrt{t}$ yields

$$\frac{d\theta}{dt} = \frac{d\theta}{du}\frac{1}{\sqrt{t}} = \frac{2}{u}\frac{d\theta}{du},$$

$$\frac{d^2\theta}{dt^2} = \frac{d}{du}\left(\frac{2}{u}\frac{d\theta}{du}\right)\frac{du}{dt} = \frac{4}{u^2}\left(\frac{d^2\theta}{du^2} - \frac{1}{u}\frac{d\theta}{du}\right).$$

The equation becomes

$$\frac{d^2\theta}{du^2} + \frac{3}{u}\frac{d\theta}{du} + \theta = 0,$$

which is an equation derived from Bessel's equation.

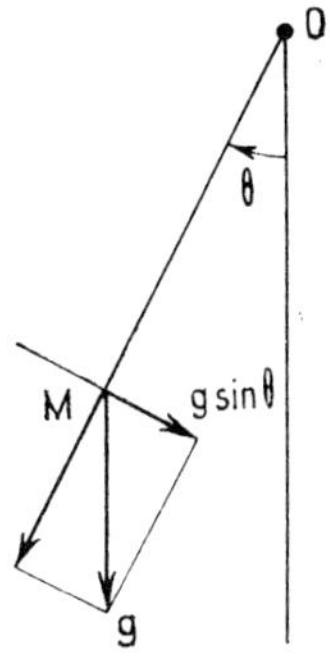

FIGURE 3

The substitution $\theta = x/u$ reduces to the Bessel equation of order 1:

$$u^2x'' + ux' + (u^2 - 1)x = 0.$$

A particular solution of this equation is the Bessel function $J_1(t)$; hence a particular solution of (1) is

$$\theta_1(t) = \frac{J_1(2\sqrt{t})}{\sqrt{t}}. \tag{2}$$

The power-series expansion of $J_1(t)$ enables us to make precise some of the properties of $\theta_1(t)$:

$$\begin{aligned}\theta_1(t) &= \frac{1}{\sqrt{t}}(\sqrt{t})\left[1 - \frac{(\sqrt{t})^2}{2} + \cdots + \frac{(-1)^n}{n!(n+1)!}(\sqrt{t})^{2n} + \cdots\right], \\ &= 1 - \frac{t}{2} + \cdots + \frac{(-1)^n}{n!(n+1)!}t^n + \cdots .\end{aligned} \tag{3}$$

The function $\theta_1(t)$ is an entire function. In particular, $\theta_1(t)$ is continuous at $t = 0$, and $\theta_1(0) = 1$.

B. *The general solution of the differential equation.* We set

$$\theta(t) = z(t)\theta_1(t).$$

Then

$$\theta(t) = z(t)\theta_1(t), \qquad \theta'(t) = z'\theta_1 + z\theta_1', \qquad \theta''(t) = z''\theta_1 + 2z'\theta_1' + z\theta_1''.$$

When we substitute these values of θ and its derivatives in (1), remembering that θ_1 is a solution of (1), we obtain

$$2z'\theta_1 + tz''\theta_1 + 2tz'\theta_1' = 0$$

or

$$\frac{z''}{z'} = -2\frac{\theta_1 + t\theta_1'}{t\theta_1} = -\frac{2}{t} - \frac{2\theta_1'}{\theta_1}.$$

We have

$$\log z' = -2\log t - 2\log\theta_1 + K, \qquad z' = \frac{C}{t^2\theta_1^{\,2}},$$

where C and K are constants. Therefore,

$$z(t) = C\int_{t_0}^{t}\frac{dv}{v^2\theta_1^{\,2}(v)} + A \qquad (t_0 > 0)$$

which yields a particular solution of (1) for a fixed *nonzero* value of t_0. Therefore, the general solution of (1) is

$$\theta(t) = C\theta_1(t)\int_{t_0}^{t}\frac{d\nu}{\nu^2\theta_1^{\,2}(\nu)} + A\theta_1(t). \tag{4}$$

As t tends to zero, $\theta_1(t)$ tends to 1. Since the integral diverges, for nonzero values of C, the solution (4) of the differential equation is unbounded in a neighborhood of zero. Thus the only bounded solutions are of the form

$$\theta(t) = A\theta_1(t).$$

They are continuous and

$$\theta(0) = A.$$

With the aid of a table of functions, we can easily construct the curve for $\theta(t)$. It has the appearance of an attenuated sinusoid.

In particular, if x is sufficiently great, the zeros of $J_1(x)$ lie very close to the points of the form

$$\frac{\pi}{4} + n\pi \qquad (n \text{ an integer}).$$

Therefore, the zeros of $J_i(2\sqrt{t})$ are approximately at the points

$$\frac{1}{4}\left(\frac{\pi}{4} + n\pi\right)^2.$$

The distance between two consecutive zeros varies linearly with the order n. It follows that, as the pendulum lengthens, the period of oscillations increases as a linear function of the time, that is, of the length, this rule becoming more and more nearly exact as the length of the string becomes greater.

13. The Laplace transforms of $J_0(t)$ and $J_0(\sqrt{t})$

This exercise can be considered as a contribution to the compilation of a catalogue of Laplace transforms. It provides the primitives of the functions $p^{-1}e^{-1/4p}$ and $(1 + p^2)^{-1/2}$. In the second case, we use the classical theorem stating that the image under the Laplace transformation of the convolution of two functions is the product of the images of these functions. We recall that the convolution $f * g$ of $f(t)$ and $g(t)$ is the function defined by the integral

$$(f * g)(t) = \int_0^t f(s)g(t - s)\, ds.$$

However, there exists another kind of convolution in which the limits of integration 0 and t of the integral are replaced with $-\infty$ and $+\infty$. If we set

$$(f \circ g)(t) = \int_{-\infty}^{\infty} f(s)g(t - s)\, ds,$$

we see that

$$f * g = (fY) \circ (gY),$$

where $Y(t)$ is the function that is equal to 0 for $t < 0$ and equal to 1 for $t \geqslant 0$.

In the following problem, use the convolution $f * g$, which is well suited for a unilateral Laplace transformation.

Problems. Recall that

$$J_0(t) = 1 - \frac{t^2}{4} + \cdots + (-1)^n \frac{1}{(n!)^2} \frac{t^{2n}}{2^{2n}} + \cdots \tag{1}$$

or, in integral form,

$$J_0(t) = \frac{2}{\pi} \int_0^1 \frac{\cos \omega t}{(1 - \omega^2)^{1/2}}\, d\omega. \tag{2}$$

A. Use the series (1) to find the Laplace transform of $J_0(\sqrt{t})$.

B. Use the same procedure to find the Laplace transform of $J_0(t)$.

C. Find the result of Problem B by using the integral (2).

D. Prove the formula

$$\int_0^t J_0(s) J_0(t - s)\, ds = \sin t.$$

Solutions. **A.** From the series

$$J_0(t) = 1 - \frac{t^2}{4} + \cdots + (-1)^n \frac{1}{(n!)^2} \frac{t^{2n}}{2^{2n}} + \cdots, \tag{1}$$

we have

$$J_0(\sqrt{t}) = 1 - \frac{t}{4} + \cdots + (-1)^n \frac{1}{(n!)^2} \frac{t^n}{2^{2n}} + \cdots \qquad (t > 0). \tag{3}$$

Let us multiply (3) by e^{-pt} and integrate from 0 to $+\infty$. If this operation is legitimate, we shall have the Laplace transform of $J_0(\sqrt{t})$. On the right we need to evaluate the Laplace transform of t^n:

$$\int_0^\infty t^n e^{-pt}\, dt = \frac{n!}{p^{n+1}} \qquad (p > 0).$$

Thus we have

$$\begin{aligned}\frac{1}{p} - \frac{1}{4p^2} + \cdots + \frac{(-1)^n}{n!}\frac{1}{2^{2n}p^{n+1}} + \cdots \\ &= \frac{1}{p}\left[1 - \frac{1}{(4p)} + \cdots + \frac{(-1)^n}{n!}\frac{1}{(4p)^n} + \cdots\right] \\ &= \frac{1}{p}\, e^{-1/4p}.\end{aligned}$$

Finally,

$$\mathscr{L} J_0(\sqrt{t}) = \frac{1}{p}\, e^{-1/4p},$$

where $\mathscr{L}$ denotes the operator of the Laplace transformation.

To justify this calculation, it will be sufficient for us to show that the integral

$$\int_0^\infty R_n(t) e^{-pt}\, dt,$$

where $R_n(t)$ denotes the remainder of the series (3), tends to zero as $n \to \infty$.

Now, the series (3) converges by virtue of the theorem on alternating series. Therefore,

$$| R_{n-1}(t)| < \frac{1}{(n!)^2}\frac{t^n}{2^{2n}}$$

and

$$\left\| \int_0^\infty R_{n-1}(t) e^{-pt}\, dt \right\| < \frac{1}{(n!)^2}\frac{1}{2^{2n}}\int_0^\infty t^n e^{-pt}\, dt = \frac{1}{p^{n+1}}\frac{1}{n!}\frac{1}{2^{2n}}\,.$$

For every fixed p, this expression, which is the general term of an exponential series, tends to zero as $n \to \infty$. The conclusion then follows.

B. Let us return to the series (1). We have

$$\mathscr{L} t^{2n} = \frac{(2n)!}{p^{2n+1}}$$

from which we get, formally,

$$\begin{aligned}\mathscr{L} J_0(t) &= \frac{1}{p} - \frac{1}{2p^3} + \cdots + (-1)^n \frac{(2n)!}{(n!)^2}\frac{1}{2^{2n}p^{2n+1}} + \cdots \\ &= \frac{1}{p}\left[1 - \frac{1}{2p^2} + \cdots + (-1)^n \frac{(2n)!}{(n!)^2}\frac{1}{2^{2n}p^{2n}} + \cdots\right].\end{aligned}$$

We know that, if $|x| < 1$,

$$\frac{1}{(1+x^2)^{1/2}} = (1+x^2)^{-1/2} = 1 - \frac{1}{2}x^2 + \cdots + \frac{(-1)^n(2n)!}{(n!)^2 2^{2n}} x^{2n} + \cdots$$

Therefore, if $p > 1$, we see that

$$\mathscr{L} J_0(t) = \frac{1}{p}\frac{1}{[1+(1/p^2)]^{1/2}} = \frac{1}{(1+p^2)^{1/2}}.$$

To justify our calculation, we note that the remainder $\rho_{n-1}(t)$ of the series (1) is less in absolute value than $[1/(n!)^2](t^{2n}/2^{2n})$. Therefore,

$$\left| \int_0^\infty \rho_{n-1}(t) e^{-pt}\, dt \right| < \frac{1}{(n!)^2}\frac{1}{2^{2n}} \int_0^\infty t^{2n} e^{-pt}\, dt = \frac{(2n)!}{(n!)^2}\frac{1}{2^{2n}p^{2n+1}}.$$

To evaluate this expression, we use Stirling's formula. The value of $n!$ is asymptotically equal to $(2\pi)^{1/2} n^{n+1/2} e^{-n}$. Therefore, after some calculation, our expression becomes

$$\frac{1}{\sqrt{\pi}}\frac{1}{p^{2n+1}\sqrt{n}},$$

which tends to zero as $n \to \infty$.

C. We shall now find the Laplace transform of $J_0(t)$ by using the integral form for $J_0(t)$:

$$J_0(t) = \frac{2}{\pi}\int_0^1 \frac{\cos \omega t}{(1-\omega^2)^{1/2}}\, d\omega.$$

We have

$$\mathscr{L} J_0(t) = \int_0^\infty J_0(t) e^{-pt}\, dt = \frac{2}{\pi}\int_0^\infty e^{-pt}\, dt \int_0^1 \frac{\cos \omega t}{(1-\omega^2)^{1/2}}\, d\omega.$$

This iterated integral is equal to a convergent double integral. To see this, consider the integral of the absolute value of the integrand. This absolute value is majorized by

$$e^{-pt}\frac{1}{(1-\omega^2)^{1/2}}.$$

The convergence of the original double integral followed from the convergence of the double integral

$$\iint e^{-pt}\frac{1}{(1-\omega^2)^{1/2}}\, dt\, d\omega.$$

But since we are dealing with a positive function, it will be sufficient to find a procedure for evaluating this integral. If it exists, it is equal to

$$\int_0^\infty e^{-pt}\,dt \int_0^1 \frac{d\omega}{(1-\omega^2)^{1/2}}$$

and, conversely, the existence of this product of two integrals implies the existence of the double integrals in question. And the existence of this product is obvious.

Therefore, we may permute the integrations and write

$$\mathscr{L} J_0(t) = \frac{2}{\pi}\int_0^1 \frac{d\omega}{(1-\omega^2)^{1/2}} \int_0^\infty e^{-pt}\cos \omega t\,dt.$$

Here we see the Laplace transform of $\cos \omega t$, which is the real part of

$$\int_0^\infty e^{-pt}e^{i\omega t}\,dt = \frac{1}{p - i\omega} = \frac{p + i\omega}{p^2+\omega^2}.$$

Therefore,

$$\mathscr{L} J_0(t) = \frac{2p}{\pi}\int_0^1 \frac{d\omega}{(p^2+\omega^2)(1-\omega^2)^{1/2}}.$$

The substitution $\omega = t/(1+t^2)^{1/2}$ yields

$$\mathscr{L} J_0(t) = \frac{2p}{\pi}\int_0^\infty \frac{dt}{p^2+(p^2+1)t^2} = \frac{2}{\pi}\frac{1}{(p^2+1)^{1/2}}\left[\arctan\left(\left(\frac{p^2+1}{p^2}\right)^{1/2} t\right)\right]_0^\infty$$

$$= \frac{1}{(p^2+1)^{1/2}}.$$

Here we have the result found in Problem B, without, however, the restriction on its validity: p may take any positive value.

Instead of making a change of real variable, we can also evaluate $\mathscr{L} J_0(t)$ by integrating the function

$$f(z) = \frac{1}{(p^2+z^2)(1-z^2)^{1/2}}$$

over the curve γ shown in Figure 4 in the complex plane. In the interior of γ, the function $f(z)$ is uniform and it has two poles ip and $-ip$. We can make precise the determination by assigning to $(1-z^2)^{1/2}$ the value $+1$ at the point $z = 0$, considered as being above the cut $[-1, +1]$.

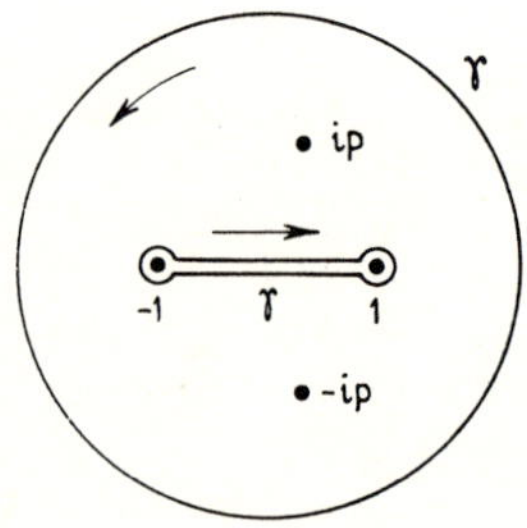

FIGURE 4

The integrals over the three circles vanish in the limit. On both sides of the cut, we find

$$2\int_{-1}^{1}\frac{d\omega}{(p^2+\omega^2)(1-\omega^2)^{1/2}}=4\int_{0}^{1}\frac{d\omega}{(p^2+\omega^2)(1-\omega^2)^{1/2}}$$

and this integral is equal to the product obtained by multiplying the two residues corresponding to the two simple poles by $2i\pi$. These residues are the values of $1/2\omega(1-\omega^2)^{1/2}$ at the points $\omega = ip$ and $\omega = -ip$. We easily see that $(1-\omega^2)^{1/2}$ assumes the value $(1+p^2)^{1/2}$ at the point ip and the opposite value at the point $-ip$. Thus

$$4\int_{0}^{1}\frac{d\omega}{(p^2+\omega^2)(1-\omega^2)^{1/2}}=2i\pi\times\frac{2}{2ip(1+p^2)^{1/2}}=\frac{2\pi}{p(1+p^2)^{1/2}},$$

so that

$$\mathscr{L}J_0(t)=\frac{1}{(1+p^2)^{1/2}}.$$

D. *An application.* We have

$$\begin{aligned}\mathscr{L}\int_0^t J_0(s)J_0(t-s)\,ds &= \mathscr{L}(J_0 * J_0)\\ &= \mathscr{L}(J_0)\times\mathscr{L}(J_0)=\frac{1}{1+p^2},\end{aligned}$$

where the asterisk denotes a convolution. Now, $1/(1+p^2)=\mathscr{L}\sin t$. It follows from the uniqueness of the Laplace transform that

$$\int_0^t J_0(s)J_0(t-s)\,ds=\sin t.$$

14. The Laplace transform of Bessel functions of integral order

This exercise is a continuation and generalization of the preceding one It uses another formula for Bessel functions of integral order:

$$J_n(t) = \frac{1}{2\pi}\int_{-\pi}^{\pi} e^{it\sin\theta} e^{-in\theta}\, d\theta.$$

This formula is equivalent to the Fourier-series expansion

$$e^{it\sin\theta} = \sum_{n=-\infty}^{\infty} J_n(t) e^{in\theta}.$$

Here, $e^{it\sin\theta}$ is the generating function of the Bessel functions $J_n(t)$.

Problems. A. Suppose that a and b are two positive numbers and that n is a nonnegative integer. Evaluate the integral

$$I = \int_0^{2\pi} \frac{e^{-in\theta}}{a - ib\sin\theta}\, d\theta.$$

B. Use the result to show that

$$\int_0^{\infty} e^{-ax} J_n(bx)\, dx = \frac{1}{(a^2+b^2)^{1/2}} \left[\frac{(a^2+b^2)^{1/2} - a}{b}\right]^n.$$

What is the Laplace transform of the Bessel function $J_n(x)$?

Solutions. **A.** Evaluation of the integral I. Let us set

$$e^{-i\theta} = z.$$

Then

$$\sin\theta = \frac{e^{i\theta} - e^{-i\theta}}{2i} = \frac{1}{2i}\left(\frac{1}{z} - z\right).$$

The function that we wish to integrate becomes

$$\frac{z^n}{[a - (b/2)][(1 - z^2)/z]} = \frac{2z^{n+1}}{bz^2 + 2az - b}.$$

As θ varies from 0 to 2π, z describes the circle γ: $|z| = 1$ (in the negative direction) in the complex plane. On that circle,

$$d\theta = \frac{i\, dz}{z}.$$

Therefore, the integral that we are evaluating takes the form of an integral over a closed curve in the complex plane:

$$I = 2i \int_\gamma \frac{z^n \, dz}{bz^2 + 2az - b}$$

and we can evaluate it by using the residue theorem.

The function $\varphi(z) = z^n/(bz^2 + 2az - b)$ has two simple poles only one of which is in the interior of γ:

$$z_0 = \frac{(a^2 + b^2)^{1/2} - a}{b}.$$

The residue ρ at this pole is equal to the value of $z^n/(2bz + 2a)$ at the point $z = z_0$:

$$\rho = \frac{1}{2(a^2 + b^2)^{1/2}} \left[\frac{(a^2 + b^2)^{1/2} - a}{b}\right]^n.$$

The residue theorem yields

$$\int_\gamma \varphi(z) \, dz = -2i\pi\rho$$

(the minus sign being due to the fact that γ is described in the negative sense). Consequently,

$$\boxed{I = \frac{2\pi}{(a^2 + b^2)^{1/2}} \left[\frac{(a^2 + b^2)^{1/2} - a}{b}\right]^n.}$$

B. *The Laplace transform of $J_n(t)$.* One of the integral representations of the Bessel function of order n is

$$J_n(t) = \frac{1}{2\pi} \int_{-\pi}^{\pi} e^{it\sin\theta} e^{-in\theta} \, d\theta.$$

Thus we have

$$\int_0^\infty e^{-ax} J_n(bx) \, dx = \frac{1}{2\pi} \int_0^\infty e^{-ax} \left[\int_{-\pi}^{\pi} e^{ibx\sin\theta} e^{-in\theta} \, d\theta\right] dx.$$

The double integral

$$\iint_\Delta e^{-ax} e^{ibx\sin\theta} e^{-in\theta} \, d\theta \, dx$$

is convergent in the region

$$-\pi \leqslant \theta \leqslant \pi, \qquad x \geqslant 0$$

since the integral of the absolute value of the integrand is

$$\iint_{\Delta} e^{-ax}\, d\theta\, dx = \int_0^{\infty} e^{-ax}\, dx \int_{-\pi}^{\pi} d\theta.$$

Therefore, we may permute the two integrations

$$\int_0^{\infty} e^{-ax} J_n(bx)\, dx = \frac{1}{2\pi}\int_{-\pi}^{\pi} e^{-in\theta}\, d\theta \int_0^{\infty} e^{-ax} e^{ibx\sin\theta}\, dx$$

$$= \frac{1}{2\pi}\int_{-\pi}^{\pi} e^{-in\theta}\, \frac{1}{a - ib\sin\theta}\, d\theta.$$

Therefore,

$$\int_0^{\infty} e^{-ax} J_n(bx)\, dx = \frac{1}{(a^2+b^2)^{1/2}}\left[\frac{(a^2+b^2)^{1/2} - a}{b}\right]^n.$$

If we choose $b = 1$ and write p instead of a, in accordance with the usual notations involving the Laplace transformation $\mathscr{L}$, we see that

$$\mathscr{L} J_n(x) = \frac{1}{(1+p^2)^{1/2}}\,[(1+p^2)^{1/2} - p]^n.$$

In particular (see Exercise 13),

$$\mathscr{L} J_0(x) = \frac{1}{(1+p^2)^{1/2}}\,.$$

15. Solution of a linear differential equation

Problems. The present problem is to find the solution of the differential equation

$$x'' - 2x' + x = \frac{1}{\sqrt{(\pi t)^{1/2}}}, \tag{1}$$

that vanishes along with its derivative at $t = 0$.

A. Solve the problem by an elementary method.

B. Solve the problem by the use of operational calculus. Let $X(p)$ denote the Laplace transform of $x(t)$. Verify that

$$X(p) = \frac{1}{(p-1)^2\sqrt{p}}\,.$$

Find the primitive of $X(p)$ by applying the convolution product formula.

Solutions. **A.** *An elementary solution.* We apply the method of variation of parameters. Two linearly independent solutions of the corresponding homogeneous equation are

$$x_1 = e^t \qquad \text{and} \qquad x_2 = te^t.$$

Therefore, we set

$$x = u_1 e^t + u_2 te^t$$

and we impose on the unknown functions u_1 and u_2 the condition

$$x' = u_1 x_1' + u_2 x_2'. \tag{2}$$

When we differentiate and make use of (2), we have

$$u_1'e^t + u_2'te^t = 0. \tag{3}$$

When we differentiate (3) and make use of the original differential equation, we get

$$u_1' + u_2'(t+1)e^t = \frac{1}{\sqrt{(\pi t)^{1/2}}}.$$

Thus u_1 and u_2 are solutions of the linear system

$$u_1' + tu_2' = 0, \qquad u_1' + (t+1)u_2' = \frac{e^{-t}}{(\pi t)^{1/2}}. \tag{4}$$

Thus we have

$$u_1' = -\left(\frac{t}{\pi}\right)^{1/2} e^{-t} \quad (t \geqslant 0), \qquad u_2' = \frac{e^{-t}}{(\pi t)^{1/2}},$$

from which we get

$$u_1 = -\int_0^t \left(\frac{u}{\pi}\right)^{1/2} e^{-u}\,du + K_1, \qquad u_2 = \int_0^t \frac{1}{(\pi u)^{1/2}} e^{-u}\,du + K_2.$$

Therefore, the general solution of the differential equation is

$$x(t) = -e^t \int_0^t \left(\frac{u}{\pi}\right)^{1/2} e^{-u}\,du + te^t \int_0^t \frac{1}{(\pi u)^{1/2}} e^{-u}\,du + K_1 e^t + K_2 te^t.$$

The solution that vanishes and has a vanishing derivative at $t = 0$ corresponds to $K_1 = K_2 = 0$:

$$x(t) = -e^t \int_0^t \left(\frac{u}{\pi}\right)^{1/2} e^{-u}\,du + te^t \int_0^t \frac{1}{(\pi u)^{1/2}} e^{-u}\,du. \tag{5}$$

Integration by parts in the first integral yields

$$x(t) = \left(\frac{t}{\pi}\right)^{1/2} + \left(t - \tfrac{1}{2}\right) e^t \int_0^t \frac{1}{(\pi u)^{1/2}} e^{-u}\, du.$$

By making the change of variable $u = v^2$, this last integral becomes

$$\int_0^{\sqrt{t}} \frac{2}{\sqrt{\pi}} e^{-v^2}\, dv = F(\sqrt{t}),$$

where

$$F(t) = \frac{2}{\sqrt{\pi}} \int_0^t e^{-v^2}\, dv$$

is the error function.

Finally,

$$x(t) = \left(\frac{t}{\pi}\right)^{1/2} + \left(t - \tfrac{1}{2}\right) e^t F(\sqrt{t}).$$

B. *Use of the Laplace transformation.* If $\mathscr{L}$ is the Laplace transformation operator, we set

$$X(p) = \mathscr{L}x(t).$$

Since $x'(0) = 0$ and $x''(0) = 0$, we have

$$pX(p) = \mathscr{L}x'(t), \qquad p^2X(p) = \mathscr{L}x''(t)$$

and

$$\frac{1}{\sqrt{p}} = \mathscr{L}\frac{1}{(\pi t)^{1/2}}.$$

The differential equation is thus transformed into an algebraic equation:

$$p^2X - 2pX + X = \frac{1}{\sqrt{p}},$$

where

$$X = \frac{1}{(p-1)^2\sqrt{p}}.$$

Now we know that

$$\frac{1}{(p-1)^2} = \mathscr{L}(te^t), \qquad \frac{1}{\sqrt{p}} = \mathscr{L}\frac{1}{(\pi t)^{1/2}}.$$

Thus the primitive of X is the convolution of te^t and $1/(\pi t)^{1/2}$, that is,

$$x(t) = \int_0^t \frac{1}{(\pi s)^{1/2}} (t-s)e^{t-s}\, ds.$$

We can write $x(t)$ in the form

$$x(t) = te^t \int_0^t \frac{e^{-s}}{(\pi s)^{1/2}}\, ds - e^t \int_0^t \left(\frac{s}{\pi}\right)^{1/2} e^{-s}\, ds,$$

so that we again have (5).

16. Study of a singular integral equation

In this exercise we shall study a linear integral equation of the Volterra type, that is, one with variable limits of integration. It is "singular" because the integrand is unbounded. We first solve it by making a transformation that does not seem very intuitive. The method of the Laplace transformation enables us to redo the calculation in a more natural manner. Difficulties arise if we wish to apply the method of differentiation under the integral sign to the result. They are analogous to those encountered in Exercise 5 of Chapter 2 and they are overcome in the same way.

Problems. Let $f(x)$ be a function that is continuous for $0 \leqslant x \leqslant h$, where h is a positive constant. Let α denote a positive number in the open interval (0, 1). Find a function $\varphi(x)$ that is continuous for $0 \leqslant x \leqslant h$ such that

$$\int_0^x \frac{\varphi(t)}{(x-t)^\alpha}\, dt = f(x). \tag{1}$$

A. Multiply (1) by $1/(y-x)^{1-\alpha}$ and integrate from 0 to y and then reverse the order of integration.

B. Show, either by integrating by parts or making change of variable, that it is possible, in a sense to be made precise, to reverse the orders of integration and differentiation in the result.

C. Find these results by using the Laplace transformation.

Solutions. **A.** Given a function $f(x)$ that is continuous for $0 \leqslant x \leqslant h$, we seek a continuous function $\varphi(t)$ such that the integral

$$\int_0^x \frac{\varphi(t)}{(x-t)^\alpha}\, dt,$$

where α is a given number between 0 and 1, converges to $f(x)$.

For this, we introduce a parameter y lying between 0 and h and we form the iterated integral

$$\int_0^y \frac{dx}{(y-x)^{1-\alpha}} \int_0^x \frac{\varphi(t)}{(x-t)^\alpha}\, dt = \int_0^y \frac{f(x)}{(y-x)^{1-\alpha}}\, dx.$$

This iterated integral can be represented by a double integral over the triangle T shown in Figure 5. Thus we have

$$\iint_T \frac{\varphi(t)\, dx\, dt}{(y-x)^{1-\alpha}(x-t)^\alpha} = \int_0^y \frac{f(x)}{(y-x)^{1-\alpha}}\, dx.$$

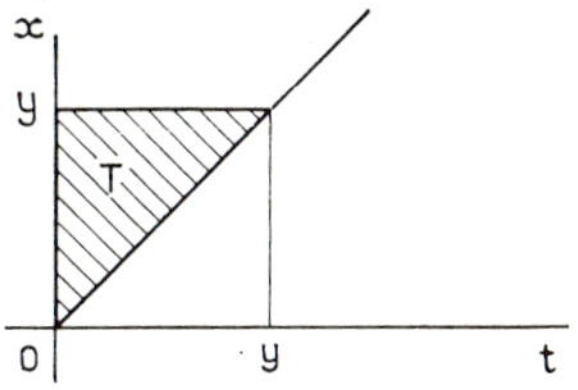

FIGURE 5

This is an improper double integral. For it to exist, it is necessary and sufficient that replacement of $\varphi(t)$ with $|\varphi(t)|$ yields a convergent integral, or, what amounts to the same thing, that it converge when $\varphi(t)$ is a positive function. A sufficient condition for this is that the calculation that we are about to perform be valid in this last case, which we shall verify.

We evaluate the double integral by integrating first with respect to x, then with respect to t. We obtain

$$I = \int_0^y \varphi(t)\, dt \int_t^y \frac{dx}{(y-x)^{1-\alpha}(x-t)^\alpha}.$$

If we set $x = t + u(y - t)$, the integral

$$\int_t^y \frac{dx}{(y-x)^{1-\alpha}(x-t)^\alpha}$$

becomes

$$\int_0^1 \frac{du}{(1-u)^{1-\alpha}u^\alpha} = B(\alpha, 1-\alpha) = \Gamma(\alpha)\Gamma(1-\alpha) = \frac{\pi}{\sin \alpha\pi},$$

from which we get

$$\int_0^y \varphi(t)\, dt = \frac{\sin \alpha\pi}{\pi} \int_0^y \frac{f(x)}{(y-x)^{1-\alpha}}\, dx.$$

Since $\varphi(t)$ is continuous, we do obtain a convergent integral when we replace $\varphi(t)$ with $|\varphi(t)|$, which justifies the operations. The function $\int_0^u \varphi(t)\,dt$ is differentiable, so that

$$\varphi(y) = \frac{\sin \alpha\pi}{\pi} \frac{d}{dy} \int_0^y \frac{f(x)}{(y-x)^{1-\alpha}}\,dx,$$

or, by making a change of notation,

$$\varphi(x) = \frac{\sin \alpha\pi}{\pi} \frac{d}{dx} \int_0^x \frac{f(t)\,dt}{(x-t)^{1-\alpha}}\,.$$

If $\alpha = \frac{1}{2}$, we have, in particular,

$$\int_0^x \frac{\varphi(t)}{(x-t)^{1/2}}\,dt = f(x), \qquad \varphi(x) = \frac{1}{\pi}\frac{d}{dx}\int_0^x \frac{f(t)}{(x-t)^{1/2}}\,dt.$$

Example. $f(t) = t^{\beta-1}$, where $\beta > 1$.

$$\begin{aligned}
\varphi(x) &= \frac{\sin \alpha\pi}{\pi} \frac{d}{dx} \int_0^x t^{\beta-1}(x-t)^{\alpha-1}\,dt \\
&= \frac{\sin \alpha\pi}{\pi} \frac{d}{dx} [x^{\alpha+\beta-1} B(\alpha, \beta)] \\
&= \frac{\sin \alpha\pi}{\pi} (\alpha + \beta - 1) x^{\alpha+\beta-2} \frac{\Gamma(\alpha)\Gamma(\beta)}{\Gamma(\alpha+\beta)} \\
&= \frac{\sin \alpha\pi}{\pi} \frac{\Gamma(\alpha)\Gamma(\beta)}{\Gamma(\alpha+\beta-1)} x^{\alpha+\beta-2} \\
&= \frac{\Gamma(\beta)}{\Gamma(1-\alpha)\Gamma(\alpha+\beta-1)} x^{\alpha+\beta-2}.
\end{aligned}$$

B. If $f(x)$ has a continuous derivative, we may consider applying a method of differentiating under the integral sign for calculating

$$\frac{d}{dx}\int_0^x \frac{f(t)}{(x-t)^{1-\alpha}}\,dt.$$

But this method cannot be applied directly. The integral obtained by formal differentiation under the integral sign diverges, and the rule for differentiation under the integral sign cannot be applied, since one cannot replace t by x in the denominator. We can get around this difficulty by various artifices.

(a) If we integrate by parts, we get

$$\int_0^x \frac{f(t)}{(x-t)^{1-\alpha}}\,dt = \Big[-\frac{1}{\alpha}(x-t)^\alpha f(t)\Big]_0^x + \frac{1}{\alpha}\int_0^x (x-t)^\alpha f'(t)\,dt$$

$$= \frac{1}{\alpha}x^\alpha f(0) + \frac{1}{\alpha}\int_0^x (x-t)^\alpha f'(t)\,dt,$$

so that

$$\frac{d}{dx}\left(\int_0^x \frac{f(x)}{(x-t)^{1-\alpha}}\,dt\right) = \frac{f(0)}{x^{1-\alpha}} + \int_0^x \frac{f'(t)}{(x-t)^{1-\alpha}}\,dt.$$

(b) By making the change of variable $t = xu$, we get

$$\int_0^x \frac{f(t)}{(x-t)^{1-\alpha}}\,dt = \int_0^1 \frac{f(xu)x\,du}{x^{1-\alpha}(1-u)^{1-\alpha}} = x^\alpha \int_0^1 \frac{f(xu)\,du}{(1-u)^{1-\alpha}},$$

$$\frac{d}{dx}\left[\int_0^x \frac{f(t)}{(x-t)^{1-\alpha}}\,dt\right] = \frac{\alpha}{x^{1-\alpha}}\int_0^1 \frac{f(xu)\,du}{(1-u)^{1-\alpha}} + x^\alpha \int_0^1 f'(xu)\frac{u\,du}{(1-u)^{1-\alpha}}$$

$$= \frac{\alpha}{x}\int_0^x \frac{f(t)\,dt}{(x-t)^{1-\alpha}} + \frac{1}{x}\int_0^x \frac{tf'(t)\,dt}{(x-t)^{1-\alpha}}$$

$$= \frac{1}{x}\int_0^x \frac{\alpha f(t) + tf'(t)}{(x-t)^{1-\alpha}}\,dt.$$

By integrating by parts, we can shift easily from one form of the solution to the other.

C. The calculation in Problem A can be avoided if we are familiar with the properties of the Laplace transformation. The integral

$$\int_0^x \frac{\varphi(t)}{(x-t)^\alpha}\,dt$$

is the convolution of $\varphi(t)$ and $t^{-\alpha}$. If $\mathscr{L}$ denotes the operator that assigns to a function $g(t)$ the Laplace transform

$$\mathscr{L}g = \int_0^\infty e^{-pt}g(t)\,dt,$$

we know that the operator $\mathscr{L}$ assigns to the convolution product $\varphi * t^{-\alpha}$ the ordinary product $\mathscr{L}\varphi \cdot \mathscr{L}t^{-\alpha}$. Thus we have

$$\mathscr{L}\varphi = \frac{\mathscr{L}f}{\mathscr{L}t^{-\alpha}}.$$

Then, if $s > 0$,

$$\mathscr{L}t^{s-1} = \int_0^\infty e^{-pt}t^{s-1}\,dt = p^{-s}\Gamma(s).$$

Therefore, for $s = 1 - \alpha$ (where $0 < \alpha < 1$),

$$\mathscr{L}\varphi = \frac{1}{\Gamma(1-\alpha)} p^{1-\alpha} \mathscr{L}f.$$

We need to find the primitive of the product of two functions. But, since $1 - \alpha$ is positive, there is no function of which $p^{1-\alpha}$ is the primitive. We are obliged to write

$$\mathscr{L}\varphi = \frac{1}{\Gamma(1-\alpha)} p \cdot [p^{-\alpha} \cdot \mathscr{L}f].$$

The primitive of $p^{-\alpha}\mathscr{L}f$ is

$$\frac{1}{\Gamma(\alpha)} x^{\alpha-1} * f(x) = \frac{1}{\Gamma(\alpha)} \int_0^x (x-t)^{\alpha-1} f(t)\, dt.$$

This function vanishes for $x = 0$. Thus, multiplication by p is the transform of the operation of differentiation with respect to x. Finally,

$$\varphi(x) = \frac{1}{\Gamma(\alpha)\Gamma(1-\alpha)} \frac{d}{dx} \int_0^x (x-t)^{\alpha-1} f(t)\, dt.$$

This is the same result as that found in Problem A, since

$$\Gamma(\alpha)\Gamma(1-\alpha) = \frac{\pi}{\sin \alpha\pi}.$$

17. Solution of a Laplace differential equation

Problems. A. Find a particular solution y_1 of the differential equation

$$xy'' + 2(\alpha x + 1)y' + (\alpha^2 x + 2\alpha)y = 0, \tag{1}$$

where α is a real constant, by using the Laplace transformation in the complex plane.

B. Solve (1) by making the transformation $y = y_1 z$.

C. Solve the nonhomogeneous equation

$$xy'' + 2(\alpha x + 1)y' + (\alpha^2 x + 2\alpha)y = g(x),$$

where $g(x)$ is a given function.

Solutions. **A.** *The finding of a solution.* Equation (1) is a Laplace differential equation, that is, a linear homogeueous differential equation the coefficients in which there are first-degree polynomials. We can apply the method of operational calculus to it; that is, we shall try to find a function $\varphi(p)$ of the complex variable p and a contour C in the complex plane such that the function

$$y(x) = \int_C e^{px}\varphi(p)\, dp$$

is a solution of (1).

Assuming that we can differentiate under the integral sign, we easily find expressions for $y'(x)$ and $y''(x)$. The values of xy, xy', and xy'' are easily found by integrating by parts. For example,

$$xy = \int_C xe^{px}\varphi(p)\, dp = [e^{px}\varphi(p)]_C - \int_C e^{px}\varphi'(p)\, dp.$$

Expressions for xy' and xy'' are found from xy by replacing $\varphi(p)$ by $p\varphi(p)$ and $p^2\varphi(p)$, respectively. When we substitute these values in (1), we obtain

$$[e^{px}(p^2 + 2\alpha p + \alpha^2)\varphi(p)]_C + \int_C e^{px}\{2(p + \alpha)\varphi(p) - [(p^2 + 2\alpha p + \alpha^2)\varphi(p)]'\}\, dp. \tag{2}$$

The second term vanishes if

$$[(p^2 + 2\alpha p + \alpha^2)\varphi(p)]' = 2(p + \alpha)\varphi(p),$$

that is,

$$\frac{[(p + \alpha)^2\varphi(p)]'}{(p + \alpha)^2\varphi(p)} = \frac{2}{p + \alpha}.$$

By integrating, we obtain

$$\log (p + \alpha)^2\varphi(p) = \log (p + \alpha)^2 + K.$$

Since the desired solution is defined up to a constant factor, we may take $K = 0$. Then

$$\varphi(p) = 1.$$

For the first term of (2) to vanish also, it is necessary that

$$[e^{px}(p + \alpha)^2]_C = 0.$$

This will be the case if we take as the curve C that portion of the real axis between $-\infty$ and $-\alpha$ since, assuming $x > 0$, we have at the end points

$$e^{px} = 0 \qquad \text{for} \quad p = -\infty \text{ (on the real axis)},$$
$$p + \alpha = 0 \qquad \text{for} \quad p = -\alpha.$$

The corresponding solution of (1) is

$$y_1(t) = \int_{-\infty}^{-\alpha} e^{px}\, dp = \left(\frac{e^{px}}{x}\right)_{-\infty}^{-\alpha} = \frac{1}{x} e^{-\alpha x}. \tag{3}$$

We can verify directly that this is a solution of (1) even if x is negative.

B. *Solution of* (1). We set

$$y = y_1 z.$$

We have

$$y' = y_1' z + z' y_1, \qquad y'' = y_1'' z + 2z' y_1' + z'' y_1.$$

Since y_1 is already a solution of the equation, we have

$$2(\alpha x + 1) z' y_1 + x(2z' y_1' + z'' y_1) = 0,$$

$$\frac{z''}{z'} = -2\left(\frac{y_1'}{y_1} + \alpha + \frac{1}{x}\right).$$

From this, we get

$$\log z' = -2(\log y_1 + \alpha x + \log x) + K.$$

We may take $K = 0$. From (3) we find

$$\log z' = 0, \qquad z' = 1.$$

Therefore, a new solution is

$$z = x,$$

that is,

$$y_2 = e^{-\alpha x}.$$

The general solution of (1) is

$$y(x) = A\,\frac{e^{-\alpha x}}{x} + Be^{-\alpha x}.$$

C. *Solution of the nonhomogeneous equation.* We shall use the method of variation of parameters. We know two linearly independent solutions of (1):

$$y_1(x) = \frac{e^{-\alpha x}}{x}, \qquad y_2(x) = e^{-\alpha x}.$$

We set

$$y = u_1 y_1 + u_2 y_2$$

and we impose on the unknown functions $u_1(x)$ and $u_2(x)$ the condition

$$y' = u_1 y_1' + u_2 y_2'.$$

Thus u_1 and u_2 are solutions of the system of differential equations

$$u_1' y_1 + u_2' y_2 = 0,$$

$$u_1' y_1' + u_2' y_2' = \frac{g(x)}{x},$$

$$u_1' + u_2' x = 0,$$

$$u_1' \left(\frac{1}{x} + \alpha\right) + u_2' \alpha x = -g(x) e^{\alpha x}.$$

From this we conclude that

$$u_1' = -x g(x) e^{\alpha x}, \qquad u_1 = -\int_{x_1}^{x} u g(u) e^{\alpha u}\, du,$$

and hence,

$$u_2' = g(x) e^{\alpha x}, \qquad u_2 = \int_{x_2}^{x} g(u) e^{\alpha u}\, du.$$

The general solution of the nonhomogeneous equation is

$$y(x) = -\frac{e^{-\alpha x}}{x} \int_{x_1}^{x} u g(u) e^{\alpha u}\, du + e^{-\alpha x} \int_{x_2}^{x} g(u) e^{\alpha u}\, du.$$

It depends on the two constants x_1 and x_2.

18. Legendre polynomials

The differential equation

$$(1 - t^2) \frac{d^2 x}{dt^2} - 2t \frac{dx}{dt} + n(n + 1)x = 0$$

is not a Laplace equation, since one of the coefficients is a polynomial of degree 2. Therefore, it is not possible to solve it by a Laplace transformation in the complex plane as we have done in connection with Hermite polynomials and on other occasions (Exercises 7 and 17 of this chapter).

However, there are other integral transformations than the Laplace transformation that can be used for more complicated equations than the Laplace equations. An example is the transformation

$$x(t) = \int_C (t-u)^\alpha f(u)\,du.$$

We shall again find the properties that we know for Legendre polynomials.

Problems. Consider the differential equation

$$(1-t^2)\frac{d^2x}{dt^2} - 2t\frac{dx}{dt} + n(n+1)x = 0, \tag{1}$$

where n is a nonnegative integer.

A. To solve this equation, we set

$$x(t) = \int_C (t-u)^\alpha f(u)\,du,$$

where C is a curve in the complex plane u, $f(u)$ is an analytic function, and α is a real number. Evaluate dx/dt and d^2x/dt^2. Show that, if $x(t)$ is a solution of (1), then $f(u)$ is a solution of the differential equation

$$\frac{d^2}{d^2u}[(1-u^2)f] - 2\alpha\frac{d}{du}(uf) + [n(n+1) - \alpha(\alpha+1)]f = 0, \tag{2}$$

in which we must impose certain conditions on the limits.

Simplify this equation by taking $\alpha = -n-1$. Show that $x(t)$ can then be put in the form

$$x(t) = K\int_C \frac{(u^2-1)^n}{(t-u)^{n+1}}\,du, \tag{3}$$

where K is an arbitrary constant.

B. Show that certain closed contours can be chosen for C. Then show that the function

$$P_n(t) = \frac{1}{2^n n!}\frac{d^n}{dt^n}(t^2-1)^n$$

is a solution of (1). Note that $P_n(t)$ is a polynomial.

C. Use the residue theorem and the expression for the polynomial P_n with a convenient value for the constant K to evaluate the sum of the series

$$\sum_{n=0}^{\infty} \lambda^n P_n(t),$$

where λ is a parameter.

D. Show with the aid of the differential equation (1), and without using the results of the preceding problems, that the function

$$h(\lambda, t) = \sum_{n=0}^{\infty} \lambda^n P_n(t)$$

is a harmonic function of the variables

$$\lambda \sin \theta \cos \varphi, \qquad \lambda \sin \theta \sin \varphi, \qquad \lambda \cos \theta,$$

where $t = \cos \theta$. The expression for the Laplacian in spherical coordinates is

$$\Delta h = \frac{1}{\lambda^2} \frac{\partial}{\partial \lambda} \left(\lambda^2 \frac{\partial h}{\partial \lambda} \right) + \frac{1}{\lambda^2} \frac{1}{\sin \theta} \frac{\partial}{\partial \theta} \left(\sin \theta \frac{\partial h}{\partial \theta} \right) + \frac{1}{\lambda^2 \sin^2 \theta} \frac{\partial^2 h}{\partial \varphi^2}.$$

Assume the legitimacy of term-by-term differentiation with respect to the different variables.

Solutions. **A.** *Transformation of the equation.* We use a generalization of the method of the Laplace transformation. We set

$$x(t) = \int_C (t - u)^\alpha f(u) \, du, \tag{4}$$

where the kernel $(t - u)^\alpha$ replaces the kernel e^{tu} of the Laplace integral.

Assuming that we may differentiate under the integral sign, we have

$$\frac{dx}{dt} = \int_C \alpha (t - u)^{\alpha - 1} f(u) \, du. \tag{5}$$

Integrating by parts, we obtain

$$\frac{dx}{dt} = - [(t - u)^\alpha f(u)] + \int_C (t - u)^\alpha \frac{df}{du} \, du. \tag{6}$$

Similarly, we obtain

$$\frac{d^2x}{dt^2} = - \left[\alpha (t - u)^{\alpha - 1} f(u) + (t - u)^\alpha \frac{df}{du} \right]_C + \int_C (t - u)^\alpha \frac{d^2 f}{du^2} \, du. \tag{7}$$

To get an expression for $t\,dx/dt$, we rewrite (5) in the form

$$t\frac{dx}{dt} = \int_C \alpha(t-u)(t-u)^{\alpha-1}f(u)\,du + \int_C \alpha u(t-u)^{\alpha-1}f(u)\,du.$$

Then we transform the second integral by an integration by parts:

$$t\frac{dx}{dt} = \int_C (t-u)^\alpha \alpha f\,du - [(t-u)^\alpha uf]_C \quad \int_C (t-u)^\alpha \frac{d}{du}(uf)\,du,$$

that is,

$$t\frac{dx}{dt} = -[(t-u)^\alpha uf]_C + \int_C (t-u)^\alpha \left[\alpha f + \frac{d}{du}(uf)\right] du. \tag{8}$$

Proceeding similarly for $t^2\,d^2x/dt^2$, we obtain

$$t^2\frac{d^2x}{dt^2} = -\left[2(\alpha-1)(t-u)^\alpha uf + \alpha(t-u)^{\alpha-1}u^2f + (t-u)^\alpha\frac{d}{du}(u^2f)\right]_C$$
$$+ \int_C (t-u)^\alpha\left\{\frac{d^2}{du^2}(u^2f) + 2(\alpha-1)\frac{d}{du}(uf) + \alpha(\alpha-1)f\right\} du. \tag{9}$$

When we substitute these expressions in (1), we obtain

$$\int_C (t-u)^\alpha\left\{\frac{d^2}{du^2}[(1-u^2)f] - 2\alpha\frac{d}{du}(uf) - \alpha(\alpha+1)f + n(n+1)f\right\} du$$
$$-\left[(t-u)^{\alpha-1}\alpha(1-u^2)f - (t-u)^\alpha\left(2\alpha uf - \frac{d}{du}[(1-u^2)f]\right)\right]_C = 0. \tag{10}$$

For this relation to be satisfied, it will be sufficient for the two terms to vanish separately. For the first to vanish, it will be sufficient for $f(u)$ to be a solution of the differential equation

$$\frac{d^2}{du^2}[(1-u^2]f - 2\alpha\frac{d}{du}(uf) + [n(n+1) - \alpha(\alpha+1)]f = 0. \tag{11}$$

If we choose $\alpha = -n-1$, this equation is simplified to

$$\frac{d^2}{du^2}[(1-u^2)f] + 2(n+1)\frac{d}{du}(uf) = 0. \tag{12}$$

Integrating once, we obtain

$$\frac{d}{du}[(1-u^2)f] + 2(n+1)uf = A, \tag{13}$$

where A is a constant. If we choose $A = 0$, this equation is easily integrated. Separating the variables, we have

$$\frac{(d/du)[(u^2-1)f]}{(u^2-1)f} = (n+1)\frac{2u}{u^2-1} = (n+1)\frac{(d/du)(u^2-1)}{u^2-1},$$

and hence,

$$(u^2-1)f = K(u^2-1)^{n+1}.$$

Therefore,

$$f(u) = K(u^2-1)^n, \tag{14}$$

where K is an arbitrary constant. The corresponding expression for $x(t)$ is

$$x(t) = K\int_C \frac{(u^2-1)^n}{(t-u)^{n+1}}\,du. \tag{3}$$

B. *Polynomial solutions of* (1). Since n is an integer, the function $f(u)$ given by (14) is a single-valued function of the complex variable u. If we take for C a closed contour, the bracketed expression in (10) is not identically zero. But then, for the function $x(t)$ given by formula (3) not to vanish identically, it is necessary that this contour encircle the pole $u = t$.

The formula giving the successive derivatives of the holomorphic function $(u^2-1)^n$ then shows immediately that

$$x(t) = (-1)^{n+1}\frac{2i\pi}{n!}K\frac{d^n}{dt^n}(t^2-1)^n.$$

The function thus obtained is a polynomial of degree n. We can verify easily that it is indeed a solution of (1), which justifies *a posteriori* the differentiations under the integral sign that we made in Problem A.

To obtain the Legendre polynomial $P_n(t)$ itself, we need to take

$$K = \frac{(-1)^{n+1}}{2i\pi\cdot 2^n}.$$

C. *The generating function of the Legendre polynomials.* We have

$$\lambda^n P_n(t) = \frac{1}{2i\pi}\int_C \frac{\lambda^n(u^2-1)^n}{2^n(u-t)^{n+1}}\,du. \tag{15}$$

Let us take for the curve C the circle $|u| = 1$ over which we integrate once in the counterclockwise direction.

Let us assign to t a fixed value such that $|t| < 1$. As u ranges over C, we have

$$|u - t| > 1 - |t|,$$

with

$$\left|\frac{1 - u^2}{2}\right| \leqslant 1.$$

Therefore,

$$\left|\frac{\lambda^n(u^2 - 1)^n}{2^n(u - t)^n}\right| < \frac{|\lambda|^n}{(1 - |t|)^n}.$$

This is smaller than 1 if $|\lambda| < 1 - |t|$.

The geometric series $\Sigma[\lambda^n(u^2 - 1)^n/2^n(u - t)^{n+1}]$ can then be integrated term by term over C. Thus

$$\sum_{n=0}^{\infty} \lambda^n P_n(t) = \frac{1}{2i\pi}\int_C \sum_{n=0}^{\infty} \frac{\lambda^n(u^2 - 1)^n}{2^n(u - t)^{n+1}}\, du = \frac{1}{2i\pi}\int_C \frac{du}{u - t - (\lambda/2)(u^2 - 1)}.$$

Let u_1 and u_2 be poles of the rational function

$$\frac{1}{u - t - (\lambda/2)(u^2 - 1)}.$$

These are continuous functions of λ. For $\lambda = 0$, one of these poles, let us say u_1, is equal to t. If λ is sufficiently small, the pole u_1 will remain close to t, hence in the interior of C. Similarly, the second pole remains outside C. Under these conditions, only the pole u_1 is interior to C and, according to the residue theorem, the last integral is equal to $1/(1 - \lambda u_1)$.

Now, for small values of λ, there exists a determination of the radical $(1 - 2\lambda t + \lambda^2)^{1/2}$ that assumes the value 1 for $\lambda = 0$. If we let $(1 - 2\lambda t + \lambda^2)^{1/2}$ denote this determination, we have

$$u_1 = \frac{1 - (1 - 2\lambda t + \lambda^2)^{1/2}}{\lambda}, \qquad 1 - \lambda u_1 = (1 - 2\lambda t + \lambda^2)^{1/2}.$$

Finally,

$$\sum_{n=0}^{\infty} \lambda^n P_n(t) = \frac{1}{(1 - 2\lambda t + \lambda^2)^{1/2}}.$$

For a fixed value of t such that $|t| < 1$, this result is proved when $|\lambda|$ is "sufficiently small." But we can now make more precise the conditions under which this expansion is valid. It is valid when $|\lambda|$ is less than the smallest of the absolute values of the zeros of $1 - 2\lambda t + \lambda^2$.

The series is then uniformly convergent and, since the two members are equal for an infinite set of values of λ (close to 0), they are equal for all values of λ for which they are holomorphic.

Remark. The above formula can be established by several other procedures. A different proof for it is given in Exercise 8 of Chapter 4.

D. We wish to show that, if $P_n(t)$ is a solution of the differential equation

$$(1 - t^2)P_n''(t) - 2tP_n'(t) + n(n + 1)P_n(t) = 0, \tag{1'}$$

the function

$$h(\lambda, t) = \sum_{n=0}^{\infty} \lambda^n P_n(t)$$

is a harmonic function relative to the point M with radius vector λ and colatitude θ, where $t = \cos\theta$. The function $h(\lambda t)$ does not depend on the longitude φ.

This result is obvious for the expression found for $h(\lambda t)$, since $1/(1 - 2\lambda t + t^2)^{1/2}$ is the reciprocal of the distance from M to the point $(0, 0, 1)$. Thus it is a harmonic function of the point M. However, we wish to prove this result without solving (1′).

To do this we write Δh in spherical coordinates. If we assume that h does not depend on φ and if we set $t = \cos\theta$, we see that

$$\lambda^2 \Delta h = \frac{\partial}{\partial\lambda}\left(\lambda^2 \frac{\partial h}{\partial\lambda}\right) + \frac{\partial}{\partial t}\left[(1 - t^2)\frac{\partial h}{\partial t}\right].$$

Now we have

$$\frac{\partial h}{\partial\lambda} = P_1 + \cdots + n\lambda^{n-1}P_n + \cdots,$$

$$\lambda^2\frac{\partial h}{\partial\lambda} = \lambda^2 P_1 + \cdots + n\lambda^{n+1}P_n + \cdots,$$

$$\frac{\partial}{\partial\lambda}\left(\lambda^2\frac{\partial h}{\partial\lambda}\right) = 2\lambda P_1 + \cdots + n(n + 1)\lambda^n P_n + \cdots.$$

If we multiply both members of the differential equation (1′) by λ^n and sum from $n = 0$ to $n = \infty$, we get

$$(1 - t^2)\frac{\partial^2 h}{\partial t^2} - 2t\frac{\partial h}{\partial t} + \frac{\partial}{\partial\lambda}\left(\lambda^2\frac{\partial h}{\partial\lambda}\right) = 0$$

or

$$\frac{\partial}{\partial t}\left[(1 - t^2)\frac{\partial h}{\partial t}\right] + \frac{\partial}{\partial\lambda}\left(\lambda^2\frac{\partial h}{\partial\lambda}\right) = 0.$$

The left member of this equation is $\lambda^2 \Delta h$. Thus $\Delta h = 0$, and h is a harmonic function.

Remark. It is easy to deduce from this result that each term $\lambda^n P_n(t)$ is a harmonic function. It is a harmonic polynomial in the coordinates of M and is homogeneous of degree n.

19. Calculus of variations (non-Euclidean straight lines)

This exercise interprets the "Euclidean line" (see Exercise 5 of Chapter 4) as the shortest path between two points.

Problem. Let $y = y(x)$ be the equation of an arc γ passing through two given points A and B of the xy-plane. Determine this arc in such a way that the integral

$$\int_\gamma \frac{ds}{y},$$

where s is an element of arc, will be maximized or minimized.

Solutions. On γ we have $ds = (1 + y'^2)^{1/2}\, dx$ and the integral in question is of the form

$$\int_{x_0}^{x_1} \frac{(1 + y'^2)^{1/2}}{y}\, dx,$$

from which we get Euler's equation,

$$\frac{\partial}{\partial y} \frac{(1 + y'^2)^{1/2}}{y} - \frac{d}{dx} \frac{\partial}{dy'} \frac{(1 + y'^2)^{1/2}}{y} = 0$$

or

$$\frac{(1 + y'^2)^{1/2}}{y^2} + \frac{d}{dx} \frac{y'}{y(1 + y'^2)^{1/2}} = 0.$$

This is a second-order equation and it cannot be immediately integrated. It is more convenient to treat x locally as a function of y and to start with the integral

$$\int \frac{(1 + x'^2)^{1/2}}{y}\, dy.$$

We obtain the same Euler equation in a more tractable form. Specifically, it can be written

$$\frac{\partial}{dx}\frac{(1 + x'^2)^{1/2}}{y} - \frac{d}{dy}\frac{\partial}{dx'}\frac{(1 + x'^2)^{1/2}}{y} = 0.$$

But the first term vanishes, so that we immediately have a first integral:

$$\frac{\partial}{\partial x'}\frac{(1 + x'^2)^{1/2}}{y} = C$$

or

$$\frac{x'}{(1 + x'^2)^{1/2}} = Cy.$$

Let us set

$$x' = \tan \varphi.$$

We then have

$$Cy = \sin \varphi, \qquad C\, dy = \cos \varphi \, d\varphi,$$

from which we get

$$dx = \tan \varphi \, dy = \frac{1}{C} \sin \varphi \, d\varphi,$$

or, finally,

$$x = x_0 - \frac{1}{C} \cos \varphi, \qquad y = \frac{1}{C} \sin \varphi.$$

The solution curves are circles centered on the x-axis. The problem regarding the limits is determined. There exists, in general, a solution in the form of a circle passing through two given points A and B (except when the segment AB is perpendicular to the x-axis, which is contrary to the hypotheses of a calculus-of-variations problem).

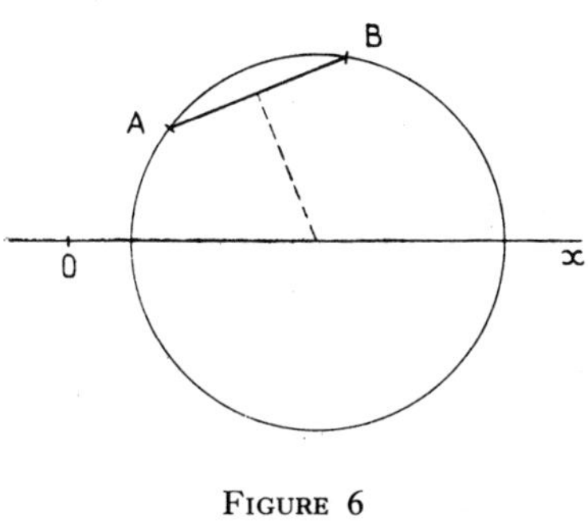

FIGURE 6

The integral

$$\int \frac{ds}{y} = \int \frac{d\varphi}{\sin \varphi}$$

is meaningful only if A and B lie on the same side of the x-axis. Then the arc representing the solution is that one of the two arcs AB that does not touch the x-axis.

20. A refraction problem

Light rays can be considered as the trajectories of a moving particle that takes the smallest amount of time to go from one point to another, its speed (the absolute value of the velocity vector) constituting a given scalar field over the space in question. This field is constant in every homogeneous region.

Problems. A. Consider a domain $\mathscr{D}$ in n-dimensional space. Suppose that $\mathscr{D}$ is partitioned into two subregions by a surface S. Let A denote a point in one of these subregions and B a point in the other. For two given positive numbers n_1 and n_2, find on S a point M such that the expression

$$n_1 MA + n_2 MB$$

(where MA and MB are the distances from M to the points A and B) will be minimized.

Assume that this problem has a solution in the interior of $\mathscr{D}$. Consider the case in which $\mathscr{D}$ is the entire space and S is a plane.

B. Let $\varphi(x, y, z)$ denote a given continuous function in three-dimensional space. Suppose that φ has continuous first derivatives. Out of all the curves C that connect two points A and B and that possess the regularity properties usually assumed in the calculus of variations, find the one that minimizes the integral

$$\int_C \varphi(x, y, z)\, ds,$$

where s denotes an element of arc on C.

Write the differential equations for C. Study the particular case in which φ is independent of z. Show that, in this case, C is a plane curve. Consider the case in which φ is dependent only on z and has a constant value n_1 for $z > z_1$ and a constant value n_2 for $z < z_2$, where z_2 and z_1 ($> z_2$) are two given numbers.

Solutions. **A.** We denote by a_i the coordinates of A, b_i the coordinates of B, x_i the coordinates of the point M that we are seeking on the surface S, and $f(x_1, ..., x_n) = 0$ the equation for S.

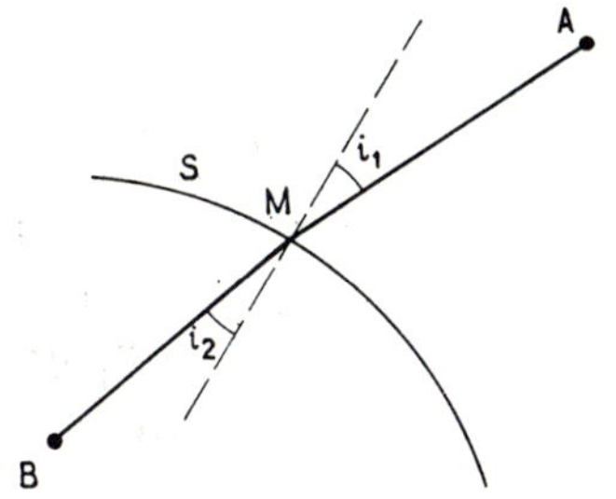

FIGURE 7

We are looking for a point M on S such that

$$n_1 MA + n_2 MB$$

will be minimized, that is,

$$n_1[\sum (x_i - a_i)^2]^{1/2} + n_2[\sum (x_i - b_i)^2]^{1/2},$$

with

$$f(x_i) = 0.$$

We introduce Lagrange's multiplier λ and we set the partial derivatives of

$$n_1[\sum (x_i - a_i)^2]^{1/2} + n_2[\sum (x_i - b_i)^2]^{1/2} + \lambda f(x_i)$$

equal to zero. If we set $d_1 = MA$ and $d_2 = MB$, we find the n equations

$$\frac{n_1}{d_1}(x_i - a_i) + \frac{n_2}{d_2}(x_i - b_i) + \lambda \frac{\partial f}{\partial x_i} = 0, \tag{1}$$

which, with $f(x_i) = 0$, theoretically determine $x_1, ..., x_n$ and λ. We define

$$p = \left[\sum \left(\frac{\partial f}{\partial x_i}\right)^2\right]^{1/2}.$$

If we denote by i_1 and i_2 the angles that MA and MB make with the normal to S at M, we have

$$\cos i_1 = \frac{1}{p\,d_1} \sum (x_i - a_i) \frac{\partial f}{\partial x_i}, \qquad \cos i_2 = \frac{1}{p\,d_2} \sum (x_i - b_i) \frac{\partial f}{\partial x_i}.$$

Finally, let us set

$$\sum (x_i - a_i)(x_i - b_i) = h.$$

With the aid of the multipliers $x_i - a_i$, $x_i - b_i$, and $\partial f/\partial x_i$, we derive from equations (1) the following three equations:

$$\begin{aligned} n_1 d_1 + \frac{n_2}{d_2} h + \lambda p\, d_1 \cos i_1 &= 0, \\ \frac{n_1}{d_1} h + n_2 d_2 + \lambda p\, d_2 \cos i_2 &= 0, \\ n_1 \cos i_1 + n_2 \cos i_2 + \lambda p &= 0. \end{aligned}$$

Elimination of h and λ in these three equations yields

$$n_1^2 - n_2^2 = n_1^2 \cos^2 i_1 - n_2^2 \cos^2 i_2$$

or

$$n_1^2 \sin^2 i_1 = n_2^2 \sin^2 i_2. \tag{2}$$

Equations (1) express the fact that the plane defined by the points A, M, and B contains the normal to S at M. Equation (2) may be written

$$n_1 \sin i_1 = \pm n_2 \sin i_2.$$

To determine the sign, we simplify the problem by considering a plane in the neighborhood of the solution and locally replacing with a straight line the intersection of S and this plane.

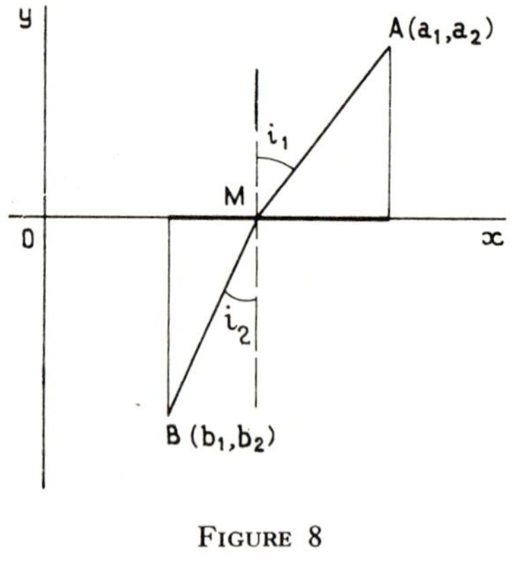

FIGURE 8

We take this line as the x-axis. Now let us find the extrema of

$$n_1[(x - a_1)^2 + a_2^2]^{1/2} + n_2[(x - b_1)^2 + b_2^2]^{1/2}$$

(notations as Figure 8), with $a_2 > 0$ and $b_2 < 0$.

Here we have only one parameter. When we set the derivative of the above function equal to zero, we obtain

$$\frac{n_1}{d_2}(x - a_1) + \frac{n_2}{d_2}(x - b_1) = 0.$$

Here $x - a_1$ and $x - b_1$ are of opposite signs. The point M is in the interior of the projection of the segment AB onto the x-axis. The angles i_1 and i_2 can be defined without ambiguity as angles lying between 0 and $\pi/2$. Then we have

$$n_1 \sin i_1 = n_2 \sin i_2 ,$$

which is the familiar law of refraction.

B. Suppose now that we are dealing with three-dimensional space. Light takes the smallest possible time to go from a point A to another point B. However, in a nonhomogeneous isotropic medium, the speed V is a point function. Let us suppose here that this function is continuous. The time required for light to travel between two points is given by an integral of the form

$$\int \frac{ds}{V} = \int \varphi(x, y, z)\, ds,$$

where $\varphi(x, y, z)$ is the variable index of refraction.

We seek y and z as a function of x in such a way as to minimize the above integral over an arc connecting two fixed points A and B. Thus we need to set up Euler's differential system for the function

$$\varphi(x, y, z)(1 + y'^2 + z'^2)^{1/2}.$$

This system is

$$\frac{\partial\varphi}{\partial y}(1 + y'^2 + z'^2)^{1/2} - \frac{d}{dx}\frac{\varphi y'}{(1 + y'^2 + z'^2)^{1/2}} = 0,$$

$$\frac{\partial\varphi}{\partial z}(1 + y'^2 + z'^2)^{1/2} - \frac{d}{dx}\frac{\varphi z'}{(1 + y'^2 + z'^2)^{1/2}} = 0.$$

Let us interpret these equations. We have $ds = (1 + y'^2 + z'^2)^{1/2}\, dx$. Let us denote by α, β, and γ the direction cosines of the tangent T oriented in the direction of the light ray C. We have

$$\frac{\partial\varphi}{\partial y} = \frac{d}{ds}(\varphi\beta), \qquad \frac{\partial\varphi}{\partial z} = \frac{d}{ds}(\varphi\gamma).$$

Similarly, if we took y or z as the independent variable instead of x, we would have

$$\frac{\partial\varphi}{\partial x} = \frac{d}{ds}(\varphi\alpha).$$

Differentiation produces the derivatives

$$\frac{d\varphi}{ds} = \alpha\frac{\partial\varphi}{\partial x} + \beta\frac{\partial\varphi}{\partial y} + \gamma\frac{\partial\varphi}{\partial z}$$

and $d\alpha/ds$. In accordance with Frenet's formulas,

$$\frac{d\alpha}{ds} = \frac{\alpha_1}{R}, \qquad \frac{d\beta}{ds} = \frac{\beta_1}{R}, \qquad \frac{d\gamma}{ds} = \frac{\gamma_1}{R},$$

where α_1, β_1, and γ_1 are the direction cosines of the principal normal to the curve C and R is the radius of curvature. Thus we have

$$\begin{aligned}
\frac{\partial\varphi}{\partial x} &= \alpha\left(\alpha\frac{\partial\varphi}{\partial x} + \beta\frac{\partial\varphi}{\partial y} + \gamma\frac{\partial\varphi}{\partial z}\right) + \varphi\frac{\alpha_1}{R},\\
\frac{\partial\varphi}{\partial y} &= \beta\left(\alpha\frac{\partial\varphi}{\partial x} + \beta\frac{\partial\varphi}{\partial y} + \gamma\frac{\partial\varphi}{\partial z}\right) + \varphi\frac{\beta_1}{R},\\
\frac{\partial\varphi}{\partial z} &= \gamma\left(\alpha\frac{\partial\varphi}{\partial x} + \beta\frac{\partial\varphi}{\partial y} + \gamma\frac{\partial\varphi}{\partial z}\right) + \varphi\frac{\gamma_1}{R}.
\end{aligned} \tag{3}$$

These formulas express the fact that the oscillating plane to the curve C at a point M contains the normal to the surface $\varphi = \text{const}$ that passes through M.

Let us denote by i the angle between this normal and the tangent to C. We have

$$p \sin i = \alpha_1\frac{\partial\varphi}{\partial x} + \beta_1\frac{\partial\varphi}{\partial y} + \gamma_1\frac{\partial\varphi}{\partial z}$$

where

$$p = \left[\left(\frac{\partial\varphi}{\partial x}\right)^2 + \left(\frac{\partial\varphi}{\partial y}\right)^2 + \left(\frac{\partial\varphi}{\partial z}\right)^2\right]^{1/2}.$$

From equations (3) we now conclude that

$$p \sin i = \frac{\varphi}{R},$$

which is the law of refraction in a nonhomogeneous medium.

Since $\alpha = dx/ds$, the differential equations for C become

$$\varphi \frac{d^2x}{ds^2} + \frac{\partial\varphi}{dx}\left(\frac{dx}{ds}\right)^2 + \frac{\partial\varphi}{\partial y}\frac{dx}{ds}\frac{dy}{ds} + \frac{\partial\varphi}{\partial z}\frac{dx}{ds}\frac{dz}{ds} = \frac{\partial\varphi}{\partial x}$$

and two analogous equations.

By introducing notations with indices, we write these equations in the form

$$\varphi \frac{d^2x_i}{ds^2} + \sum_k \frac{\partial\varphi}{\partial x_k}\frac{dx_i}{ds}\frac{dx_k}{ds} = \frac{\partial\varphi}{dx_i}. \tag{4}$$

These equations are compatible with

$$\Sigma\left(\frac{dx_i}{ds}\right)^2 = 1.$$

They are the differential equations for light rays in nonhomogeneous media.

Example. Suppose that φ does not depend on z. Then

$$\varphi x'' + \varphi' x' z' = 0, \qquad \varphi y'' + \varphi' y' z' = 0, \qquad \varphi z'' + \varphi' z'^2 = \varphi',$$

where the symbols x'... denote derivatives with respect to s.

From the first two equations, we derive

$$x''y' - y''x' = 0,$$

so that $y' = Kx'$, where K is a constant, and $y = Kx + K_1$.

The trajectory C is in a "vertical" plane. Let us take this plane for the xz-plane, which means making $y = 0$. It follows that

$$\varphi x'' + \varphi' x' z' = 0, \qquad \varphi z'' + \varphi' z'^2 = \varphi'.$$

The second of these equations is a differential equation involving z, z', and z'', which are functions of s. The second then provides x. It also has the first integral

$$\varphi \frac{dx}{ds} = \text{const.}$$

In this case we can also solve the problem by seeking directly the extreme values of the integral

$$\int \varphi(z)(1 + x'^2\, dz)^{1/2},$$

where x' this time represents the derivative with respect to z.

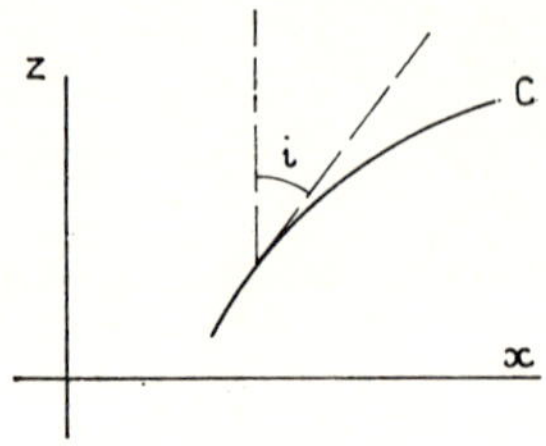

FIGURE 9

Euler's equation has the first integral

$$\frac{\varphi(z)x'}{(1 + x'^2)^{1/2}} = \text{const.}$$

Let i denote the angle between the trajectory and the z-axis [normal to the curves $\varphi(z) = \text{const}$]. We then have $x' = \tan i$ or

$$\frac{x'}{(1 + x'^2)^{1/2}} = \sin i,$$

from which we get

$$\varphi(z) \sin i = \text{const.}$$

Suppose that, for $z > z_1$, $\varphi(z)$ has a constant value n_1 and that, for $z < z_2$, it has a constant value n_2. Outside the region $z_2 < z < z_1$,

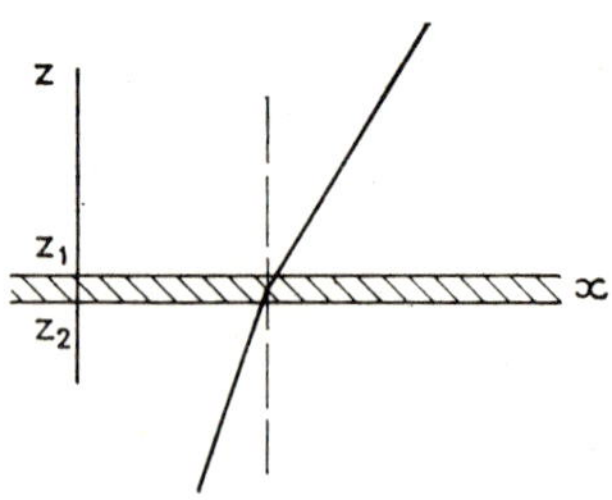

FIGURE 10

the curve C is locally constituted by a line segment. But, of course, we have two distinct lines in the upper and lower portions of the plane. These lines make angles i_1 and i_2, respectively, with the z-axis, and we have

$$n_1 \sin i_1 = n_2 \sin i_2 .$$

We may assume that this result remains valid in the limit as $z_1 - z_2$ tends to zero, although the function $\varphi(z)$ cannot then be continuous and satisfy the hypotheses made on Euler's equation. We again find the familiar law of refraction, which we proved directly in Problem A.

CHAPTER 6

Partial Differential Equations

This chapter contains two kinds of exercises. First, some of the exercises are on first-order partial differential equations, linear and nonlinear. The techniques used in connection with these problems always reduce to finding first integrals of ordinary differential systems, and it is possible to obtain each particular solution from the "general solution," which is actually written out explicitly. Second, there are exercises on second-order linear partial differential equations. These exercises have little to do with general theories. They do not even require familiarity with the elementary properties of characteristics. We have attempted primarily to show, by means of simple examples, how it is possible to start with initial and boundary conditions and construct elementary solutions satisfying some of these conditions and, if possible, to find by means of a linear combination of such solutions, a solution satisfying all the given conditions. Most frequently these problems lead to boundary-value problems and eigenvalue problems for second-order linear differential equations of the type that were treated in Chapter 5. The solutions derived from these problems involve Fourier series, Fourier integrals, or series of orthogonal functions.

Primarily the exercises have to do with the heat equation (homogeneous or otherwise), the vibrating-string equation, and Laplace's equation (harmonic functions), that is, the simplest type of parabolic, hyperbolic, and elliptic equations.

The methods used consist, in general, in writing series expansions that formally constitute solutions corresponding to the given conditions. Once these expansions are formed, it is necessary to verify that they are convergent and that they satisfy the partial differential equation in question. More precisely, we need to specify the requirements that the given conditions must satisfy for this to be the case. In the following exercises we shall often ask for this verification and we accept the formal, that is, not entirely justified, results. We do this for two reasons. In the first place, justification is sometimes quite difficult and will burden unnecessarily some of the exercises, the purpose of which is of quite a different nature. In the second place, these verifications are sometimes illusory. The series obtained may well fail to be differentiable and hence fail to verify the partial differential equation in question and yet be a valid solution of the physical problem posed. Series of this type constitute generalized solutions, the derivatives of which are distributions that verify the equation in the sense of distributions. Therefore, from this point of view, it is not worthwhile to seek the conditions under which the solutions have derivatives that are functions.

The second portion of this chapter requires only slight additional preparation as far as new concepts are concerned. However, it does require skill and familiarity with the handling of the fundamental concepts of analysis to which the first five chapters have been devoted.

1. Linear partial differential equations and integrating factors

Problems. Consider the differential form

$$\omega = 2y^3\,dx + (3xy^2 - 1)\,dy$$

and the partial differential equation

$$(3xy^2 - 1)\frac{\partial \mu}{\partial x} - 2y^3\frac{\partial \mu}{\partial y} - 3\mu y^2 = 0. \tag{E}$$

A. Find all the integrating factors of ω and from them the general solution of (E).

B. Solve (E) and from it all the integrating factors of ω.

Solutions. **A.** With the differential form ω, we associate the differential equation $\omega = 0$, that is, the equation

$$2y^3\frac{dx}{dy} + 3xy^2 - 1 = 0. \tag{1}$$

This is a linear equation in x and dx/dy. The corresponding homogeneous equation may be written

$$\frac{dx}{x} = -\frac{3}{2}\frac{dy}{y},$$

from which we get

$$x = Cy^{-3/2}.$$

If C is treated as a function of y, we have

$$C = -y^{-1/2} + K.$$

The general solution is

$$x = Ky^{-3/2} - y^{-2} \qquad \text{or} \qquad xy^{3/2} + y^{-1/2} = K.$$

Therefore, the function $xy^{3/2} + y^{-1/2}$ is a first integral of the differential equation (1). If we set this function equal to a constant K, we obtain a representation of the solutions of (1). But, if $\varphi(s)$ is a differentiable function of a variable s, the function

$$u = \varphi(xy^{3/2} + y^{-1/2})$$

is also a first integral. If we differentiate u, we obtain

$$\begin{aligned} du &= \varphi' \times [y^{3/2}\,dx + (\tfrac{3}{2}\,xy^{1/2} - \tfrac{1}{2}\,y^{-3/2})\,dy] \\ &= \tfrac{1}{2}\,y^{-3/2}\,\varphi' \times [2y^3\,dx + (3xy^2 - 1)\,dy] \\ &= \tfrac{1}{2}\,y^{-3/2}\varphi' \cdot \omega. \end{aligned}$$

Thus the function

$$\tfrac{1}{2}\,y^{-3/2}\varphi'(xy^{3/2} + y^{-1/2})$$

is an integrating factor of ω, and it follows from the theory that we have all the integrating factors. By changing the notations, we see that the general expression for the integrating factors is

$$\mu = y^{-3/2}\Phi(xy^{3/2} + y^{-1/2}). \tag{2}$$

In this expression, Φ is an arbitrary function of a single variable, to which we assign the combination $xy^{3/2} + y^{-1/2}$.

The partial differential equation

$$(3xy^2 - 1)\frac{\partial\mu}{\partial x} - 2y^3\frac{\partial\mu}{\partial y} - 3\mu y^2 = 0 \tag{E}$$

may also be written

$$\frac{\partial}{\partial y}(2\mu y^3) = \frac{\partial}{\partial x}[\mu(3xy^2 - 1)].$$

In this form it expresses the fact that μ is an integrating factor of ω. The general solution of (E) is given by (2).

We can also verify that E may be put in the form

$$\frac{D(u, v)}{D(x, y)} = 0,$$

where

$$u = \mu y^{3/2} \qquad \text{and} \qquad v = xy^{3/2} + y^{-1/2},$$

which expresses the fact that a functional relationship exists between u and v.

B. To integrate (E) *a priori*, which is a first-order quasi-linear equation, we write the associated characteristic differential system

$$\frac{dx}{3xy^2 - 1} = \frac{dy}{-2y^3} = \frac{d\mu}{3\mu y^2}. \tag{3}$$

The first equation has already been solved. A first integral of it is

$$xy^{3/2} + y^{-1/2}.$$

On the other hand,

$$\frac{3dy}{2y} = \frac{d\mu}{-\mu},$$

from which we get

$$\mu y^{3/2} = \text{const.}$$

The expression $\mu y^{3/2}$ is another first integral of the system (3) and is independent of the preceding one.

The general integral of (E) is obtained by writing that these two first integrals are functionally related. If the equation expressing this relation is solved for $\mu y^{3/2}$, we have

$$\mu y^{3/2} = \Phi(xy^{3/2} + y^{-1/2}).$$

Thus μ is an integrating factor of ω.

2. First-order linear partial differential equations. Geometric solution

Problem. Find the surfaces whose normals are tangent to a given cylinder of revolution.

To work with this exercise, we write the partial differential equation for the problem. We obtain the equation

$$(qx - py)^2 = R^2(p^2 + q^2) \tag{1}$$

with the axes chosen in such a way that the equation for the cylinder will be $x^2 + y^2 = R^2$. This equation is not linear, but it would be clumsy to handle it by the general methods (finding the general solution...) because we can see immediately that it can be decomposed into two linear equations. Equation (1) is homogeneous in p and q. These equations, furthermore, are irrational and complicated to write.

In a problem of this type, it is completely useless to write the partial differential equations. It is sufficient to know that the problem depends on a first-order partial differential equation. We can then find the solution by purely geometric considerations.

Solutions. We seek to find the surfaces whose normals are tangent to a cylinder of revolution. Let M be a point in space outside the cylinder. For a ray originating at M to be tangent to the cylinder, it is necessary and sufficient that one of the two planes (which we denote by π and π') that are tangent to the cylinder pass through M.

Therefore, for a surface to meet the requirements of the problem, it is necessary and sufficient that its tangent plane at M contain a straight line MT perpendicular to one of the two planes π or π'.

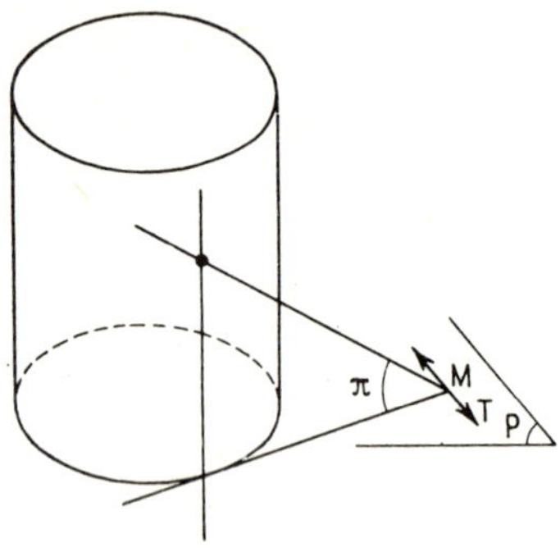

FIGURE 1

Let us consider only the plane π and the line MT. We recognize the geometric interpretation of a problem of solving a *linear* first-order partial differential equation. The characteristic curves are the curves that are tangent at M to the lines MT, that is, the curve developed on the tangent space of the circular cross sections of the cylinder. A solution representing the surface is a one-parameter locus of these curves.

Through a given curve γ there pass two solution surfaces since two involutes pass through a point. They are generated by those of the involutes of one of the families that touch γ.

3. A linear partial differential equation generalizing Riccati's equation

Problems. Consider the partial differential equation

$$P\frac{\partial z}{\partial x} + Q\frac{\partial z}{\partial y} = Az^2 + Bz + C,$$

where P, Q, A, B, and C are functions of x and y.

A. Show that the general solution is of the form

$$z = \frac{\alpha F(\varphi) + \beta}{\gamma F(\varphi) + \delta},$$

where α, β, γ, δ, and φ are known functions of x and y and where F is an arbitrary function of a single variable.

B. Show that the cross ratio of four particular solutions z_1, z_2, z_3, z_4 is a solution of the homogeneous equation

$$P\frac{\partial u}{\partial x} + Q\frac{\partial u}{\partial y} = 0.$$

Solutions. Consider the linear partial differential equation

$$P\frac{\partial z}{\partial x} + Q\frac{\partial z}{\partial y} = Az^2 + Bz + C, \tag{1}$$

where P, Q, A, B, and C are functions of x and y.

The associated differential system is

$$\frac{dx}{P} = \frac{dy}{Q} = \frac{dz}{Az^2 + Bz + C}. \tag{2}$$

We obtain a first integral of (2) by integrating the differential equation $Q\,dx - P\,dy = 0$. We write the result of this integration in the form of $\varphi(x, y) = \mu$, where φ is a function depending only on P and Q. From this we have

$$y = \psi(x, \mu).$$

Thus the equation

$$\frac{dx}{P[x, \psi(x, \mu)]} = \frac{dz}{A[x, \psi(x, \mu)]z^2 + B[x, \psi(x, \mu)]z + C[x, \psi(x, \mu)]}$$

is a Riccati differential equation whose general solution is of the form

$$z = H(\lambda),$$

where $H(\lambda)$ is a homographic function of the constant of integration λ, of which the coefficients are functions of x and μ.

Let us replace μ with $\varphi(x, y)$. Then z becomes a function

$$z = \mathscr{H}(x, y, \lambda)$$

that is homographic in λ, and we have

$$\lambda = K(x, y, z).$$

The function $K(x, y, z)$, which is homographic in z, is a first integral of Riccati's equation.

Thus the general solution of (1) is given implicitly by the relation

$$G[\varphi(x, y), K(x, y, z)] = 0,$$

where G is an arbitrary function of two variables.

This equation can be solved for K. Since K is a homographic function of z, we finally put z in the form

$$z = \frac{\alpha(x, y)F(\varphi) + \beta(x, y)}{\gamma(x, y)F(\varphi) + \delta(x, y)},$$

where F is an arbitrary function of a single variable.

Since cross ratio is conserved under homographic transformation, we have

$$(z_1, z_2, z_3, z_4) = (F_1(\varphi), F_2(\varphi), F_3(\varphi), F_4(\varphi)) = L(\varphi).$$

Now, φ is a first integral of the differential system

$$\frac{dx}{P} = \frac{dy}{Q}.$$

Thus $L(\varphi)$ is a solution of the partial differential equation

$$P\frac{\partial u}{\partial x} + Q\frac{\partial u}{\partial y} = 0.$$

The cross ratio of four solutions of the given equation is a solution of the associated homogeneous linear equation.

Remark. In the case of an ordinary Riccati differential equation

$$P(x)\frac{dz}{dx} = A(x)z^2 + B(x)z + C,$$

the cross ratio of four solutions is constant. Thus it is another solution of the homogeneous equation

$$\frac{\partial z}{\partial x} = 0.$$

4. A nonlinear partial differential equation

Let $F(x, y, z, p, q) = 0$ be a first-order nonlinear partial differential equation in two independent variables x and y. Let us set

$$p = \frac{\partial z}{\partial x}, \qquad q = \frac{\partial z}{\partial y}, \qquad \frac{\partial F}{\partial x} = F_x, \ldots, \frac{\partial F}{\partial q} = F_q.$$

With this equation, we associate in five-dimensional space the characteristic differential system

$$\frac{dx}{F_p} = \frac{dy}{F_q} = \frac{dz}{pF_p + qF_q} = \frac{-dp}{F_x + pF_z} = \frac{-dq}{F_y + qF_z}.$$

Let $G(x, y, z, p, q)$ be a first integral of this system in which either p or q does actually appear. We know that the system of partial differential equations

$$F = 0, \qquad G = \lambda,$$

where λ is a constant, is completely integrable. This means that it has solutions depending not just on λ but also on a constant μ and that they can be found by integrating an exact differential. These solutions constitute a complete integral (an integral depending on two parameters) of the given equation, and the general solution is obtained as the envelope of a one-parameter family of surfaces taken from the complete integral. (This is the method of Lagrange and Charpit.)

The two-parameter envelope of the complete integral constitutes a singular solution.

Problems. **A.** Let M be a point on a surface S, let P be the projection of M onto the z-axis, and let N be the projection of P onto the normal to S at M.

Find the surfaces S for which MN has a constant length a.

B. Which of these surfaces S are surfaces of revolution about the z-axis?

Solutions. **A.** The condition of the problem may be written

$$(\mathbf{MN}, \mathbf{MP}) = a \, \| \mathbf{MN} \| ,$$

where

$$(px + qy)^2 = a^2(1 + p^2 + q^2).$$

We need to integrate a first-order partial differential equation. Let us seek a complete solution of this equation by using the method of Lagrange and Charpit.

Out of all the equations appearing in the differential system, let us keep the following one:

$$\frac{dp}{F_x + pF_z} = \frac{dq}{F_y + qF_z},$$

which can be written simply

$$\frac{dp}{p} = \frac{dq}{q}.$$

Thus p/q is a first integral of the characteristic system and we are led to seek solutions of the compatible system of partial differential equations

$$(px + qy)^2 = a^2(1 + p^2 + q^2), \qquad p = \lambda q.$$

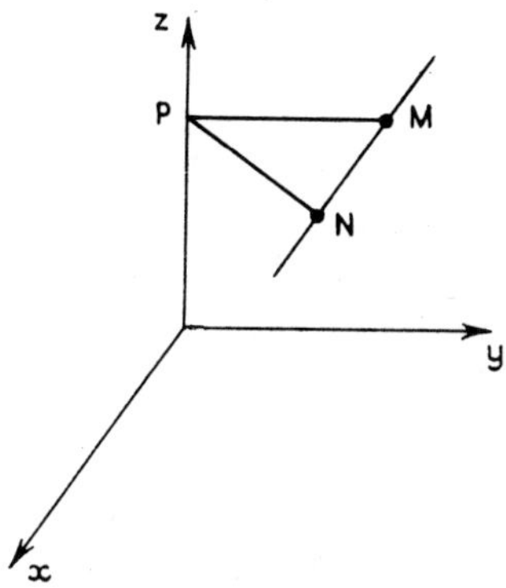

FIGURE 2

If we solve for p and q, we get

$$p = \frac{\lambda a}{[(\lambda x + y)^2 - a^2(\lambda^2 + 1)]^{1/2}}, \qquad q = \frac{a}{[(\lambda x + y)^2 - a^2(\lambda^2 + 1)]^{1/2}}.$$

Thus we need to integrate the differential form

$$dz = \frac{a(\lambda\, dx + dy)}{[(\lambda x + y)^2 - a^2(\lambda^2 + 1)]^{1/2}} = \frac{ad(\lambda x + y)}{[(\lambda x + y)^2 - a^2(\lambda^2 + 1)]^{1/2}}.$$

We find

$$z = a \cosh^{-1} \frac{\lambda x + y}{a(\lambda^2 + 1)^{1/2}} + \mu,$$

where λ and μ are constants.

The complete integral can also be written

$$a \cosh\left(\frac{z - \mu}{a}\right) = \frac{\lambda x + y}{(\lambda^2 + 1)^{1/2}}$$

or

$$a \cosh\left(\frac{z - \mu}{a}\right) = x \cos\varphi + y \sin\varphi \qquad \text{with} \quad \lambda = \cot\varphi.$$

To find the general solution we treat μ as a function of φ and seek the envelope of the one-parameter family of surfaces thus extracted from the complete integral.

To find the envelope we differentiate with respect to φ. Thus we obtain the two equations

$$\sinh\left(\frac{z-\mu}{a}\right)\mu' = x\sin\varphi - y\cos\varphi,$$
$$\frac{1}{a}\cosh\left(\frac{z-\mu}{a}\right) = x\cos\varphi + y\sin\varphi.$$

Solving for x and y, we obtain

$$x = \mu'(\varphi)\sin\varphi\sinh\frac{z-\mu}{a} + \frac{1}{a}\cos\varphi\cosh\frac{z-\mu}{a},$$
$$y = -\mu'(\varphi)\cos\varphi\sinh\frac{z-\mu}{a} + \frac{1}{a}\sin\varphi\cosh\frac{z-\mu}{a}.$$

These equations, to which the equation $z = z$ is conventionally added, represent parametrically the solution surfaces with the aid of the parameters φ and z.

Remark. A variation in the above calculations consists in solving the equation $p = \lambda q$. The general solution is of the form

$$z = f(\lambda x + y),$$

where f is a function of a single argument. The function f needs to be determined so that this solution will also verify the given equation. This amounts to finding solution surfaces that are cylinders whose generators are horizontal and parallel to the straight line $\lambda x + y = 0$.

B. If any of the solutions of the problem posed are represented by surfaces of revolution about the z-axis, they can be found as solutions common to the given equation

$$(px + qy)^2 = a^2(1 + p^2 + q^2)$$

and the equation

$$py - qx = 0,$$

which represents the surfaces of revolution about the z-axis.

It is not obvious that these two equations have common solutions, but this can be verified. However, it is instructive to make it precise by returning to the characteristic differential system. This system is written as follows:

$$\frac{dx}{x(px+qy) - a^2p} = \frac{dy}{y(px+qy) - a^2q} = \frac{dz}{(px+qy)^2 - a^2(p^2+q^2)}$$
$$= \frac{-dp}{p(px+qy)} = \frac{-dq}{q(px+qy)}.$$

One can verify immediately that it admits the integrable combination

$$p\,dy + y\,dp - q\,dx - x\,dq = 0.$$

The equations

$$(px + qy)^2 = a^2(1 + p^2 + q^2) \qquad \text{and} \qquad py - qx = H$$

thus have common solutions no matter what the value of the constant h. This is true, in particular, for $h = 0$, in which case the solutions represent surfaces of revolution. Then

$$p = \frac{ax}{[(x^2 + y^2)(x^2 + y^2 - a^2)]^{1/2}}, \qquad q = \frac{ay}{[(x^2 + y^2)(x^2 + y^2 - a^2)]^{1/2}},$$

from which we get

$$dz = p\,dx + q\,dy = \frac{ar\,dr}{[r^2(r^2 - a^2)]^{1/2}} = \frac{a\,dr}{[r^2 - a^2]^{1/2}}, \qquad r = (x^2 + y^2)^{1/2}$$

and

$$z = a \cosh^{-1} \frac{(x^2 + y^2)^{1/2}}{a} + \text{const.}$$

Remark 1. The fact that the solutions of the problem posed include surfaces of revolution is obvious from a geometric point of view. Let γ denote the meridian of a surface of revolution about the z-axis. Since the triangle PMN is in the plane of the meridian, we are led to a plane geometry problem. For a given line Oz, find the curves γ such that, if P is the projection of a point M on γ onto Oz and N is the projection of P onto the normal to γ at M, the segment MN has a constant length. Solution of this problem leads to an ordinary differential equation.

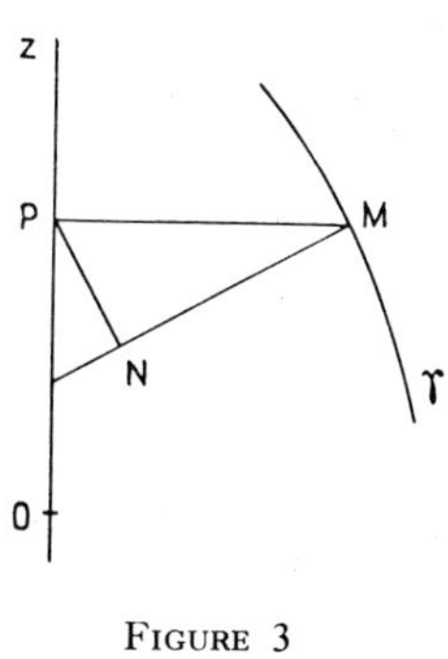

FIGURE 3

Remark 2. The cylinder of revolution of radius a about the z-axis is not a solution. This should not be surprising, since its equation is not of the form $z = f(x, y)$. However, it is clear that it is, from a geometric point of view, a solution of the problem posed. It is a singular solution, obtained by seeking the *two*-parameter envelope of the surfaces of the complete solution. We derive independently the equation

$$a \cosh\left(\frac{z-\mu}{a}\right) = x \cos \varphi + y \sin \varphi$$

in μ and φ. We obtain

$$\sinh\left(\frac{z-\mu}{a}\right) = 0 \qquad \text{or} \qquad z = \mu$$

and

$$-x \sin \varphi + y \cos \varphi = 0.$$

Thus we have

$$x \cos \varphi + y \sin \varphi = a \cosh 0 = a,$$

from which, by elimination of φ, we obtain

$$x^2 + y^2 = a^2.$$

This is indeed the equation of the cylinder referred to.

5. The heat-flow equation. Elementary and periodic solutions

In the following exercises we shall treat various problems connected with the heat-flow equation

$$\frac{\partial y}{\partial t} = a^2 \frac{\partial^2 y}{\partial x^2}.$$

In the first exercise we shall study the eigenvalue problem corresponding to the search for periodic solutions in x. We are led to represent y by a series that, for $t = 0$, reduces to a Fourier series.

In the following exercises we shall introduce more complicated boundary conditions, which lead to representations by means of Fourier integrals or series of orthogonal functions.

Problems. Consider the heat-flow equation

$$\frac{\partial y}{\partial t} = a^2 \frac{\partial^2 y}{\partial x^2}, \tag{E}$$

where a is a given real constant.

A. Find solutions of the form

$$y = f(x)g(t).$$

B. Set

$$\exp(\omega x + a^2\omega^2 t) = \sum_{n=0}^{\infty} \frac{\omega^n}{n!} V_n(x, t).$$

Show that the functions $V_n(x, t)$ are solutions of (E) and that, for $t < 0$, they can be expressed in terms of Hermite polynomials $P_n(s)$ defined by

$$\exp(-u^2 + 2us) = \sum_{n=0}^{\infty} \frac{1}{n!} u^n P_n(s)$$

(see Exercise 9 of Chapter 5).

C. Find the solutions y of (E) that are periodic functions of x with period 2π that remain bounded as $t \to \infty$ and that, for $t = 0$, reduce to a given function $y_0(x)$.

D. Carry out the calculations when $y_0(x) = u_1$ for $0 < x < \alpha$ and $y_0(x) = u_2$ for $\alpha < x < 2\pi$, where u_1 and u_2 are given constants. What is the limit of $y(x, t)$ as $t \to \infty$?

(*N.B.*: The second question is independent of the following two.)

Solutions. **A.** We have

$$fg' = a^2 f'' g \qquad \text{or} \qquad \frac{1}{a^2} \cdot \frac{g'}{g} = \frac{f''}{f}.$$

The left-hand member of this last equation is a function of t and the right-hand member is a function of x. They can be identically equal only by having a common constant value. If this constant is negative, we have

$$f'' + \omega^2 f = 0, \qquad \frac{g'}{g} = -a^2\omega^2.$$

From this we get the elementary solutions

$$f = e^{i\omega x}, \qquad g = \exp(-a^2\omega^2 t)$$

and

$$y = \exp(i\omega x - a^2\omega^2 t).$$

In terms of real quantities, we have the two solutions

$$y = \exp(-a^2\omega^2 t) \cos \omega x \qquad \text{and} \qquad y = \exp(-a^2\omega^2 t) \sin \omega x.$$

If the constant is positive, we have

$$f'' - \omega^2 f = 0, \qquad \frac{g'}{g} = a^2\omega^2,$$

from which we get

$$y = \exp(\omega x + a^2\omega^2 t).$$

B. Every series of the form

$$\sum c_n \exp(i\omega_n x - a^2\omega_n{}^2 t)$$

that is twice differentiable with respect to x and once differentiable with respect to t is a solution of the heat-flow equation. [An example is the function θ (Exercise 19 of Chapter 4).]

We shall go into greater detail with this type of solution in the following problems. Here we shall examine a family of solutions that is associated with the elementary solution $\exp(\omega x + a^2\omega^2 t)$. This function, treated as a function of the complex variable ω, is holomorphic and hence it can be expanded in a power series in any circle. We write

$$\exp(\omega x + a^2\omega^2 t) = \sum_{n=0}^{\infty} \frac{\omega^n}{n!} V_n(x, t)$$

and then confine ourselves to real values of ω. Let A denote the operator

$$\frac{\partial}{\partial t} - a^2 \frac{\partial^2}{\partial x^2}.$$

By hypothesis, we have

$$A\left(\sum_{n=0}^{\infty} \frac{\omega^n}{n!} V_n(x, t)\right) = 0.$$

Assuming that term-by-term differentiation of this series with respect to x or t is permissible (see Exercise 9 of Chapter 5), we have

$$\sum_{n=0}^{\infty} \frac{\omega^n}{n!} AV_n(x, t) = 0.$$

Since the limit of this power series is zero no matter what the value of the variable ω, all the coefficients must be zero and hence $AV_n(x, t) = 0$. For every value of n, the function $V_n(x, t)$ is a solution of the heat-flow equation.

Now we have defined the Hermite polynomials $P_n(s)$ by

$$\exp(-u^2 + 2us) = \sum_{n=0}^{\infty} \frac{1}{n!} u^n P_n(s).$$

Let us set

$$u^2 = -a^2\omega^2 t, \qquad 2us = \omega x,$$

or

$$u = a\omega \sqrt{-t}, \qquad s = \frac{x}{2a\sqrt{-t}}$$

and let us assume that $t < 0$. We see that

$$\exp(a^2\omega^2 t + \omega x) = \sum \frac{1}{n!} (-t)^{n/2} a^n \omega^n P_n \left(\frac{x}{2a\sqrt{-t}}\right).$$

Therefore,

$$V_n(x, t) = a^n(-t)^{n/2} P_n \left(\frac{x}{2a\sqrt{-t}}\right),$$

which shows the relation between the functions V_n and the Hermite polynomials.

Examples.

$$P_2(s) = 4s^2 - 2, \qquad V_2(x, t) = x^2 + 2a^2 t,$$

$$P_3(s) = 8s^3 - 12s, \qquad V_3(x, t) = x^3 + 6xa^2 t.$$

The polynomials $x^2 + 2a^2t$ and $x^3 + 6xa^2t$ are solutions of the equation

$$\frac{\partial y}{\partial t} = a^2 \frac{\partial^2 y}{\partial x^2}.$$

C. We wish $y(x, t)$ to be a periodic function of x with period 2π that remains bounded at $t \to +\infty$. Thus we need to choose certain elementary solutions

$$\exp(-a^2\omega^2 t) \cos \omega x \qquad \text{and} \qquad \exp(-a^2\omega^2 t) \sin \omega x.$$

We then combine the chosen solutions into a single unique solution of the form

$$y = \exp(-a^2\omega^2 t)(A \cos \omega x + B \sin \omega x),$$

which depends on two new constants A and B, still to be determined.

The function $f(x)$ is to be periodic with period 2π, which requires that ω have integral values n. To each value of n, we assign values A_n and B_n of A and B. Thus we construct a solution that is formally represented by the series

$$y = \sum_{n=0}^{\infty} \exp(-a^2n^2t)(A_n \cos nx + B_n \sin nx). \tag{1}$$

From the initial conditions, y is equal to a given function $y_0(x)$ for $t = 0$:

$$y_0(x) = \sum_{n=0}^{\infty} (A_n \cos nx + B_n \sin nx).$$

Thus we evaluate the A_n and B_n as Fourier coefficients of the function $y_0(x)$ on the interval $[0, 2\pi]$:

$$A_n = \frac{1}{\pi}\int_0^{2\pi} y_0(s) \cos ns\, ds, \qquad B_n = \frac{1}{\pi}\int_0^{2\pi} y_0(s) \sin ns\, ds, \qquad A_0 = \frac{1}{2\pi}\int_0^{2\pi} y_0(s)\, ds,$$

$$A_n \cos nx + B_n \sin nx = \frac{1}{\pi}\int_0^{2\pi} y_0(s) \cos n(x - s)\, ds.$$

Therefore, we have

$$y = \frac{1}{2\pi}\int_0^{2\pi} y_0(s)\, ds + \frac{1}{\pi}\sum_{n=1}^{\infty} \exp(-a^2n^2t) \int_0^{2\pi} y_0(s) \cos n(x - s)\, ds.$$

The series whose general term is

$$\frac{1}{\pi} \exp(-a^2n^2t) y_0(s) \cos n(x - s)$$

is majorized for all $t > 0$ by the series whose general term is $(M/\pi) \exp(-a^2n^2t)$, where M is the least upper bound of $|y_0(s)|$. Therefore, it is uniformly convergent with respect to s and we may integrate term by term. If we set

$$h(t, x) = \frac{1}{\pi}\left[\frac{1}{2} + \sum_{n=1}^{\infty} \exp(-a^2n^2t) \cos nx\right],$$

we have

$$y = \int_0^{2\pi} h(t, x - s) y_0(s)\, ds.$$

The function $y(x, t)$ is indeed a solution of the heat-flow equation if the series (1) can be term-wise differentiated once with respect to t and twice with respect to x, that is, if the series whose general term is

$$n^2 \exp(-a^2n^2t) \int_0^{2\pi} y_0(s) \cos n(x - s)\, ds$$

is uniformly convergent with respect to t and x. Now, the latter series is majorized by the series whose general term is

$$2\pi M n^2 \exp(-a^2n^2t),$$

which in turn is majorized, for all $t \geqslant t_0 > 0$, by the convergent series whose general term is

$$2\pi M n^2 \exp(-a^2n^2t_0).$$

The uniform convergence follows from this.

D. *An example.*

$$y_0(x) = \begin{cases} u_1 & \text{if } 0 < x < \alpha, \\ u_2 & \text{if } \alpha < x < 2\pi. \end{cases}$$

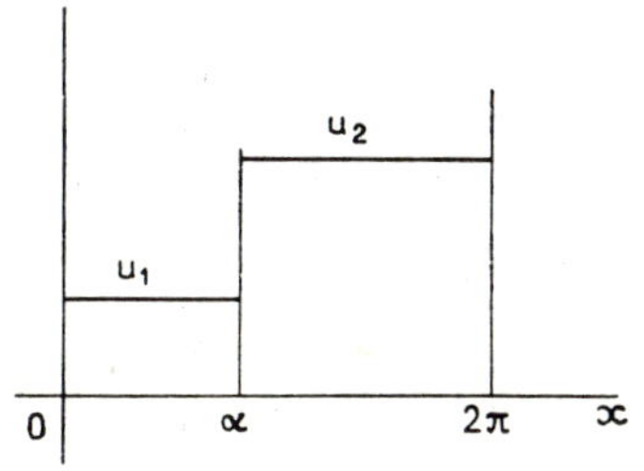

FIGURE 4

Here we have

$$A_n = \frac{u_1}{\pi} \int_0^{\alpha} \cos ns\, ds + \frac{u_2}{\pi} \int_{\alpha}^{2\pi} \cos ns\, ds,$$

$$B_n = \frac{u_1}{\pi} \int_0^{\alpha} \sin ns\, ds + \frac{u_2}{\pi} \int_{\alpha}^{2\pi} \sin ns\, ds,$$

or

$$A_n = \frac{u_1 - u_2}{\pi} \frac{\sin n\alpha}{n}, \qquad A_0 = \frac{\alpha u_1}{2\pi} + (2\pi - \alpha) \frac{u_2}{2\pi},$$

$$B_n = \frac{u_1 - u_2}{n\pi} (1 - \cos n\alpha),$$

$$A_n \cos nx + B_n \sin nx = 2 \frac{u_1 - u_2}{n\pi} \sin \frac{n\alpha}{2} \cos n \left(x - \frac{\alpha}{2}\right).$$

From this we get the following expression for y:

$$y = \alpha \frac{u_1}{2\pi} + (2\pi - \alpha) \frac{u_2}{2\pi} + \frac{2(u_1 - u_2)}{\pi} \sum_{n=1}^{\infty} \frac{\exp(-a^2 n^2 t)}{n} \sin \frac{n\alpha}{2} \cos n \left(x - \frac{\alpha}{2}\right).$$

Since the series is uniformly convergent in the interval $t \geqslant t_0 > 0$, we can take the limit as $t \to \infty$ term by term. All the terms tend to zero except the first and y tends to $(\alpha u_1/2\pi) + (2\pi - \alpha)(u_2/2\pi)$, which is the mean value of the function $y_0(x)$ that represents the initial conditions.

6. The heat-flow equation with boundary conditions

Here we shall treat an eigenvalue problem analogous to that in Exercise 5 but one that leads to more complicated boundary conditions and to series analogous to Fourier series but which are not periodic.

Problems. Consider the partial differential equation

$$\frac{\partial y}{\partial t} = a^2 \frac{\partial^2 y}{\partial x^2}. \tag{E}$$

We shall seek a solution, defined for $0 \leqslant x \leqslant l$, $t > 0$, such that

(1) $\partial y/\partial t + Ky = 0$ for $x = l$, t arbitrary (K a given constant).

(2) $\partial y/\partial x = 0$ for $x = 0$, t arbitrary.

(3) $\lim_{t\to\infty} y(x, t) = 0$ for arbitrary x.

(4) $y(x, 0) = f(x)$, where $f(n)$ is a given function defined for $0 \leqslant x \leqslant l$.

A. Find solutions of the form

$$y = e^{-a^2\omega^2 t} \cos \omega x$$

that verify conditions (1), (2), and (3) and show that ω is a solution of the equation

$$\omega \tan \omega l = K.$$

B. Let $\omega_1, \omega_2, \ldots, \omega_n, \ldots$ be the positive roots of the above equation. Show that the functions

$$g_n(x) = \cos \omega_n x$$

constitute an orthogonal series on the interval $[0, l]$ and evaluate their norm.

C. Exhibit (without proof of the legitimacy of the calculations) the solution of (E) that verifies conditions (1), (2), (3), and (4) by means of a series of the form

$$\sum_n A_n \exp(-a^2\omega_n{}^2 t) \cos \omega_n x.$$

Evaluate the coefficients A_n with the aid of $f(x)$. Treat the particular cases in which $K = 0$ and $K = \infty$.

Solutions. **A.** We have seen (Exercise 5) that the heat-flow equation is satisfied by

$$y(x, t) = \exp(-a^2\omega^2 t)(A \cos \omega x + B \sin \omega x),$$

where A, B, and ω are real constants.

Let us write the boundary conditions:

$$\frac{\partial y}{\partial x} = \exp(-a^2\omega^2 t)(-\omega A \sin \omega x + \omega B \cos \omega x).$$

The condition

$$\frac{\partial y}{\partial x}(0, t) = 0$$

implies that $B = 0$. Therefore,

$$\frac{\partial y}{\partial x} + Ku = \exp(-a^2\omega^2 t)(KA \cos \omega x - \omega A \sin \omega x).$$

The condition

$$\left(\frac{\partial y}{\partial x} + Ky\right)(l, t) = \exp(-a^2\omega^2 t)(KA \cos \omega l - \omega A \sin \omega l) = 0$$

implies either that $A = 0$ [which is an inadmissible solution since it would imply that $y(x, t) = 0$] or that $\omega \tan \omega l = K$.

The function

$$y(x, t) = A \exp(-a^2\omega^2 t) \cos \omega x \qquad \text{where} \quad \omega \tan \omega l = K$$

is an elementary solution of the heat-flow equation that verifies the boundary conditions. Furthermore, it tends to zero as $t \to \infty$, which justifies *a posteriori* the choice of a real positive constant ω^2.

B. *Orthogonality of the eigenfunction solutions.* The equation $\omega \tan \omega l = K$ may be written

$$\tan \omega l = \frac{Kl}{\omega l}$$

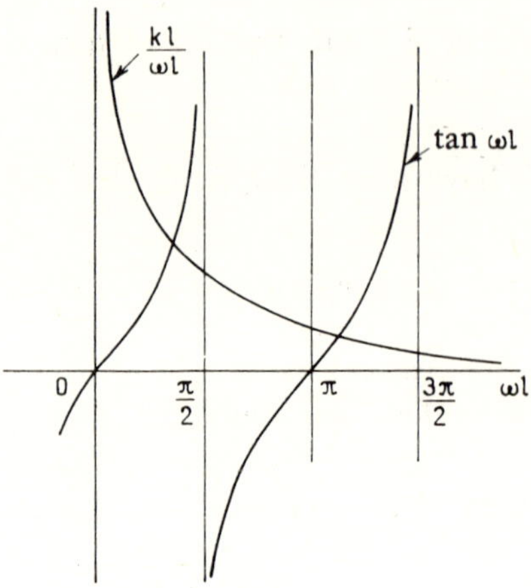

FIGURE 5

We can see from the graph that this equation has an infinite number of solutions $\omega_1, \omega_2, ..., \omega_n, ...$ lying between 0 and $\pi/2l$, between π/l and $3\pi/2l$, etc.

These are the eigenvalues of the boundary-value problem that we are considering. This problem consists in finding solutions $\cos \omega x$ of the differential equation

$$g''(x) + \omega^2 g(x) = 0$$

such that

$$g'(x) = 0 \qquad \text{for} \quad x = 0$$

and

$$g'(x) + Kg(x) = 0 \qquad \text{for} \quad x = l.$$

These conditions on the boundaries define a vector space Ω of functions on the interval $[0, l]$. We already know that the eigenfunctions are orthogonal in this space. To show this consider two distinct eigenvalues ω_q and ω_p corresponding to two eigenfunctions g_q and g_p. We have

$$g_q'' + \omega_q^2 g_q = 0, \qquad g_p'' + \omega_p^2 g_p = 0,$$

from which we get

$$g_q g_p'' - g_q'' g_p + (\omega_p^2 - \omega_q^2) g_q g_p = 0.$$

If we integrate from 0 to l, noting that

$$g_q g_p'' - g_q'' g_p = \frac{d}{dx}(g_q g_p' - g_q' g_p),$$

we get

$$[g_q g_p' - g_q' g_p]_0^l + (\omega_p^2 - \omega_q^2) \int_0^l g_q g_p \, dx = 0.$$

From the boundary conditions, the bracketed expression vanishes at the lower limit, $x = 0$. Similarly, for $x = l$, we have

$$g'(l) = -Kg(l),$$

from which we get

$$g_q(l) g_p'(l) - g_q'(l) g_p(l) = -Kg_q(l) g_p(l) + Kg_q(l) g_p(l) = 0.$$

Therefore, the entire bracketed expression vanishes, and, since $\omega_p^2 - \omega_q^2 \neq 0$, we have

$$\int_0^l g_q g_p \, dx = 0.$$

Thus the functions g_q and g_p are indeed orthogonal on the interval $[0, l]$.

This result can be verified by an elementary calculation on integrals by using the expression for $g(x)$. We define

$$I = \int_0^l \cos \omega_p x \cos \omega_q x \, dx \qquad (p \neq q).$$

Then

$$2I = \int_0^l [\cos(\omega_p - \omega_q)x + \cos(\omega_p + \omega_q)x] \, dx$$

$$= \frac{\sin(\omega_p - \omega_q)l}{\omega_p - \omega_q} + \frac{\sin(\omega_p + \omega_q)l}{\omega_p + \omega_q} = \frac{2\omega_p \tan \omega_p l - 2\omega_q \tan \omega_q l}{\omega_p^2 - \omega_q^2}.$$

Since the value of $\omega_p \tan \omega_p l$ is independent of p, we see that $I = 0$. Thus the sequence of functions $\cos \omega_n(x)$ is orthogonal on $[0, l]$.

Finally, let us evaluate the norm of $\cos \omega_n x$:

$$2 \| \cos \omega_n x \|^2 = 2 \int_0^l \cos^2 \omega_n x \, dx = l + \frac{\sin 2\omega_n l}{2\omega_n} = l + \frac{2 \tan \omega_n l}{2\omega_n(1 + \tan^2 \omega_n l)}$$

$$= l + \frac{K}{K^2 + \omega_n^2},$$

$$\| \cos \omega_n x \| = \frac{1}{\sqrt{2}} \left(l + \frac{K}{K^2 + \omega_n^2} \right)^{1/2}.$$

The orthonormal sequence corresponding to the orthogonal sequence $\cos \omega_n x$ is the sequence whose general term is $\cos \omega_n x / \| \cos \omega_n x \|$, that is,

$$\left(\frac{\sqrt{2} \cos \omega_n x}{\{l + [K/(K^2 + \omega_2{}^n)]\}^{1/2}} \right).$$

C. Let us assume that the function $y(x, t)$ can be represented by a series of the form

$$y = \sum_{n=0}^{\infty} A_n \exp(-a^2 \omega_n{}^2 t) \cos \omega_n x. \tag{1}$$

Let us now write the initial condition $y = f(x)$ for $t = 0$. We have

$$\sum_{n=0}^{\infty} A_n \cos \omega_n x = f(x) \qquad (0 < x < l).$$

The function $f(x)$ is thus the sum of a trigonometric series. But, *a priori*, the ω_n are not multiples of a single pulsation ω_0. Nothing allows us to conclude that $\omega_n = n\omega_0$ or, more generally, that there exists an integer $\varphi(n)$, dependent on n, such that $\omega_n = \omega_0 \varphi(n)$. The series (1) is probably not a periodic Fourier series.

Without seeking to prove the correctness of this hypothesis, we shall show how relation (1) enables us to calculate the coefficients A_n. For simplicity of notation let us set $\cos \omega_n x = g_n(x)$ and let us write

$$\sum_{n=0}^{\infty} A_n g_n = f.$$

Let us take the scalar product (in the sense of the functional space Ω) of each member of this equation with g_p. Let us assume that this operation can be done termwise. Because of the orthogonality of the g_n, there remains only

$$A_p(g_p, g_p) = (f, g_p) \qquad \text{or} \qquad A_p = \frac{(f, g_p)}{\| g_p \|^2}.$$

The norm $\| g_p \|$ was calculated in Problem B. The scalar product in the numerator is

$$(f, g_p) = \int_0^l f(x) \cos \omega_p x \, dx.$$

Thus the A_p are determined.

We then have

$$y = \sum_{n=0}^{\infty} A_n \exp(-a^2\omega_n^2 t) \cos \omega_n x \qquad (t > 0).$$

Let us show that y is once differentiable with respect to t and twice with respect to x. It will then follow that y is indeed a solution of the heat-flow equation.

The whole problem consists in showing that the series

$$\sum_n \omega_n^2 A_n \exp(-a^2\omega_n^2 t) \cos \omega_n x$$

converges uniformly. If $t \geqslant t_0 > 0$, this series is majorized by the series

$$\sum_n \omega_n^2 \mid A_n \mid \exp(-a^2\omega_n^2 t_0).$$

According to Schwarz' inequality,

$$\mid A_n \mid = \frac{|(f, g_n)|}{\| g_n \|^2} \leqslant \frac{\| f \| \, \| g_n \|}{\| g_n \|^2} = \frac{\| f \|}{\| g_n \|}.$$

Now,

$$\| g_n \| = \frac{1}{\sqrt{2}} \left(l + \frac{K}{K^2 + \omega_n^2} \right)^{1/2} > \frac{1}{\sqrt{2}} \frac{\sqrt{K}}{(K^2 + \omega_n^2)^{1/2}}.$$

We can verify that

$$\omega_n^2 \mid A_n \mid \leqslant \| f \| \frac{\omega_n^2}{\| g_n \|} \leqslant \left(\frac{2}{K} \right)^{1/2} \omega_n^2 (K^2 + \omega_n^2)^{1/2} \| f \|.$$

The general term of the majorizing series tends to zero in the same way as

$$\omega_n^3 \exp(-a^2\omega_n^2 t_0).$$

Since $\omega_n > (n = 1)\pi/l$, the exponential assures convergence of this series, and it follows that the series obtained by differentiating y converges uniformly.

Particular cases. If $K = 0$, we have $\tan \omega l = 0$. In this case, $\omega_n = n\pi/l$, and the series (1) is an ordinary periodic Fourier series. We have an analogous conclusion if $K = \infty$.

Remark. The majorization $\| A_n \| \leqslant \| f \| / \| g_n \|$ by Schwarz' inequality is sufficient to prove the differentiability of y. However, this procedure is clumsy. It shows only that $\| A_n \|$ is bounded, since $\| g_n \|$ tends a limit as $n \to \infty$. Now, we know that the integral

$$\int_0^l f(x) \cos \omega_n x \, dx$$

tends to zero as $\omega_n \to \infty$, that is, $A_n \to 0$. This is not yet enough to show that the series $\Sigma A_n \cos \omega_n x$ is convergent. Just as for ordinary Fourier series, we would need to make precise the properties of the function $f(x)$.

7. The nonhomogeneous heat-flow equation

Consider the heat-flow equation

$$\frac{\partial y}{\partial t} = a^2 \frac{\partial^2 y}{\partial x^2} \qquad (a > 0).$$

We know that this equation possesses exactly one solution that, for $t = 0$, reduces to a function $f(x)$ which is defined for all x and absolutely integrable from $-\infty$ to $+\infty$.

This solution is

$$y = \frac{1}{2a(\pi t)^{1/2}} \int_{-\infty}^{\infty} \exp\left[-\frac{(x-s)^2}{4a^2 t}\right] f(s) \, ds. \tag{1}$$

Of course, we may replace the value $t = 0$ with an arbitrary value t', writing $t - t'$ in place of t.

We shall now study the nonhomogeneous heat-flow equation

$$\frac{\partial y}{\partial t} = a^2 \frac{\partial^2 y}{\partial x^2} + \varphi(x, t)$$

and show that the solution that vanishes for $t = 0$ can be represented by means of a formula analogous to (1). We shall derive from this formula the solution that, for $t = 0$, reduces to the given function $f(x)$.

Problem. Consider the nonhomogeneous heat-flow equation

$$\frac{\partial y}{\partial t} = a^2 \frac{\partial^2 y}{dx^2} + \varphi(x, t), \tag{1}$$

where $\varphi(x, t)$ is a given function that is absolutely integrable from $x = -\infty$ to $+\infty$ for every fixed value of t.

Show that the solution y_0 of this equation that vanishes for $t = 0$ is equal to

$$y_0 = \int_0^t u_0(x, t, t')\, dt',$$

where u_0 is that solution of the homogeneous equation $\partial u/\partial t = a^2 \partial^2 u/\partial x^2$ that, for $t = t'$, reduces to $\varphi(x, t')$. Then find the solution of (1) that, for $t = 0$, reduces to a given function $f(x)$.

Solution. Let $u_0(x, t, t')$ be that solution of the homogeneous equation

$$\frac{\partial u}{\partial t} = a^2 \frac{\partial^2 u}{\partial x^2}$$

that reduces to $\varphi(x, t')$ for $t = t'$. The existence and uniqueness of this solution are assumed known. Consider the function

$$y_0(x, t) = \int_0^t u_0(x, t, t')\, dt'.$$

Since $u_0(x, t, t')$ is continuous, we have

$$\frac{\partial y_0}{\partial t} = u_0(x, t, t) + \int_0^t \frac{\partial u_0}{\partial t}(x, t, t')\, dt' = \varphi(x, t) + \int_0^t \frac{\partial u_0}{\partial t}(x, t, t')\, dt'.$$

We can also differentiate under the integral sign with respect to x:

$$a^2 \frac{\partial^2 y_0}{\partial x^2} = \int_0^l a^2 \frac{\partial^2 u_0}{\partial x^2}(x, t, t')\, dt' = \int_0^t \frac{\partial u_0}{\partial t}(x, t, t')\, dt'.$$

Thus we have two expressions for

$$\int_0^t \frac{\partial u_0}{\partial t}(x, t, t')\, dt',$$

namely,

$$\frac{\partial y_0}{\partial t} - \varphi(x, t) \qquad \text{and} \qquad a^2 \frac{\partial^2 y_0}{\partial x^2}.$$

Therefore,

$$\frac{\partial y_0}{\partial t} = a^2 \frac{\partial^2 y_0}{\partial x^2} + \varphi(x, t).$$

The function y_0 is indeed a solution of the nonhomogeneous heat-flow equation, and this solution vanishes for $t = 0$.

Since we know that

$$u_0(x, t, t') = \frac{1}{2a[\pi(t - t')]^{1/2}} \int_{-\infty}^{\infty} \exp\left[-\frac{(x-s)^2}{4a^2(t-t')}\right] \varphi(s, t')\, ds,$$

the solution y_0 is precisely determined.

That solution y of the nonhomogeneous equation that assumes nonzero values $f(x)$ for $t = 0$ is then easily constructed. We have

$$\frac{\partial y}{\partial t} = a^2 \frac{\partial^2 y}{\partial x^2} + \varphi(x, t), \qquad \frac{\partial y_0}{\partial t} = a^2 \frac{\partial^2 y_0}{\partial x^2} + \varphi(x, t).$$

Thus $y - y_0$ is a solution of the homogeneous equation

$$\frac{\partial}{\partial t}(y - y_0) = a^2 \frac{\partial^2}{\partial x^2}(y - y_0).$$

Now, at $t = 0$, the function $y - y_0$ assumes the same values $f(x)$ as does y_0. Therefore, we have

$$y - y_0 = y_1 \qquad \text{or} \qquad y = y_0 + y_1,$$

with

$$y_1 = \frac{1}{2a(\pi t)^{1/2}} \int_{-\infty}^{\infty} \exp\left[-\frac{(x-s)^2}{4a^2 t}\right] f(s)\, ds.$$

We pass from $f(s)$ to y_1 by a linear integral transformation. Let us write this transformation schematically in the form

$$y_1 = \int_{-\infty}^{\infty} K(x, s, t)\, f(s)\, ds \quad \text{where} \quad K(x, s, t) = \frac{1}{2a(\pi t)^{1/2}} \exp\left[-\frac{(x-s)^2}{4a^2 t}\right].$$

Thus we have

$$y = \int_0^t dt' \left[\int_{-\infty}^{\infty} K(x, s, t - t')\varphi(s, t')\, ds\right] + \int_{-\infty}^{\infty} K(x, s, t)\, f(s)\, ds.$$

8. The heat-flow equation. The Laplace and Fourier transformations

Here we shall treat a Cauchy problem relative to the variable x for the heat-flow equation: Find that solution that, at $x = 0$, reduces on a given segment S of the t-axis to a given function and whose derivative $\partial y/\partial x$ reduces to a second given function.

According to the Cauchy-Kowalewsky theorem, this problem has exactly one solution defined in a neighborhood of the segment S in the xt-plane.

Of course, this is valid when S is a finite segment.

In this problem we take for S the entire positive half of the t-axis. We seek to determine the solution in the entire half-plane $t > 0$. In this case we cannot give as general a theorem on the existence and uniqueness. We shall content ourselves with constructing *formally* a solution satisfying the given conditions. This solution will be represented by an integral and not a Fourier series, since the given conditions apply to an unbounded set.

Problems. A. Find the solution of the heat-flow equation

$$\frac{\partial y}{\partial t} = a^2 \frac{\partial^2 y}{\partial x^2}$$

that satisfies the following conditions:

For $x = 0$,

$$y = \begin{cases} 0 & \text{if } t < 0, \\ f(t) & \text{if } t > 0, \end{cases} \qquad \frac{\partial y}{\partial x} = \begin{cases} 0 & \text{if } t < 0, \\ g(t) & \text{if } t > 0, \end{cases}$$

where f and g are two given functions defined for $t > 0$. Assume the legitimacy of the calculations. Represent y by an integral of the form

$$y = \int_{-\infty}^{\infty} [A(\omega) \cos \omega x + B(\omega) \sin \omega x] \exp(-a^2\omega^2 t)\, d\omega,$$

where $A(\omega)$ is an even function and $B(\omega)$ an odd function. Then reduce the problem to a Laplace transform problem by setting $a^2\omega^2 = s$.

B. Carry out the calculations when $f(t) = 1/\sqrt{t}$ and $g(t) = 0$.

Solution. Consider the heat-flow equation

$$\frac{\partial y}{\partial t} = a^2 \frac{\partial^2 y}{\partial x^2}\,.$$

We seek a solution satisfying the following conditions:

For $x = 0$,

$$y = \begin{cases} 0 & \text{if } t \leqslant 0, \\ f(t) & \text{if } t > 0, \end{cases} \qquad \frac{\partial y}{\partial x} = \begin{cases} 0 & \text{if } t \leqslant 0, \\ g(t) & \text{if } t > 0. \end{cases}$$

The given conditions are boundary conditions and not initial conditions. This problem is a Cauchy problem defined on the entire half-axis $t > 0$.

Since $y = 0$ is a solution for $t < 0$, it will be sufficient to consider positive values of t.

We shall seek this solution in the form of a Fourier integral and not a Fourier series as in the preceding exercises:

$$y(x, t) = \int_{-\infty}^{\infty} [A(\omega) \cos \omega x + B(\omega) \sin \omega x] \exp(-a^2\omega^2 t)\, d\omega,$$

where $A(\omega)$ is an even function and $B(\omega)$ is an odd function. For $x = 0$, we have

$$y(0, t) = \int_{-\infty}^{\infty} A(\omega) \exp(-a^2\omega^2 t)\, d\omega = 2 \int_0^{\infty} A(\omega) \exp(-a^2\omega^2 t)\, d\omega$$

because $A(\omega)$ is even and we have

$$\frac{\partial y}{\partial x}(0, t) = \int_{-\infty}^{\infty} \omega B(\omega) \exp(-a^2\omega^2 t)\, d\omega = 2 \int_0^{\infty} \omega B(\omega) \exp(-a^2\omega^2 t)\, d\omega$$

because $B(\omega)$ is odd.

We set $a^2\omega^2 = s$. Then

$$f(t) = \int_0^{\infty} \frac{A(\sqrt{s}/a)}{a\sqrt{s}}\, e^{-st}\, ds, \qquad g(t) = \int_0^{\infty} \frac{B(\sqrt{s}/a)}{a^2}\, e^{-st}\, ds.$$

The functions $f(t)$ and $g(t)$ are the Laplace transforms of the functions

$$\frac{A(\sqrt{s}/a)}{a\sqrt{s}} \qquad \text{and} \qquad \frac{1}{a^2} B\left(\frac{\sqrt{s}}{a}\right).$$

Therefore, if F and G are the primitives of f and g, respectively, we have

$$A(\omega) = a^2\omega F(a^2\omega^2), \qquad B(\omega) = a^2 G(a^2\omega^2),$$

which determines the function y.

Example. $f(t) = 1/\sqrt{t}$, $g(t) = 0$. First of all, we have

$$B(\omega) = 0.$$

On the other hand, $1/\sqrt{t}$ is the Laplace transform of $1/\sqrt{\pi s}$. Therefore, $A(\omega) = a/\sqrt{\pi}$. From this we get the expression for $y(\pi, t)$:

$$y(x, t) = \int_{-\infty}^{\infty} \frac{a}{\sqrt{\pi}} \cos \omega x \exp(-a^2\omega^2 t)\, d\omega$$

or

$$y(x, t) = \int_{-\infty}^{\infty} \frac{a}{\sqrt{\pi}} e^{i\omega x} \exp(-a^2\omega^2 t)\, d\omega.$$

When the exponential in this last integral is written in trigonometrical form, the integral of the term containing the odd function $\sin \omega x$ vanishes.

Let us set $a^2\omega^2 t = s^2/2$. Then

$$y(x, t) = \int_{-\infty}^{+\infty} \frac{1}{(2\pi t)^{1/2}} \exp\left(is \frac{x}{a(2t)^{1/2}}\right) \exp\left(-\frac{s^2}{2}\right) ds.$$

We now set $x/a(2t)^{1/2} = z$:

$$y(x, t) = \int_{-\infty}^{+\infty} \frac{1}{(2\pi t)^{1/2}} e^{isz} \exp(-s^2/z)\, ds.$$

The integral on the right is, up to the factor $1/(2\pi t)^{1/2}$, the Fourier transform of $\exp(-s^2/2)$, that is, $(2\pi)^2 \exp(-z^2/2)$ (see Exercise 7 of Chapter 2).

Therefore,

$$g(x, t) = \frac{1}{\sqrt{t}} \exp\left(-\frac{x^2}{4a^2 t}\right) \qquad (t > 0).$$

Note that y tends to 0 as t tends to 0 through positive values. The solution that we have found can therefore be extended as the zero function for $t < 0$. The same is true of $\partial y/\partial x$.

9. Small motions of a vibrating string

Consider a material one-dimensional body that is rectilinear in its equilibrium position. If it resists bending, it is an elastic or plastic solid. For a thin elastic rod, the vibrations satisfy a partial differential equation of the form

$$\frac{\partial^2 y}{\partial t^2} + K \frac{\partial^4 y}{\partial x^4} = 0,$$

where K is a positive constant and y is the displacement from the equilibrium position of a point on the rod of abscissa x.

If the body does not resist bending, it is a thread or cord. If it is stretched, its movements obey the vibrating-string equation

$$\frac{\partial^2 y}{\partial t^2} = a^2 \frac{\partial^2 y}{\partial x^2}\,.$$

If the cord is not stretched, its motion depends essentially on the exterior forces to which it is subjected and which, in the first two cases, are often negligible as compared with the elastic forces or tensions. In this exercise we shall study the small-amplitude motions of a heavy cord fastened at one end. We shall also content ourselves with finding the elementary solutions (eigenfunctions and eigenvalues) analogous to the sinusoidal solutions of the vibrating-string problem.

The equation that we obtain is linear and hyperbolic in the relevant region $x > 0$ of the xt-plane. It is of mixed type in the entire xt-plane (elliptic for $x < 0$).

Problems. Consider a heavy homogeneous cord of length l fastened at the end A. We choose x- and y-axes in space so that the coordinates of A are $(l, 0)$. The x-axis is vertical. The displacements y of the cord from the vertical equilibrium position are made in the xy-plane. They are functions of time and of the abscissa x on the cord (for $0 \leqslant x \leqslant l$). Assume that y is small and that the angle θ between the tangent to the cord and the x-axis is small.

A. Show that the motion of the cord is given by the equation

$$\frac{1}{g}\frac{\partial^2 y}{\partial t^2} = x\frac{\partial^2 y}{\partial x^2} + \frac{\partial y}{\partial x}, \tag{1}$$

where g is the acceleration due to gravity.

B. Find the solutions of (1) that are of the form

$$y = f(x)\varphi(t)$$

and that satisfy the given boundary conditions.

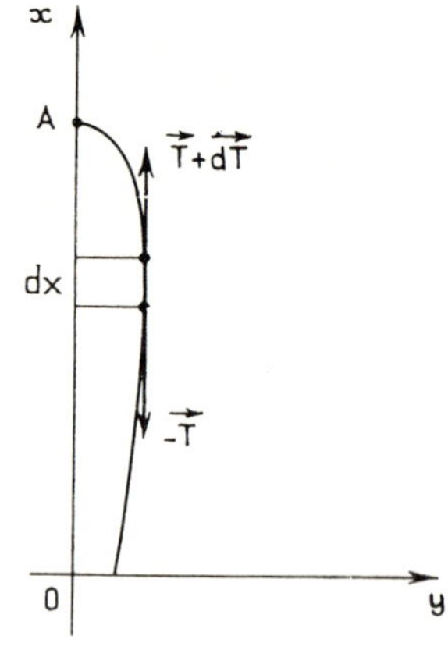

FIGURE 6

Solutions. **A.** The theorem on the kinetic moment states that the tensile forces are tangent to the cord.

The relation $\mathbf{F} = \mathbf{m}\gamma$ projected onto the x-axis provides the equation

$$-\rho\, dx\, g - T\cos\theta + (T + dT)\cos(\theta + d\theta) = 0.$$

If we make the approximation $\cos\theta = 1$, we have

$$\frac{dT}{dx} = \rho g \qquad \text{or} \qquad T = \rho g x.$$

The same relation projected onto the y-axis yields

$$(\rho\, dx)\frac{\partial^2 y}{\partial t^2} = (T + dT)\sin(\theta + d\theta) - T\sin\theta$$

$$= T\cos\theta\, d\theta + dT\sin\theta$$

as a second-order approximation.

Since the angles θ are small, we may write

$$\sin\theta = \frac{\partial y}{\partial x}, \qquad d\theta = \frac{\partial^2 y}{\partial x^2}\, dx$$

and

$$\rho\frac{\partial^2 y}{\partial t^2} = T\frac{\partial^2 y}{\partial x^2} + \frac{dT}{dx}\frac{\partial y}{\partial x}.$$

Since $T = \rho g z$, we have

$$\frac{1}{g}\frac{\partial^2 y}{\partial t^2} = x\frac{\partial^2 y}{\partial x^2} + \frac{\partial y}{\partial x}. \tag{1}$$

B. Let us seek the solutions of (1) that are of the form $y = f(x)\varphi(t)$:

$$\frac{1}{g}f(x)\varphi''(t) = x\varphi(t)f''(x) + f'(x)\varphi(t).$$

Separating the variables, we get

$$\frac{1}{g}\frac{\varphi''(t)}{\varphi(t)} = \frac{xf''(x) + f'(x)}{f(x)}.$$

The common value of these two expressions is a constant, which we must assume to be negative if we confine ourselves to small motions. We denote it by $-\omega^2$.

Then

$$\varphi(t) = A \cos \omega \sqrt{g}t + B \sin \omega \sqrt{g}t$$

and f verifies the differential equation

$$xf''(x) + f'(x) + \omega^2 f(x) = 0. \tag{2}$$

We set

$$x = \frac{u^2}{4\omega^2}, \qquad f(x) = g(u), \qquad f'(x) = g'(u) \times \frac{2\omega^2}{u},$$

$$f''(x) = \frac{4\omega^4}{u^2} g''(u) - \frac{4\omega^4}{u^3} g'(u).$$

The differential equation becomes

$$ug''(u) + g'(u) + ug(u) = 0. \tag{3}$$

We recognize this as *Bessel's equation of order zero*, the only bounded solution of which is, up to a constant factor, $J_0(u)$. Therefore,

$$f(x) = J_0(2\omega \sqrt{x})$$

and

$$y = J_0(2\omega \sqrt{x})(A \cos \omega \sqrt{g}t + B \sin \omega \sqrt{g}t).$$

Let us express the fact that the point A remains fixed for all t:

$$J_0(2\omega \sqrt{l}) = 0.$$

This equation provides values of ω that are consonant with the imposed boundary conditions. The function J_0 has infinitely many positive zeros, $x_1, x_2, \ldots, x_n, \ldots$, and we must choose ω of the form

$$\omega_n = \frac{1}{2\sqrt{l}} x_n .$$

We recall the numerical values of the first five zeros of J_0:

$$x_1 = 2.40, \qquad x_2 = 5.52, \qquad x_3 = 8.65, \qquad x_4 = 11.79, \qquad x_5 = 14.93.$$

*10. Vibrations of a string that is fastened at one end but movable at the other

Let us find the motion of a string if one end is fastened (the position and velocity are also zero for $x = 0$), if it is in its equilibrium position at an initial instant ($u = 0$ for $t = 0$, $0 \leqslant x \leqslant l$), and if the other end $x = l$ is constrained to move in a certain manner compatible with the preceding conditions.

We know that the equation

$$\frac{\partial^2 u}{\partial x^2} = \frac{1}{a^2} \frac{\partial^2 u}{\partial t^2},$$

which describes the vibration of a string, has the elementary solution

$$(A \cos a\,\omega t + B \sin a\,\omega t)(A' \cos \omega x + B' \sin \omega x),$$

where A, B, A', B', and ω are constants. This solution is compatible with the conditions

$$u = 0 \qquad \text{for} \quad t = 0 \qquad \text{and for} \quad x = 0$$

if the constants are such that it reduces to

$$u = B \sin a\omega t \sin \omega x.$$

But the condition $\partial u/\partial t = 0$ for $t = 0$ cannot then be satisfied and neither, of course, can the condition $u = \varphi(t)$ for $x = l$. We can try to let B be a function of ω and to represent u as a sum of the form

$$u = \sum_{\omega} B_{\omega} \sin a\omega t \sin \omega x$$

with

$$\sum_{\omega} \omega B_{\omega} \sin \omega x = 0, \qquad \sum B_{\omega} \sin \omega l \sin a\omega t = \varphi(t)$$

or possibly as an integral of analogous form. But the first of these two conditions seems to imply that $B_{\omega} = 0$ and to be incompatible with the second condition, which is difficult to achieve. Therefore, we shall solve the problem by other methods not involving the elementary solution.

Problems. Consider the partial differential equation

$$\frac{\partial^2 u}{\partial x^2} = \frac{1}{a^2} \frac{\partial^2 u}{\partial t^2},$$

where a is a given positive number. Find solutions u such that

For	$t = 0$:	$u = 0,$	$\dfrac{\partial u}{\partial t} = 0,$
For	$x = 0$:	$u = 0,$	
For	$x = l (l > 0 \text{ given})$:	$u = \varphi(t),$	

where $\varphi(t)$ is a given continuous differentiable function such that $\varphi(0) = 0$ and $\varphi'(0) = 0$. Employ in succession two different methods and then compare them.

A. Work the problem with a Laplace transformation relative to the variable t.

B. Represent u in the interval $0 \leqslant x < l$ by a series of the form

$$u = \sum_{n=1}^{\infty} c_n(t) \sin \frac{n\pi}{l} x, \tag{1}$$

where the $c_n(t)$ are unknown functions of t. Show that series (1) cannot be differentiated term by term with respect to x but that $\partial u/\partial x$ can be represented by a series of analogous form:

$$\frac{\partial u}{\partial x} = \frac{\gamma_0(t)}{2} + \sum_{n=1}^{\infty} \gamma_n(t) \cos \frac{n\pi}{l} x. \tag{2}$$

Evaluate the coefficient $\gamma_n(t)$ with the aid of the $c_n(t)$ and $\varphi(t)$. Show that series (2) is differentiable with respect to x (by means of the hypotheses of a general nature which we assume verified). Then show that $c_n(t)$ is a solution of a second-order nonhomogeneous linear equation.

C. Solve this differential equation taking into account the initial conditions for $t = 0$. Then derive an expression for the desired solution u.

D. Take the Laplace transform of the solution found for Problem C and find again the result of Problem A. Use the following Fourier series:

$$\frac{\sinh \omega x}{\sinh \omega l} = \frac{2\pi}{l^2} \sum_{n=1}^{\infty} (-1)^{n-1} \frac{n}{\omega^2 + (n^2\pi^2/l^2)} \sin \frac{n\pi}{l} x \qquad (-l < x < l)$$

(the proof of which can be given).

E. Discuss the motion of the string with the aid of the result of Problem A. Expand $\sinh(p/a)x/\sinh(p/a)l$ in a real exponential series and use the lag theorem, according to which if $\Phi(p)$ is the Laplace transform of $\varphi(t)$, the primitive of $e^{-p\alpha}\Phi(p)$ is zero for $t < \alpha$ and equal to $\varphi(t - \alpha)$ if $t > \alpha$.

Solutions. **A.** *Use of the Laplace transformation.* The solution $u(x, t)$ is defined for all $t > 0$ and it satisfies the conditions

$$u = 0, \quad \frac{\partial u}{\partial t} = 0 \qquad \text{for} \quad t = 0.$$

If we take the Laplace transform of this solution *with respect to time*, we find

$$U(p, x) = \int_0^\infty e^{-pt} u(x, t)\, dt = \mathscr{L}u.$$

Then

$$p^2 U = \mathscr{L}\frac{\partial^2 u}{\partial t^2}, \qquad \frac{\partial^2 U}{\partial x^2} = \mathscr{L}\frac{\partial^2 u}{dx^2}.$$

From this we derive the transformed equation

$$\frac{\partial^2 U}{\partial x^2} = \frac{p^2}{a^2} U \tag{3}$$

with $U(p, 0) = 0$ and $U(p, l) = \Phi(p)$, where Φ is the Laplace transform of φ.

Equation (3) can be solved immediately. The solution that vanishes for $x = 0$ is

$$U = A(p) \sinh \frac{p}{a} x.$$

For $x = l$, we have

$$A(p) \sinh \frac{p}{a} l = \Phi(p).$$

If we then solve for $A(p)$ and substitute into the above expression for U, we get

$$U(p, x) = \Phi(p) \frac{\sinh(p/a)x}{\sinh(p/a)l}. \tag{4}$$

Thus we can in principle evaluate $u(x, t)$, since we have its Laplace transform.

B. *Use of Fourier series.* Let us seek to represent u by means of a Fourier series in x, the coefficients in which will be unknown functions of t. The sine series

$$u(x, t) = \sum_{n=1}^{\infty} c_n(t) \sin \frac{n\pi}{l} x \tag{5}$$

satisfies the condition $u = 0$ for $x = 0$. But the different terms in the series are not individually solutions of the vibrating-string equation.

Before writing that the sum satisfies the string equation and the conditions

$$u = 0, \quad \frac{\partial u}{\partial t} = 0 \quad \text{for} \quad t = 0,$$
$$u = \varphi(t) \quad \text{for} \quad x = l,$$

let us make some remarks.

(a) The series represents u only for $0 \leqslant x < l$. Its sum is not in general continued to the point $x = l$:

$$\lim_{x \to l} u(x, t) = \varphi(t) \neq 0,$$

whereas, according to (5),

$$u(x, l) = 0.$$

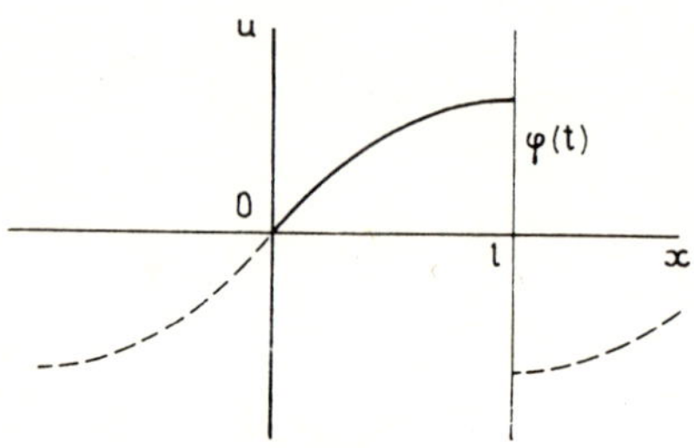

FIGURE 7

(b) In contrast with the procedure used in the case in which the string is fastened at both ends, the series (1) is not a superposition of elementary solutions of the partial differential equation. In other words, $c_n(t)$ is not given *a priori* and is not a sinusoidal function of t.

(c) Let us assume that we can differentiate the series (1) term by term with respect to t:

$$\frac{\partial^2 u}{\partial t^2} = \sum_{n=1}^{\infty} c_n''(t) \sin \frac{n\pi}{l} x.$$

(d) The function $u(x, t)$ is assumed to be continuous and differentiable for $0 \leqslant x < l$. Since it is discontinuous at the point $x = l$, the coefficients $c_n(t)$ tend to zero as $n \to \infty$ and are of the same order as $1/n$. Therefore, to obtain $\partial u/\partial x$, it is not permissible to differentiate the series (5) term by term with respect to x, because the series obtained by doing so diverges.

However, we can expand $\partial u/\partial x$ in a Fourier series (under regularity hypotheses that are easily stated). Let us write

$$\frac{\partial u}{\partial x} = \frac{\gamma_0}{2} + \sum_{n=1}^{\infty} \gamma_n \cos \frac{n\pi}{l} x \tag{6}$$

($\partial u/\partial x$ is an even function) with

$$\gamma_n = \frac{2}{l} \int_0^l \frac{\partial u}{\partial x} \cos \frac{n\pi}{l} x \, dx = \frac{2}{l} \left[u(x) \cos \frac{n\pi}{l} x \right]_0^l + \frac{2n\pi}{l^2} \int_0^l u(x) \sin \frac{n\pi}{l} x \, dx.$$

Thus γ_n is related to c_n by the formula

$$\gamma_n(t) = \frac{2(-1)^n}{l} \varphi(t) + \frac{n\pi}{l} c_n(t).$$

Therefore, we have

$$\frac{\partial u}{\partial x} = \frac{\varphi(t)}{l} + \sum_{n=1}^{\infty} \left[2 \frac{(-1)^n}{l} \varphi(t) + \frac{n\pi}{l} c_n(t) \right] \cos \frac{n\pi}{l} x.$$

Now, $\partial u/\partial x$ is a *continuous* even function at the point $x = l$. Hence we may differentiate the series for $\partial u/\partial x$ term by term, whereas this was quite impossible for the series representing u. We get

$$\frac{\partial^2 u}{\partial x^2} = \sum_{n=1}^{\infty} \frac{n\pi}{l} \left[2 \frac{(-1)^{n-1}}{l} \varphi(t) - \frac{n\pi}{l} c_n(t) \right] \sin \frac{n\pi}{l} x.$$

When we substitute the values found for $\partial^2 u/\partial t^2$ and $\partial^2 u/\partial x^2$ into the partial differential equation and equate terms with like values of n, we obtain

$$\boxed{c_n''(t) + \frac{a^2 n^2 \pi^2}{l^2} c_n(t) = 2(-1)^{n-1} a^2 \frac{n\pi}{l^2} \varphi(t).} \tag{7}$$

Thus each function $c_n(t)$ satisfies a second-order nonhomogeneous linear differential equation.

C. *Solution of* (7). The general solution of the homogeneous equation corresponding to (7) is

$$c_n(t) = \lambda \cos a \frac{n\pi}{l} t + \mu \sin a \frac{n\pi}{l} t.$$

Let us apply to λ and μ the method of variation of parameters. We need to take account of the as-yet-unused conditions

$$u = 0, \quad \frac{\partial u}{\partial t} = 0 \qquad \text{for} \quad t = 0.$$

We find

$$\lambda = 2(-1)^n \frac{a}{l} \int_0^t \varphi(s) \sin \frac{an\pi}{l} ds, \qquad \mu = 2(-1)^{n-1} \frac{a}{l} \int_0^t \varphi(s) \cos \frac{an\pi}{l} s\, ds.$$

Finally,

$$c_n(t) = \frac{2a}{l} (-1)^{n-1} \int_0^t \varphi(s) \sin \frac{an\pi}{l} (t - s)\, ds. \tag{8}$$

With these values of the $c_n(t)$, the function

$$u(x, t) = \sum_{n=1}^{\infty} c_n(t) \sin \frac{n\pi}{l} x$$

is a formal solution of the vibrating-string equation that satisfies the initial and boundary conditions.

If $\varphi(t)$ has a continuous derivative, we know that the integral that represents $c_n(t)$ tends to zero as $n \to 1$ at least as fast as does $1/n^2$. Thus the series (5) converges absolutely, uniformly with respect to x, and uniformly with respect to t in every finite interval. For it to be a solution of the vibrating-string equation, it must be twice differentiable; c_n must tend to zero at least as fast as does $1/n^3$. This will be the case if $\varphi(t)$ possesses a continuous third derivative except possibly at isolated points at which this derivative has salti.

D. *Comparison of the two methods.* Consider the periodic function of period $2l$ that is equal to $\sinh \omega x$ for $-l < x < l$. This function can be expanded in a sine series. We easily see that

$$\frac{\sinh \omega x}{\sinh \omega l} = \frac{2\pi}{l^2} \sum_{n=1}^{\infty} (-1)^{n-1} \frac{n}{\omega^2 + (n^2\pi^2/l^2)} \sin \frac{n\pi}{l} x. \tag{9}$$

Now, the coefficient c_n is, up to the factor $2a(-1)^{n-1}/l$, the convolution of $\varphi(t)$ and $\sin(an\pi t/l)$. We know that the image of $\sin \alpha t$ under the Laplace transformation is $\alpha/(p^2 + \alpha^2)$. Therefore, the Laplace transform of $\sin(an\pi t/l)$ is

$$\frac{an\pi/l}{p^2 + (a^2n^2\pi^2/l^2)} = \frac{n\pi}{al} \frac{1}{(p^2/a^2) + (n^2\pi^2/l^2)}\,.$$

Therefore,

$$\mathscr{L}c_n(t) = \left[\frac{2n\pi}{l^2}(-1)^{n-1}\frac{1}{(p^2/a^2) + (n^2\pi^2/l^2)}\right] \times \Phi$$

and

$$U = \mathscr{L}u = \left[\frac{2\pi}{l^2}\sum_{n=1}^{\infty}(-1)^{n-1}\frac{n}{(p^2/a^2) + (n^2\pi^2/l^2)}\sin\frac{n\pi}{l}x\right] \times \Phi.$$

With $\omega = p/a$, we see from (9) that

$$U = \frac{\sinh(p/a)x}{\sinh(p/a)l}\,\Phi.$$

This is the formula found in the answer to Problem A.

Remark. We have seen that

$$u(x, t) = \sum_{n=1}^{\infty} c_n(t)\sin\frac{n\pi}{l}x = \frac{2a}{l}\sum_{n=1}^{\infty}(-1)^{n-1}\int_0^t \varphi(s)\sin\frac{an\pi}{l}(t-s)\,ds \cdot \sin\frac{n\pi}{l}x.$$

If we make reasonable hypotheses on φ, the function u is represented by a convergent series. However, we cannot reverse the order of summation and integration because the series

$$\sum_{n=1}^{\infty}\sin\frac{an\pi}{l}(t-s)\sin\frac{n\pi}{l}x$$

does not converge to any function. We can only write

$$u(x, t) = \frac{2a}{l}\lim_{n\to\infty}\int_0^t \varphi(s)\left[\sum_{n=1}^{N}(-1)^n\sin\frac{an\pi}{l}(t-s)\sin\frac{n\pi}{l}x\right]ds.$$

Thus u is represented by a Dirichlet integral. This amounts to saying that the primitive of

$$\frac{\sinh(p/l)x}{\sinh(p/l)a} = \frac{2\pi}{l^2}\sum_{n=1}^{\infty}(-1)^{n-1}\frac{n}{(p^2/a^2) + (n^2\pi^2/l^2)}\sin\frac{n\pi}{l}x$$

(this series converges slowly in the ordinary sense) does not exist as a function (although it does exist as a distribution). The primitive of

$$U = \Phi(p)\frac{\sinh(p/l)x}{\sinh(p/l)a}$$

is not the convolution of $\varphi(t)$ with any function.

We can avoid these difficulties by writing

$$U = p\Phi(p)\frac{1}{p}\frac{\sinh(p/l)x}{\sinh(p/l)a},$$

which would mean transforming the expression for the c_n by integrating by parts, which is a valid procedure if φ is differentiable.

E. Let us now use expression (4) for the Laplace transform of u in a different manner. We have

$$U = \Phi(p)\exp\left[-\frac{p}{a}(l-x)\right]\left[1-\exp\left(-\frac{2p}{a}x\right)\Big/1-\exp\left(-\frac{2p}{a}l\right)\right].$$

We expand $1/[1-e^{-(2p/a)l}]$ in powers of the exponential expression:

$$\frac{1}{1-e^{-(2p/a)l}} = 1 + e^{-(2p/a)l} + \cdots + e^{-(2p/a)ln} + \cdots.$$

It follows that U is the product of $\Phi(p)$ and the series

$$e^{-(p/a)(l-x)} - e^{-(p/a)(l+x)} + \cdots + e^{-(p/a)[(2n+1)l-x]} - e^{-(p/a)[(2n+1)l+x]} + \cdots. \quad (10)$$

Thus u is the sum of convolutions of $\varphi(t)$ and the primitives of the terms in series (10).

But we know that the primitive of $e^{-p\alpha}\Phi(p)$ is equal to

$$\begin{aligned} &\varphi(t-\alpha) && \text{if } t > \alpha, \\ &0 && \text{if } t < \alpha \end{aligned}$$

[since $\varphi(t)$ is assumed zero for $t < 0$]. The primitives of the terms shown in (10) have the values

$$\begin{array}{lll}
\varphi\left(t-\dfrac{l-x}{a}\right) & \text{if } t > \dfrac{l-x}{a}, & 0 \text{ if } t < \dfrac{l-x}{a}, \\
\varphi\left(t-\dfrac{l+x}{a}\right) & \text{if } t > \dfrac{l+x}{a}, & 0 \text{ if } t < \dfrac{l+x}{a}, \\
\varphi\left(t-\dfrac{(2n+1)l-x}{a}\right) & \text{if } t > \dfrac{(2n+1)l-x}{a}, & 0 \text{ if } t < \dfrac{(2n+1)l-x}{a}, \\
\varphi\left(t-\dfrac{(2n+1)l+x}{a}\right) & \text{if } t > \dfrac{(2n+1)l+x}{a}, & 0 \text{ if } t < \dfrac{(2n+1)l+x}{a}.
\end{array}$$

Consider a fixed value of x (where $0 < x < l$). Let us examine what happens at that point. At first, for $t < (l-x)/a$, all the terms of series

(10) vanish. The string as yet does not budge. It starts to move at the instant $(l - x)/a$. For $(l - x)/a < t < (l + x)/a$, we have

$$u(x, t) = \varphi\left(t - \frac{l - x}{a}\right).$$

The motion is the same as at the end $x = l$ with a lag. When t exceeds $(l - x)/a$, a reflected motion on the fastened end $(x = 0)$ is superimposed on the initial motion. For $(l - x)/a < t < (l + x)/a$, we have

$$u(x, t) = \varphi\left(t - \frac{l - x}{a}\right) - \varphi\left(t - \frac{l + x}{a}\right).$$

For $(l + x)/a < t < (3l - x)/a$, a new term in series (10) ceases to vanish and we have

$$u(x, t) = \varphi\left(t - \frac{l - x}{a}\right) - \varphi\left(t - \frac{l + x}{a}\right) + \varphi\left(t - \frac{l - x}{a}\right),$$

and so forth.

11. Positiveness of the Green's function

In this exercise we introduce the Green's function $G(M, P)$ of a domain $\mathscr{D}$ bounded by a closed surface S. This is a function defined on the ordered pairs of points (M, P) in the interior of $\mathscr{D}$ that vanishes on S and is equal to

$$G(M, P) = \frac{1}{MP} - g(M, P),$$

where g is a function that, relative to P (or M), is harmonic and regular in $\mathscr{D}$.

We know that a function that is harmonic and regular in $\mathscr{D}$ has no maximum or minimum in the interior of $\mathscr{D}$. Its extrema are on the boundary S of $\mathscr{D}$.

We shall use these properties to show that G and its normal derivative dG/dn on S are positive. This result is of practical interest for the following reason: The formula

$$u(M) = \frac{1}{4\pi} \iint_S u(P) \frac{dG}{dn}(M, P)\, dS_p, \tag{1}$$

which solves the Dirichlet problem, shows, in particular, that, when $u(P) = 1$ on S, then

$$1 = \frac{1}{4\pi} \iint_S \frac{dG}{dn}(M, P)\, dS_p\,.$$

It follows that $(1/4\pi)\, dG(M, P)/dn$ has, relative to P, the nature of a *probability density* defined on S. Formula (1) expresses the fact that $u(M)$, where M is a point in the interior of $\mathscr{D}$, is the mean value of u on S. This mean value can be calculated with the density $(1/4\pi)\, dG/dn$.

Problems. A. Show that the Green's function $G(M, P)$ of a region $\mathscr{D}$ is positive. Use the maximum principle, noting that $G = (MP)^{-1} - g$ vanishes on the boundary surface S of $\mathscr{D}$ and is positive in a neighborhood of the point M.

B. Show that the normal derivative dG/dn of G on the surface S is positive.

C. Let s be a closed surface in the interior of S. Let G_s be the Green's function of the open region enclosed by s and let G_S be the Green's function of the open region enclosed by S. Show that $G_s \leqslant G_S$.

Solutions. **A.** We have set

$$G(M, P) = \frac{1}{MP} - g(M, P).$$

Since the solution g is regular and therefore bounded, there exists a sphere S' of radius α with center at M that is sufficiently small that $G(M, P)$ is positive in the interior of this sphere S' and on its surface.

On the other hand, $G(M, P)$ vanishes on the surface S and is regular in the region $\mathscr{D}$ constituting the complement of S' with respect to S.

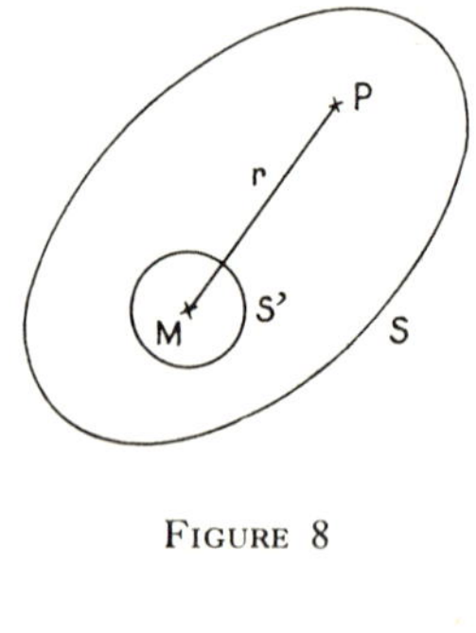

FIGURE 8

If, at some point P_0, the values of $G(M, P_0)$ were negative, there would exist a point P_1 in the interior of $\mathscr{D}$ at which G would have a minimum, which is impossible since G is harmonic and regular in the interior of $\mathscr{D}$.

Example. If S is a sphere of radius R with center at the origin, we know that

$$G = \frac{1}{r} - \frac{R}{\rho r'},$$

with

$$\rho = OM, \qquad r = MP, \qquad r' = M'P,$$

where M' is the inverse of M with respect to S. Elementary geometry shows that $G > 0$ in the interior of S.

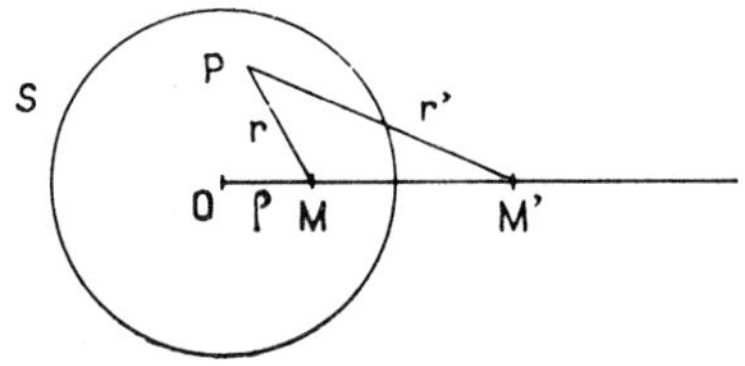

FIGURE 9

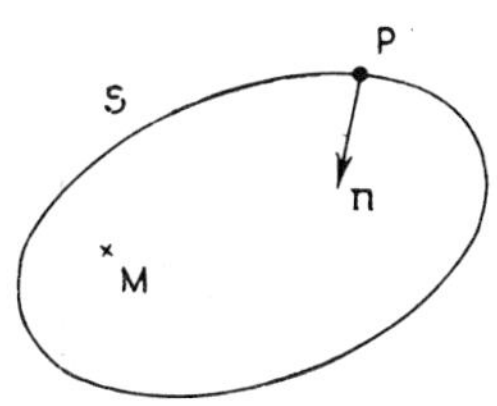

FIGURE 10

B. Consider a point P on the surface S and an inwardly directed unit normal vector to the surface. Let λ denote the abscissa of a point on the normal measured inwardly from P. The values assumed by G on the normal are of the form $H(M, \lambda)$, where M is fixed and

$$\frac{dG}{dn} = \frac{d}{d\lambda} H(M, \lambda).$$

$H(M, \lambda)$ is a function of the real variable λ that vanishes for $\lambda = 0$ and is positive for $\lambda > 0$. Therefore, its derivative is positive at the point $\lambda = 0$.

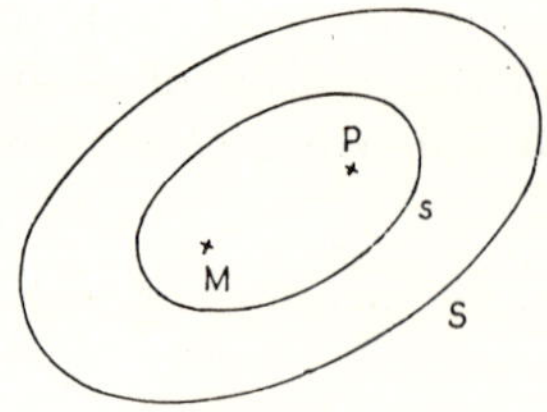

FIGURE 11

Example. If S is a sphere,

$$\frac{dG}{dn} = \frac{R^2 - \rho^2}{Rr^3}.$$

C. The function $G_S - G_s$ is harmonic and regular in the interior of s because

$$G_S - G_s = g_s - g_S.$$

If P is on s, the difference $G_S - G_s = G_S$ is a positive function by virtue of what was said above. Since this function has its minimum on s, it is positive in and on s. Therefore, $G_s < G_S$.

12. An example of the Dirichlet problem for a sphere

Problems. Find a harmonic function $u(x, y, z)$ that is regular in the interior of the sphere S of radius 1 with center at the origin and that assumes values equal to z^2 on S. Use the following two methods in succession:

A. Use the linearity properties of Laplace's equation, constructing as a preliminary second-degree harmonic polynomials.

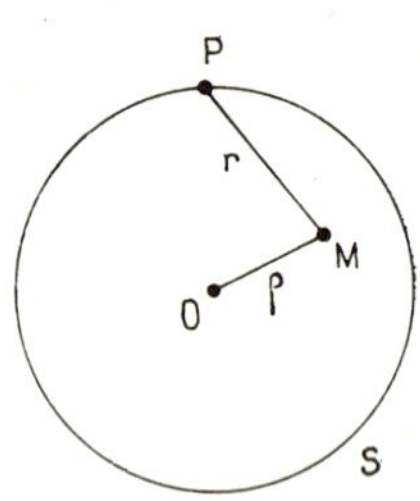

FIGURE 12

B. Use the formula

$$u(M) = \frac{1}{4\pi} \iint_S u(P) \frac{R^2 - \rho^2}{Rr^3} \, dS,$$

choosing axes that will be most convenient.

Solutions. **A.** The function z^2 is a second-degree polynomial. We shall construct the function u by using the linearity of Laplace's equation. We already know how to construct second-degree harmonic polynomials. We see immediately that $2z^2 - x^2 - y^2 = 3z^2 - (x^2 + y^2 + z^2)$ is such a polynomial. Now, on the sphere S, $x^2 + y^2 + z^2$ has a constant value R^2. Therefore, the polynomial in question assumes the value $3z^2 - R^2$ on the sphere. Thus the polynomial

$$z^2 - \frac{x^2 + y^2 + z^2}{3}$$

is harmonic and is equal to $z^2 - R^2/3$ on the sphere.

Now, the constant $R^2/3$ is harmonic. The function

$$z^2 - \frac{x^2 + y^2 + z^2}{3} + \frac{R^2}{3}$$

is harmonic and regular in the interior of S, and it assumes the values z^2 on S. Since the Dirichlet problem has a unique solution, this function is the desired solution:

$$u = \frac{2z^2 - x^2 - y^2 + R^2}{3}.$$

B. The Dirichlet problem for a sphere is "solvable" in the sense that there exists an *explicit* formula that reduces the solution to the evaluation of a double integral.

Let us try to apply this formula, which is written

$$u(M) = \frac{1}{4\pi} \iint_S u(P) \frac{R^2 - \rho^2}{Rr^3} \, dS,$$

where ρ is the distance from M to the center O of the sphere and r is the distance from M to the point P which describes the sphere. Here we have

$$4\pi u = \frac{(R^2 - \rho^2)}{R} \iint_S \frac{z^2}{r^3} \, dS.$$

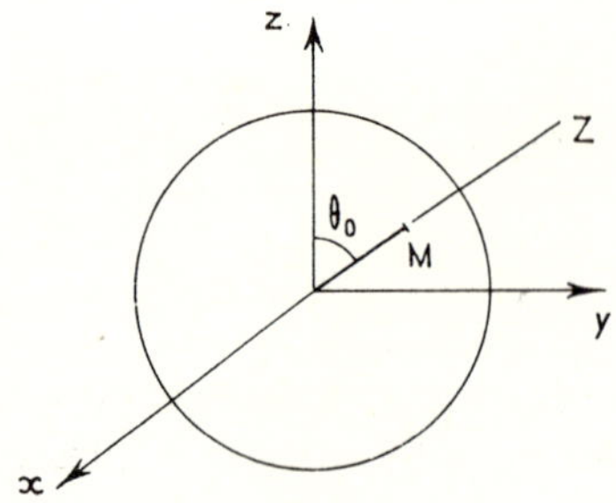

FIGURE 13

Let us shift to spherical coordinates. If we denote by x_0, y_0, and z_0 the Cartesian coordinates of M, and by R, θ, and φ the spherical coordinates of P, we have

$$z = R\cos\theta, \qquad dS = R^2 \sin\theta \, d\theta \, d\varphi,$$
$$r^2 = R^2 + \rho^2 - 2R(x_0 \sin\theta\cos\varphi + y_0 \sin\theta \sin\varphi + z_0 \cos\theta),$$

from which we get the following expression for $4\pi u$:

$$R^3(R^2 - \rho^2) \iint_S \frac{\cos^2\theta \sin\theta \, d\theta \, d\varphi}{[R^2 + \rho^2 - 2R(x_0 \sin\theta\cos\varphi + y_0 \sin\theta\sin\varphi + z_0\cos\theta)]^{3/2}}.$$

This integral may seem irreducible to elementary expressions. Actually, it can be evaluated by elementary procedures, but to do this it will be convenient to make a transformation of axes.

Let us place a Z-axis along OM in such a way that the coordinates of M are

$$0, \qquad 0, \qquad \rho > 0.$$

Then z is transformed into a linear function $\alpha X + \beta Y + \gamma Z$ of the new coordinates, where $\alpha^2 + \beta^2 + \gamma^2 = 1$. Then

$$4\pi u = \frac{R^2 - \rho^2}{R} \iint \frac{(\alpha X + \beta X + \gamma Z)^2 \, dS}{r^3},$$

with

$$X = R\sin\theta\cos\varphi, \qquad Y = R\sin\theta\sin\varphi, \qquad Z = R\cos\theta,$$
$$dS = R^2 \sin\theta \, d\theta \, d\varphi, \qquad r^2 = R^2 + \rho^2 - 2R\rho\cos\theta,$$

where θ and φ are the spherical coordinates in the new coordinate system.

Thus we have

$$\begin{aligned}(\alpha X + \beta Y + \gamma Z)^2 = {} & \alpha^2 R^2 \sin^2\theta \cos^2\varphi + \beta^2 R^2 \sin^2\theta \sin^2\varphi + \gamma^2 R^2 \cos^2\theta \\ & + 2\alpha\beta R^2 \sin^2\theta \sin\varphi \cos\varphi + 2\alpha\gamma R^2 \sin\theta \cos\theta \cos\varphi \\ & + 2\beta\gamma R^2 \sin\theta \cos\theta \sin\varphi.\end{aligned}$$

Here θ varies from 0 to π, and φ from 0 to 2π.

As r does not depend on φ, we integrate first with respect to φ:

$$\iint \frac{(\alpha X + \beta Y + \gamma Z)^2}{r^3}\, dS = R^2 \int_0^{\pi} \frac{\sin\theta\, d\theta}{r^3} \int_0^{2\pi} (\alpha X + \beta Y + \gamma Z)^2\, d\varphi.$$

Now, the integral with respect to φ is elementary. The terms corresponding to $\sin\varphi$, $\cos\varphi$, and $\sin\varphi\cos\varphi$ disappear, so that there remains

$$\pi R^2(\alpha^2 \sin^2\theta + \beta^2 \sin^2\theta + 2\gamma^2 \cos^2\theta).$$

Thus we have

$$u = \frac{R^3(R^2 - \rho^2)}{4} \int_0^{\pi} \frac{[(\alpha^2 + \beta^2)\sin^2\theta + 2\gamma^2\cos^2\theta] \sin\theta\, d\theta}{(R^2 + \rho^2 - 2R\rho\cos\theta)^{3/2}}.$$

Let us take as our variable of integration

$$s = R^2 + \rho^2 - 2R\rho\cos\theta.$$

We set

$$(\alpha^2 + \beta^2)\sin^2\theta + 2\gamma^2\cos^2\theta = \frac{As^2 + Bs + c}{4R^2\rho^2},$$

where

$$\begin{aligned} A &= 2\gamma^2 - \alpha^2 - \beta^2, \qquad B = -2(R^2 + \rho^2)(2\gamma^2 - \alpha^2 - \beta^2), \\ C &= 2\gamma^2(R^2 + \rho^2)^2 - (\alpha^2 + \beta^2)(R^2 - \rho^2)^2. \end{aligned}$$

Then we have

$$u = \frac{R^2 - \rho^2}{32\rho^3} \int_{(R-\rho)^2}^{(R+\rho)^2} (As^{1/2} + Bs^{-1/2} + Cs^{-3/2})\, ds.$$

From this we easily find

$$u = \frac{1}{8\rho^2}\left[A\left(R^2 + \frac{\rho^2}{3}\right) + B\right](R^2 - \rho^2) + \frac{C}{8\rho^2},$$

and, after some simplifications,

$$u = \frac{(\alpha^2 + \beta^2)(R^2 - \rho^2) + (R^2 + 2\rho^2)\gamma^2}{3}.$$

This is the same formula as was found in Problem A. There we had

$$u = \frac{3z^2 + R^2 - (x^2 + y^2 + z^2)}{3}.$$

We need to replace z with $\alpha X + \beta Y + \gamma Z$ and $x^2 + y^2 + z^2$ with $X^2 + Y^2 + Z^2$, and then apply the formula found with $X = Y = 0$ and $Z = \rho$. We then have

$$u = \frac{3\gamma^2\rho^2 + R^2 - \rho^2}{3} = \frac{3\gamma^2\rho^2 + (R^2 - \rho^2)(\alpha^2 + \beta^2 + \gamma^2)}{3}$$
$$= \frac{(\alpha^2 + \beta^2)(R^2 - \rho^2) + (R^2 + 2\rho^2)\gamma^2}{3},$$

which is what we expected.

13. Poisson's equation

Let $\mathscr{D}$ denote a region bounded by a closed surface S. Let $G(M, P)$ denote the Green's function of that region. Search for a solution of the Dirichlet problem involves a linear integral transformation the kernel of which is dG/dn. If u_1 is a regular harmonic function in $\mathscr{D}$ that assumes values h on S, we have

$$u_1(M) = \frac{1}{4\pi} \iint_S \frac{dG}{dn}\, h\, dS.$$

The following question arises: What does the integral transformation in which G (rather than dG/dn) is the kernel represent? Since G is defined in the interior of $\mathscr{D}$, it is a matter of characterizing the triple integral

$$\frac{1}{4\pi} \iiint_{\mathscr{D}} Gf\, ds,$$

where f is a given function in $\mathscr{D}$. We see that it is a solution of the nonhomogeneous equation

$$\Delta u = f$$

(Poisson's equation).

Problems. **A.** Consider the equation

$$\Delta u = f(x, y, z). \tag{1}$$

Find a solution to this equation that is defined and regular in a domain $\mathscr{D}$ and that vanishes on the boundary surface S of $\mathscr{D}$. Show that

$$4\pi u(M) = \iiint_{\mathscr{D}} \frac{f}{r}\, dx\, dy\, dz - \iint_{S} \frac{1}{r} \frac{du}{dn}\, dS,$$

where r is the distance on the reference point M to the point P whose coordinates are x, y, and z. From this, show that, if $G(M, P)$ is the Green's function of the region $\mathscr{D}$, then

$$4\pi u(M) = \iiint_{\mathscr{D}} Gf\, dx\, dy\, dz.$$

B. Find the solution of (1) that assumes given values $h(x, y, z)$ on the surface S.

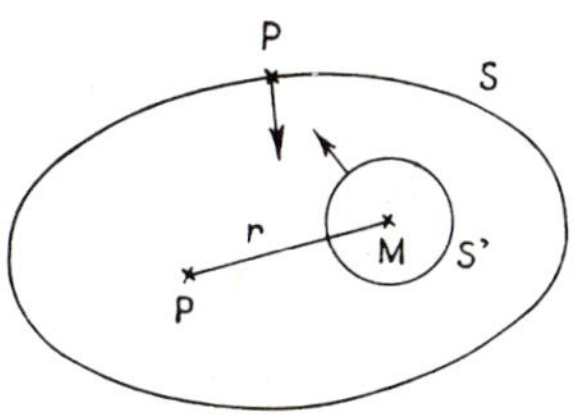

FIGURE 14

Solutions. **A.** We recall Green's formula,

$$\iiint_{\mathscr{D}} (u\Delta v - v\Delta u)\, dx\, dy\, dz = \iint_{S} \left(u \frac{dv}{dn} - v \frac{du}{dn}\right) dS.$$

Let us apply this formula, taking the unknown function for u and the function $1/r$, where r denotes the distance from the fixed point $M(a, b, c)$ to the variable point P, for v. Green's formula can be applied in a domain in which these two functions are regular. Let $\mathscr{D}'$ denote the complement with respect to $\mathscr{D}$ of a sphere S' of radius ϵ with center at M. Since $\Delta u = f$ and $u = 0$ on S by hypothesis, we have

$$-\iiint_{\mathscr{D}'} \frac{f}{r}\, dx\, dy\, dz = -\iint_{S} \frac{1}{r} \frac{du}{dn}\, dS + \iint_{S'} \left(u \frac{d(1/r)}{dn} - \frac{1}{r} \frac{du}{dn}\right) dS'.$$

In this formula the normals are directed into the interior of $\mathscr{D}'$, that is, toward the interior of S and toward the exterior of S'.

On S', r is constant (equal to ϵ) and $dr^{-1}/dn = dr^{-1}/dr = -r^{-2}$. Now,

$$\iint_{S'} \frac{1}{r}\frac{du}{dn}\,dS' = \frac{1}{r}\iint_{S'} \frac{du}{dn}\,dS'.$$

From the mean-value theorem, $\iint_{S'} (du/dn)\,dS'$ is the product of a quantity bounded by

$$\iint_{S'} dS' = 4\pi r^2 = 4\pi\epsilon^2.$$

Division by r gives us $4\pi\epsilon$, which tends to zero as ϵ tends to zero.

$$\iint_{S'} u\,\frac{d(1/r)}{dn}\,dS' = -\frac{1}{r^2}\iint_{S'} u\,dS' = -4\pi\,\frac{1}{4\pi r^2}\iint_{S'} u\,dS'.$$

The expression on the right is -4π times the mean value of u on the sphere S'. As $\epsilon \to 0$, it tends to the value of u at the center, namely, $u(M)$. Thus we have

$$4\pi u(M) = \iiint_{\mathscr{D}} \frac{f}{r}\,dx\,dy\,dz - \iint_S \frac{1}{r}\frac{du}{dn}\,dS. \tag{1}$$

But this formula does not solve the problem posed, since knowledge of $u(M)$ in the interior of $\mathscr{D}$ requires knowledge of du/dn on S, which was not given.

Therefore, let us introduce the regular harmonic function g that assumes values $1/r$ on S and let us again apply Green's formula, just as before taking the unknown function as u but this time taking the function g as v. We get

$$\iiint_{\mathscr{D}} fg\,dx\,dy\,dz = \iint_S \frac{1}{r}\frac{du}{dn}\,dS. \tag{2}$$

If we eliminate du/dn between (1) and (2), we obtain

$$4\pi u(M) = \iiint_{\mathscr{D}} f\left(\frac{1}{r} - g\right) dx\,dy\,dz = \iiint_{\mathscr{D}} Gf\,dx\,dy\,dz,$$

where $G = r^{-1} - g$ is the Green's function of the domain $\mathscr{D}$.

B. Let us seek solutions of the equation $\Delta u = f$ such that u assumes values $h(x, y, z)$ on the surface S. We denote by u_0 the solution, found in Problem A, that vanishes on the sphere. We have

$$\Delta u = f, \qquad \Delta u_0 = f.$$

Therefore,

$$\Delta(u - u_0) = 0.$$

The difference $u - u_0$ is a harmonic function u_1 which assumes the values h on S. We construct it by solving a Dirichlet problem. Thus we have

$$u = u_0 + u_1 .$$

We know that if G is the Green's function, then

$$4\pi u_1(M) = \iint_S \frac{dG}{dn} h \, dS.$$

Thus we have

$$4\pi u = \iiint_{\mathscr{D}} Gf \, dx \, dy \, dz + \iint_S \frac{dG}{dn} h \, dS.$$

Index

S

T

U

V

Z